Science and Politics in the International Environment

Science and Politics in the International Environment

Edited by
Neil E. Harrison and
Gary C. Bryner

ROWMAN & LITTLEFIELD PUBLISHERS, INC.
Lanham • Boulder • New York • Toronto • Oxford

ROWMAN & LITTLEFIELD PUBLISHERS, INC.

Published in the United States of America
by Rowman & Littlefield Publishers, Inc.
A wholly owned subsidiary of The Rowman & Littlefield Publishing Group, Inc.
4501 Forbes Boulevard, Suite 200, Lanham, MD 20706
www.rowmanlittlefield.com

P.O. Box 317, Oxford OX2 9RU, UK

British Library Cataloguing in Publication Information Available

Library of Congress Cataloging-in-Publication Data Available

ISBN 0-7425-2019-6 (cloth : alk. paper)
ISBN 0-7425-2020-X (pbk. : alk. paper)

Printed in the United States of America

♾ ™The paper used in this publication meets the minimum requirements of American National Standard for Information Sciences—Permanence of Paper for Printed Library Materials, ANSI/NISO Z39.48-1992.

Contents

Figures, Boxes, and Tables

FIGURES

BOXES

TABLES

Abbreviations and Glossary

AES	Atmospheric Environment Service
AEURHYC	Ecological Association of Users of the Hardy and Colorado Rivers
AOCs	Areas of concern
AOSIS	Alliance of Small Island States
APIOS	Acid Precipitation in Ontario Study
AQC	Air Quality Committee (United States–Canada)
AQLN	Ames Quality of Life Network
BGH	Bovine growth hormone, a protein composed of approximately 190 amino acid residues, which influences distribution of feed to bodily functions like growth and lactation
BH	Bovine hormones
BRCG	Bilateral Research Consultation Group on the Long-Range Transport of Air Pollutants
BSE	Bovine spongiform encephalopathy, or "mad cow disease." Scientists believe that BSE may be causally linked to a new variant of Creutzfeldt-Jakob Disease (vCJD) in humans
BST	Bovine somatotropin, a protein hormone that affects distribution of feed to bodily functions like growth and lactation in cattle. Hormones are secretions released by specific glands or other tissues into the bloodstream in minute amounts that stimulate a specific physiological activity
CANSAP	Canadian Network for Sampling Precipitation

CAP	Common Agricultural Policy (EU)
CBD	Convention of Biological Diversity
CCD	Convention to Combat Desertification
CCSBT	Commission for the Conservation of the Southern Bluefin Tuna
CDM	Clean development mechanism
CEDO	Center for the Study of Deserts and Oceans
CEDO	Intercultural Center for the Study of Deserts and Oceans (Puerto Peñasco, Sonora, Mexico)
CES	Committee on Earth Sciences
CIBUAEA	Centro de Investigaciones en Biotecnología de la Universidad Autónoma del Estado de Morelos
CIFOR	International Center for Forest Research
CIPO	Canadian Institute of Public Opinion
CIRVA	International Committee for the Recovery of the Vaquita
CJD	Creutzfeldt-Jakob Disease
COLEF	Colegio de la Frontera Norte
COP1–7	Conference of the Parties (FCCC) (first through seventh)
COTECOCA	Comisión Téchnica Consultiva de Coeficientes de Agostadero
CPR	Common-pool resources, limited resources that, without governance arrangements, are public and from which access cannot naturally be restricted
CPUE	Catch per unit of effort
CRCI	Costa Rica–Canada Initiative
CRIEPI	Central Research Institute of the Electric Power Industry
CSCE	Conference on Security and Cooperation in Europe (Helsinki, 1975)
DDE	Dichlorodiphenyltrichoroethane
DDT	Dichlorodiphenyldichloroethylene
DEF	Dioxin emissions factor
DGIII	Directorate General for Industry
EANET	East Asian Acid Deposition Monitoring Network
EC	European Commission, the administrative body of the EU

EDs	Endocrine disruptors, substances, including rBST, that affect the functioning of organs and glands that secrete hormones
EEZ	Exclusive economic zone
EFP	Experimental fishing program
EFPWG	Experimental Fishing Program Working Group
EMEP	European Monitoring and Evaluation Programme
EP	European Parliament
EPA	Environmental Protection Agency (United States)
ESA	Endangered Species Act (United States)
EU	European Union, a unique international "club" that has central governance organizations including the EC and the EP
FAO	Food and Agriculture Organization of the United Nations
FAR	First assessment report
FCCC	Framework Convention on Climate Change (1992)
FDA	Food and Drug Administration (United States)
FMCN	"Fondo Mexicano," the Mexican Fund for the Conservation of Nature (Fondo Mexicano Conservación Nacional)
FRA	Forest Resources Assessment
Frames	Structures of meaning that scientists and policy actors use to make sense of and assign importance to issues. They generally resist change. Information and experience are interpreted and evaluated through frames.
FSA	Fish Stocks Agreement
FWCC	First World Climate Conference
GARP	Global Atmospheric Research Program
GCC	Global Climate Coalition
GEPRP	Global Environmental Protection Research Plan
GHGs	Greenhouse gases, including carbon dioxide, methane, stratospheric ozone, chlorofluorocarbons, and halons
HELCOM	Helsinki Commission
IBSFC	International Baltic Sea Fisheries Commission
ICJ	International Court of Justice
ICP	International cooperative programs (acid rain)
ICSU	International Council of Scientific Unions

IFF	Intergovernmental Forum on Forests
IGBP	International Geosphere-Biosphere Program
IIASA	International Institute for Applied Systems Analysts
IJC	International Joint Commission
INC	Intergovernmental Negotiating Committee for a Framework Convention on Climate Change
Inco	International Limited, formerly International Nickel Company of Canada, Limited
IOMAC	Indian Ocean Marine Affairs Cooperation
IPCC	Intergovernmental Panel on Climate Change
IPF	Intergovernmental Panel on Forests
ITFF	Interagency Task Force on Forests
ITLOS	International Tribunal for the Law of the Sea
ITQs	Individual transferable quotas
ITTA	International Tropical Trade Agreement
IUCN	International Union for the Conservation of Nature
IWC	International Whaling Commission
JCP	Joint Comprehensive Environmental Action Programme
JECFA	Joint FAO/WHO Expert Committee on Food Additives
LaMP	Lakewide Management Plan
LCP	Lakeshore Capacity Project
LRTAP	Convention on Long-Range Transboundary Air Pollution (1979)
MAB	Man and the Biosphere
MAFF	Ministry of Agriculture, Food, and Fisheries (UK)
MARC	Meeting of acidification research coordinators
MITI	Ministry of International Trade and Industry (Japan)
mmt	Million metric tons
MMTCE	Million metric tons of carbon equivalent
MOE	Ministry of Environment (Canada)
MOI	Memorandum of Intent
NAAEC	North American Agreement on Environmental Cooperation, often refered to as the "Environmental Side Agreement" to NAFTA

NACEC	North American Commission on Environmental Cooperation
NADP	National Atmospheric Deposition Program
NAFTA	North American Free Trade Agreement
NAPAP	National Acid Precipitation Assessment Program
NGO	Nongovernmental organization
NO_x	Nitrogen oxides
OECD	Organization for Economic Cooperation and Development
OIE	Office International des Epizootes
PAMP	Postapproval monitoring program for Prosilac
PCDD	Polychlorinated dibenzo-p-dioxins
PCDF	Polychlorinated dibenzofurans
PESCA	National Fisheries Institute of Mexico (now part of SEMARNAT)
PITF	Programme Implementation Task Force
POPs	Persistent organic pollutants
ppm	Parts per million
RAP	Remedial action plan
rBST	Recombinant bovine somatotropin, a genetically engineered bovine hormone
SALS	Sudbury Area Lakes Study
SAR	Second assessment report
SBT	Southern Bluefin tuna
SCF	Scientific Committee for Food
SCVPH	Scientific Committee on Veterinary Measures relating to Public Health
Se	Selenium
SEMARNAT	Ministry of Environment and Natural Resources (Secretaría de Medio Ambiente, Recursos Naturales). Formerly SEMARNAP
SES	Sudbury Environmental Study
SNSF	Acid Precipitation Effects on Forests and Fish Program
SO_2	Sulphur dioxide
StCF	Standing Committee for Foodstuffs

SVC	Scientific Veterinary Committee
TAC	Total allowable catch
TAR	Third assessment report
TCDD	tetrachlorodibenzo-p-dioxin
TEQ	Toxic equivalent quotient
tph	Tons per hour
UN	United Nations
UNCED	United Nations Conference on Environment and Development ("Rio Summit")
UNCLOS	United Nations Convention on the Law of the Sea
UNECE	United Nations Economic Commission for Europe
UNEP	United Nations Environment Program
UNESCO	United Nations Economic, Scientific, and Cultural Organization
UNFF	United Nations Forum on Forests
vCJD	A new variant of Creutzfeldt-Jacob disease in humans that some scientists believe is contracted via consumption of beef from cattle afflicted with BSE
VOC	Volatile organic compounds
VPA	Virtual population analysis
WCED	World Commission on Environment and Development
WGI	Working Group I of the IPCC—Science
WHO	World Health Organization
WMO	World Meteorological Organization
WTO	World Trade Organization
WWW	World Weather Watch

1

Thinking about Science and Politics

Neil E. Harrison and Gary C. Bryner

Science and politics define environmental policy making. Environmental problems invite scientific research. They invariably are complex and rarely self-evident; they require empirical research to map their limits and effects, and scientific knowledge is an important foundation for possible solutions. But environmental issues also are highly emotive, attracting extremists from across the political spectrum, even in the rarified air of international politics. Thus, this book addresses two questions:

1. How do science and politics relate in international environmental issues?
2. How does the interplay of science and politics influence international environmental policy?

We use case studies to explore the interaction of science and politics across a wide range of international environmental issues. The motivation for this project is our belief that the way science and politics interact can greatly influence international environmental policy. Beyond our belief that the interaction of science and politics is important, underresearched, and oversimplified, we are agnostic about theories and research methods. Thus, we have designed the case studies to raise questions rather than promote a narrow range of specific solutions or find evidence supporting any particular theory.

The case studies are rich in data, but light on theory, and are designed to leave readers to form their own conclusions on the connections between events in science and politics and the causes of policy. As much as is practical, the case study authors have avoided theoretical bias in the presentation of data so that readers can connect the dots and create general theories about the relationship of science and politics in international environmental issues.

This book is designed to be useful to two distinct audiences: environmental studies and policy students, and scholars of international environmental politics and policy. Our foremost objective has been to design a teaching text for environmental studies programs and environmental-policy courses. As is more fully explained in the appendix, case studies encourage and support interdisciplinary teaching in environmental studies by explicitly addressing the interaction between science and politics, which is at the core of all environmental studies programs.

In international environmental issues, it is vital that science and politics communicate and that each respect the skills and methods of the other. Believing that both science and politics are important and related causes of international environmental policy, we reject the common assumption that a natural scientist with no formal training in politics or policy can understand politics, because, as a citizen, she is a natural participant in politics. It is important that natural scientists recognize the subtleties and complexities of the political realm. Just as most policy makers are indebted to some "dead scribbler," those who are not students of politics rely on assumptions and models of politics and policy making long discarded by specialists. Similarly, it should not be assumed that a social scientist without formal training in the natural sciences must fail, and therefore should not try, to understand the limitations and recommendations of science. In the case studies that follow, the descriptions of the developments in natural science are designed to be complete and accurate while comprehensible to the nonspecialist. This book equips students from both scientific and political backgrounds with the tools to better communicate with each other and to interact in more synergistic ways. They will be better prepared to positively infuse policy with good science. Students with training in the natural sciences will better appreciate the problems of policy making and learn how to respond to the needs of policy makers. Students with a social science or policy background will gain a greater understanding of the scientific method and how natural scientists think and are constrained in their interpretation of data. They will recognize how to use science to improve policy making.

The cases are designed to encourage and energize classroom debate. Some of the chapters are collaborations between scholars from natural and social science disciplines. A fish biologist and a legal scholar should have more to say (and more interesting things to say) to a broader audience than either writing alone. Every chapter describes in detail the issues and events in science and politics, providing, in interdisciplinary courses, a rich interaction among students trained in various disciplines. Every case is open ended, allowing almost endless debate about the causes and meanings of events.

Although the teaching objective was uppermost in our minds, the case studies have lessons for scholars in international environmental politics and policy. The "inductive" approach adopted (at our request) by the authors of the case studies produces extensive data, much of it primary, on a range of international environmental issues. The case studies are all original; many address issues not analyzed before or not in as much detail. As we show in the concluding chapter, these case studies challenge accepted ideas about how science and politics relate to each other and how international environmental policy emerges from that interaction.

The remainder of the present chapter serves four purposes: First, it discusses what we mean by "science" and, second, what we mean by "politics." Third, it summarizes some of the ways in which these two practices are believed to interact and influence each other. The final section describes the layout of the book and briefly describes each chapter.

SCIENCE

Science is the application of certain formal methods founded upon rationality and empirical observation to gain knowledge and understanding of the world around us. Scientists formally believe that they are using rationality and empiricism in a search for knowledge or truth about the world. Their goal, they believe, is knowledge of the natural world with no concern for the uses or value of that knowledge. Knowledge is an intrinsic good, valued for itself, not for its uses. Philosophies of science agree, in principle, with this perception of the ultimate purpose of science, but they disagree about the essential nature of the activity.

The many theories of the philosophy of science may be summarized in three broad categories: (1) a formal process of creating and testing hypotheses with empirical methods, (2) a social process by which professional scientists agree on what is "normal" or accepted science, and (3) a challenge to the ability of science to satisfy the expectations of rationality and objectivity. The hypothesis-testing view of science is well represented by Karl Popper, a distinguished Austrian philosopher who argued that scientific theories cannot be verified or proven, only refuted or disproved.[1] The search for truth is an energetic give-and-take that is based on the assumption that individually we only possess bits of knowledge, but that by collectively proposing hypotheses, testing them, refuting them, and formulating new ones, we can move closer to an understanding of the truth. He called this inductive process of conjecture and refutation critical rationalism and argued it could be applied to questions of metaphysics, as well as science, and that even abstract issues of ethics and morality could be rationally examined and criticized. His method involves three steps: (1) propose empirically testable theories and hypotheses that examine their main elements, (2) find data and arguments to refute them, and (3) propose refined theories and hypotheses and repeat the process. Popper argued that this process is at the heart of the search for knowledge.

Science as social interaction reflects the work of Thomas Kuhn, an American philosopher of science who emphasized the importance of the paradigms or worldviews that scientists rely on to shape their search for scientific knowledge.[2] Existing paradigms determine the scope of research and place constraints on the theories and hypotheses that are tested, the methods used, and the data collected. Real breakthroughs in scientific understanding occur when scientists become convinced that the prevailing paradigms are insufficient to explain observations and structure knowledge, and they devise a new paradigm. Paradigms facilitate communication among scientists and encourage the accumulation of knowledge, but can also limit inquiry and perpetuate outmoded ways of thinking. Breaking through existing paradigms to form new ones requires entrepreneurial energy and courage, and these efforts are often initially derided and rejected by most scientists. Science is not a neutral enterprise, but is largely shaped by convention and consensus.

A more critical view of science and the scientific method is found in the writings of Paul Feyerabend, also an Austrian, who was originally a colleague of Popper, but eventually became a great skeptic of scientific inquiry and research. Like other critical theorists who deconstruct scientific research, Feyerabend argued that no methodology for seeking truth was satisfactory.[3] He believed the great thinkers such as Galileo, who founded the modern scientific revolution and the scientific method, used rhetoric, propaganda, and other tricks to support their positions. Feyerabend argued that personal whims, social factors, and other nonscientific factors played a much more decisive role in scientific research than empirical data. The search for knowledge is so complex that no general rules can guide us. Feyerabend did not advocate anarchy in the search for truth or abandoning the effort altogether; rather, he saw the goal as freeing people from abstract concepts such as truth and reality that narrow their vision and limit their possibilities for understanding the world. The urge to explain things systematically, objectively, and scientifically is a strong one, he concluded, but one that should be resisted as impossible and inconsistent with human freedom and progress.

In principle, all knowledge is equal. In practice, differentiation among types of knowledge and disciplines is the product of scientific fashion, the vagaries of research funding, and the technological or other uses to which scientific knowledge may be applied. Disciplinary boundaries may obstruct open and effective discourse and prevent knowledge accumulation. Nowhere is this more evident than in the study of environmental issues in which the natural and social sciences are invariably interdependent. The essence of environmental issues is the interaction between human and ecological systems. The environment is only a "problem" when this interaction threatens the production of ecosystem services for the support of human society.

POLITICS

Science is about knowledge, but politics is about values: what is valued, how it is valued, and who gets valued things. Politics has been defined as the "authoritative allocation of valued things" through distributive conflict; as the process of deciding who gets what, when, and how; as conflict through which the participating groups define themselves and their interests; and as the deliberate and self-conscious creation of the structures and norms governing our collective lives.[4] Politics is the process of collectively selecting which values—valued things—to pursue and how to organize society to achieve them: collective wealth or equality or a clean environment; greater security through unilateral military strength or political and economic interdependence among states; centralization or decentralization of policy making; employer profitability or employee job security; social safety net or personal responsibility for health and pensions; and so on. Politics is an unending process because there is no absolute measure of values and all can be contested.

Political philosophy is the art of constructing "rational" value systems for ordering collective life. Libertarianism, for example, primarily values "self-ownership," which extends to material possessions owned: what is owned and who owns are essentially the same, and property is an extension of the person. This philosophy supports a "negative" freedom: freedom from external control and an absolute right to own and defend private property. Property ownership would be moral if the original distribution of property were consistent with treating all members of society equally and thereafter the property were justly conveyed. An original distribution is usually considered just if no member of society is worse off as a result of the initial unequal enclosure of property previously held in common for the benefit of the community. Any enclosure of common land meets this standard if "there is enough, and as good, left in common" for other members of the community, which was possible in seventeenth-century America, but is now usually impractical.[5] If, after the initial distribution of property, the market fairly compensated others through wages for their loss of property rights, current property rights would be moral. But this has not been the case because property rights themselves assign disproportionate market power to property owners. An initial enclosure of common land by force would breach this standard. In libertarianism, policy defends property rights and protects individuals from coercion by minimizing government and its rights.

Liberalism is built on a "positive" freedom: the freedom to determine oneself and pursue personal ambitions and dreams. This necessarily implies the ability to pursue personal goals (though not the guarantee of success) and, therefore, equal access to necessary resources such as education and employment. Inequalities in distribution of wealth, liberty, and opportunity are acceptable if they draw out useful energy and talents that benefit everyone and particularly the least advantaged.[6] As complete individual freedom breaks down the collective bonds that constitute community and society, some form of self-restraint is desirable. Adam Smith suggested that merely considering the ideas, interests, and needs of others might be sufficient.[7] The rights of individuals are of predominant concern, and individuals create values; collective values are in some sense collectively negotiated. Liberal policy defends and enhances individual rights and progressively enhances opportunity, especially for the least privileged.

In recent years liberalism has come under criticism as encouraging an individualism that is corrosive of the quality of collective life.[8] From this critique arises communitarianism, which places a higher value on the community than the individual. It argues that values derive not from individual, but collective, choices and commitments and that personal values reflect community values. Individuals are socially constructed—they receive their identity from membership in the community—and individual lives are given meaning and value in the community. This is the political theory of the common good, where the role of policy is to protect and enhance community life.

Once political philosophy and political science were synonymous. In order to pursue scientific study of political behavior, political science has borrowed various assumptions and methods from economics. The principal such assumption is rational self-interest. Rational choice theories approximate the conventional wisdom, at least in the United States: the political realm, like the market, is where self-interested individuals and groups compete for advantage. Unfortunately, if all actors in the process believe that the rules of the game reward selfish behavior, there is little incentive to cooperate for a collective benefit.

Political behavior is founded on values. Libertarians value freedom from outside influence on their bodies or property; liberals give precedence to the value of freedom to self-actualize; communitarians value the community as a creator of values, not necessarily over the individuals, but as a value in itself and for its members. Political science's rational choice assumes that interests—what an actor (an individual or group) wants—drive behavior without the actor examining where these interests come from. But interests are a subset of values: interests reflect a value of self-preference. While it may be commonly assumed that politics is the forum in which individuals and groups seek their personal interests, those interests need not be selfish or material. It is rational to choose to be altruistic, to help others, to donate labor to charity, if the reward in satisfaction is sufficiently valued. In environmental issues, some actors value self-interest and seek individual gain, others value nature and hope to regulate human invasions of ecosystems, and others value negotiation and compromise, collective agreement, and forestalling catastrophe. Politics is the forum in which collective values are formed out of myriad individual values.

SCIENCE, POLITICS, AND POLICY

Conventional wisdom and many policy textbooks assume that science drives policy by defining the problem and identifying alternative solutions from which policy makers select.[9] In this rational policy model, the policy process advances in an orderly fashion from problem definition to identification of alternatives to selection of an optimal policy. This is rarely the case and, because of scientific uncertainty and value complexity, it is almost never the case with environmental issues.

The process that led to banning substances that deplete the stratospheric ozone layer is often held up as a model for international environmental policy making.[10] Anthropogenic ozone depletion was first identified in two technical papers that hypothesized that chlorine emitted from human processes in various chemical compounds was absorbed by interaction with ozone in the upper atmosphere. But the ozone issue is not a case of science objectively defining problems. First, the estimates of ozone destruction varied widely and settled at much lower levels than originally projected. Second, in 1985 new empirical evidence called into question all the models of atmospheric physics and chemistry that had previously been relied on. Even when science thought it knew the problem, in reality it did not. Finally, data, even knowledge, is not the problem. A problem is something that needs to be solved. Even if science could define the rate of the destruction of stratospheric ozone, the nations of the world would still have to agree that it was a problem and define it in terms that were amenable to a politically practical solution. The international policy that regulates ozone-depleting substances was constructed over fifteen years of international negotiations (that continue today) with the aid of technological innovation, technology transfers, and development assistance to developing countries, and plenty of horse-trading.

Critics of the rational policy model argue that policy is much messier. Policy may occur almost accidentally, as a window of opportunity opens by the fortuitous confluence of three relatively independent streams of events: problems, policies, and politics.[11] More radical critics of the rational model argue that problems are never solved, but remain works in progress that act as focal points for interactions in the political community where maintaining relationships and interactions is more important than objectively defining and solving problems.[12] Politics is a continuing process, a way of life, not a list of rational solutions to definite problems.

All three principal models of international environmental policy making reflect this subtler and less orderly view of the use of science in politics.

Epistemic Communities

Since the mid-1980s, it has become fashionable in international environmental politics to model science and politics as so enmeshed that they can barely be distinguished.[13] This was the earliest of these approaches. An epistemic community usually comprises technical experts, often scientists, who interpret uncertain technical data through their personal and collective principled and causal beliefs. Their scientific status gives them an "authoritative claim to policy-relevant knowledge within [a] domain or issue-area." But their shared "set of normative and principled beliefs, which provide a value-based rationale for social action of community members" allows them to collectively construct knowledge out of uncertainty.[14] If community members have direct access to policy makers (as policy advisors, for example), their influence on policy is magnified. If several members of the community have political influence in states concerned with the environmental issue on which they advise, they can encourage international convergence of state policies around a single solution.

There are three significant problems with the epistemic community model. First, the model does not explain the linkage between membership in the community and influence over policy. Domestic policy making is more varied and complex than the model suggests. In some instances individuals, because of their unique talents or position, may significantly influence policy. But however influential a scientist may be, he or she shares the policy-making process with nonscientists, who also have personal beliefs that color their view of the world and drive their political behavior. If the beliefs and values of scientists can influence policy, the beliefs of policy makers can also. The epistemic community model may explain why some scientists act as they do, but it says little about the preferences and behaviors of political members of the policy-making cadre. Third, there is rarely any overt evidence of direct coordination of activities among groups of scientists toward a political goal. Where there is scientific uncertainty, an epistemic community is imputed from the parallel actions of several scientists who appear to have a common set of values. Such a community is more virtual than real, with more theoretical value than political influence.

Discursive Practices

This model builds upon and criticizes epistemic communities. Ideas in vogue have "disciplinary power," and even science is interpreted within a discursive context. The nature of environmental debate has changed radically since the early 1970s. As the discourse about the environment changed, what could be thought, what problems were observed and solutions conceived changed as well. The concept of environmental justice, for example, would have made no sense to scientists or policy makers forty years ago.

As used here, discourses are mixtures of facts and values in broad "sets of linguistic practices embedded in networks of social relations and tied to narratives about the construction of the world."[15] They conscribe and direct the discussion of science and politics within an overarching collection of ideas, norms, and beliefs. They determine "not only what can be said, but what can be thought," and so constitute a diffuse form of power through the interpretation of reality and the choice of social values and preferred social futures.[16]

Knowledge brokers help to construct discourses. They "serve as channels for discourse," framing knowledge and translating scientific information into policy-relevant knowledge for decision makers, "often clothing bare facts with social [and political] meaning." Knowledge brokers' ability to influence the policy process comes from the "plausibility of their interpretation, the loudness of their voices, and the political milieu [context] in which they act."[17]

But this approach also has its weaknesses. Although a discourse can be defined, specification of its constituents—values, ideas, individuals as representatives of organizations or as themselves, and so on—is highly subjective and may include both material and nonmaterial factors. For example, what is the relative importance of ideas, structures, and agents in discourse construction? How important is context, such as the existence of an unpredicted Antarctic ozone hole? Data that comes to hand may be pressed into service as (subjective) "evidence" of the existence or effect of a discourse. Second, links between science, the beliefs of individuals, and state policy can be plausible, but difficult or impossible to verify. Ideas are characterized as important causes, but can only be correlated post hoc with observed effects.[18] Finally, this approach demands a hermeneutical epistemology in which everything may be of importance. Divining what is important among myriad individuals, events, beliefs, and actions is as structured and "scientific," in the normal sense, as examining chickens' entrails to forecast the weather. The connections between science, politics, knowledge, and policy cannot be segregated and independently examined: everything affects everything and cause-and-effect relations are buried in a mass of detail.

Mutual Construction

Discursive constraints are more ethereal than the evident influence of institutions on the practice of science. The need for specific policy-relevant information sets research goals and influences methods, affects the interpretation of scientific knowledge and, thus, the range of potential policies. Science for policy produces a specialized type of knowledge outside the exclusive control of technical experts and directly or indirectly responsive to policy makers' needs. Demands by policy makers for scientific knowledge that is directly relevant to fashioning a policy (that, for example, predicts the future effect of alternate policies) encourage scientists to adapt their work to political concepts such as point estimates of future ecosystem states (e.g., stratospheric ozone concentrations) and policy efficiency.[19] But there is a trade-off: policy makers must simultaneously accept "a particular body of knowledge and methods as the only available option." Science becomes not the search for truth, but an artifact of material, institutional, and organizational structures scattered across the realm of "regulatory science."

This explanation of the interaction of science and politics also has shortcomings. First, the significant influence of regulatory science institutions is imputed, rather than directly demonstrated. Counterfactual arguments override empirical evidence.[20] Choice of a policy-relevant research goal does not guarantee "bad" science, but it may delay full understanding of the relevant ecosystem processes. Second, scientists usually design, lead, and define the rules of the institutions that direct science in the service of policy. This shows that scientists recognize that politics has specific information needs and is not evidence of a dominance of politics over scientific goals and methods. Third, the assumed dominance of material structures, institutions, and organizations understates the influence of individuals and ideas. In contrast with the epistemic communities model, the values and choices of individuals are notable by their absence, and in contrast with the discursive practice model, the force of ideas is overlooked. Finally, most international environmental issues create their own science. In such rapidly developing, new scientific areas, structures and procedures are appropriately malleable to increase opportunities for new ideas to stimulate research and knowledge.

Though these theories of the relationship between science and politics in international environmental issues diverge, they agree on a number of points. First, there is no singular, unidirectional relationship between science and politics. While science may first identify the issue, science and politics together usually frame the policy problem. Science and politics develop together, often in parallel, and interact recursively. The student of politics would freely accept this almost dialectical relationship; most natural scientists would probably reject it.

Second, not only rational interests or objective rules influence science and politics. Ideas, individuals and their personal preferences, groups and their values (including self-interest), the structures of processes and institutions, accidental events, and changes in collective expectations also affect them. Science and politics are more complex and less monolithic than commonly supposed; when they interact, this complexity is only multiplied.

Third, science and politics, individually and in their interaction, are affected by context. The science and politics of each environmental issue are different. The actors, issues, values and interests challenged, scientific disciplines concerned, and knowledge available vary widely. Cultural differences in trust of science and respect for politics may, therefore, have an effect. Although science is an international activity with standardized rules of conduct, countries differ in their reliance on scientific data for policy making.

The complexity of interaction between science and politics means that no single theory about science, politics, or their interaction is likely to explain satisfactorily the full range of observed events or the outcome of the process. However, practicing application of the various theories outlined above can achieve two goals. First, the strengths and weaknesses of the theories will become clear. Second, building from their strengths, it should encourage creative modification of these theories. How would stronger theorizing about domestic policy formation improve epistemic community theory? How could the discursive practices approach, in which ideas are conditioned by the discursive structure of the moment, be improved by elaboration of the psychology of independent thought?[21] In the concluding chapter, we isolate the most important factors to be considered in exploring how science and politics interact in international environmental issues and how that interplay influences policy. We offer three new theories or analytical approaches that are tentatively supported by the evidence in the case studies.

PLAN OF THE BOOK

The book is divided into five parts, each of which collects two or more cases that have salient factors in common. The first part compares two regional issues: the desert ecosystems that straddle the United States–Mexico border and food-security issues in Europe. Both sets of issues must be considered within the wider context of relations between neighboring countries and regional arrangements.

Brusca and Bryner examine the scientific and political cooperation between the United States and Mexico over the desert ecosystems that straddle their border. They detail the damage from overuse of fragile desert and marine ecosystems and the cultural and political complexities of organizing effective cooperation in their protection. Contrary to conventional wisdom, the centralized government of Mexico, which historically has been less than sympathetic to the environment and indigenous people, adopted bioreserve protection before its usually more environmentally aware neighbor. This case opens up many questions about the interaction of local, national, and transnational politics, the potential influence of local community organizations and international nongovernmental organizations (NGOs), and the possibilities for environmental preservation among rapid economic growth.

European food security was much in the news in the 1990s as bovine spongiform encephalopathy (BSE), a new disease, crossed the species barrier from cattle to humans and the United States began exporting large quantities of genetically modified foods to Europe. The issue of bovine hormones arrived on the political agenda at almost the same time as BSE, but was handled very differently. In chapter 3 Brown tells a lurid tale of bureaucratic obstruction, ideology, and institutional blindness that allowed a new disease to terrorize a nation. Many of the same actors played a very different game when challenged by U.S. companies wanting to sell hormone-enhanced milk and beef in the EU.

The second part addresses global environmental issues through two studies of the case of climate change. In chapter 4 Soroos recounts the scientific and political issues that make this issue uniquely complicated. His chronology of the scientific and political developments is an excellent introduction to climate change and raises many questions about which the public is generally unaware.

The next chapter looks in more detail at the political response to major developments in science. Harrison shows that science has failed to communicate the dangers of climate change. He also shows that scientists have struggled to make their reports useful to policy makers, but that political actors have taken progressively less account of them. He concludes the chapter with an analysis of common explanations and theories of the climate-change policy process and finds them all inadequate.

The third part comprises two very different case studies connected by fragmented science and deaf policy makers that together defeated the precautionary principle. Chapter 6 tells the story of uncoordinated research into acid deposition in Ontario, the results of which were ignored by provincial bureaucrats and failed to capture public attention. Munton closely follows the many twists and turns that only culminated in stirrings of public interest in acid deposition when Toronto's vacation lakes seemed threatened. Research implicated U.S. sources in the damage, and the provincial environment minister rapidly changed policy and authorized substantial further research. Acid deposition had been observed in Ontario for over seventy years by the time Canada and the United States agreed on regulations to reduce acid rain.

In chapter 7 Dimitrov shows how the international system failed to create an international institution to globally regulate forests. Despite much research and extensive discussion of a global treaty, deforestation remains a national preserve. Countries with equatorial forests wished to preserve their sovereign right to resource exploitation. Developed countries were unwilling to fund an effective institution, and many wished to exploit their own forest resources without let or hindrance from international oversight. Both developed and developing countries focused on international effects ignoring the extensive data available on the local and international causes of deforestation.

The fourth part compares four very different cases in which ideas and culture dominated scientific method and directed interpretation of science or the design of international treaties. In chapter 8 Wilkening compares the scientific and political responses in Europe, North America, and Japan. In each region the science and politics progressed at different rates and in distinct ways. Pressed by the Nordic countries and then by Germany, Europe acted first, while Canada ignored the science and the United States ignored the issue. Japan's late, but surprisingly rapid, response resulted from culturally determined links between science and politics not seen elsewhere.

Firestone and Polacheck's case study in chapter 9 of the Southern Bluefin Tuna follows efforts to block Japan's use of rules for experimental fishing to increase its catch beyond limits imposed by a regional agreement. Examination of the scientific and legal maneuverings reveals substantial political content and raises questions about the possible effectiveness of international agreements to govern common-pool resources. The case study also shows how sovereign countries may interpret detailed rules to their advantage, particularly if provisions in regional agreements clash with those in international agreements.

The case in chapter 10 revolves around the small town of Ames, Iowa, USA. This is an example of the cause side of the international equation, because the town's incinerator was charged with creating a plume of dioxin that threatened the health of indigenous groups in Arctic Canada. Public interest plays a peripheral role or is discussed in several cases, but here the public is central to the debate about arcane science and engineering. Although NGOs played a minimal role, this study's conclusion on how the public understands and makes use of scientific information suggests ways in which activists might pressure governments to act on environmental science.

The final case study by Eflin in chapter 11 compares ecosystem-management in the Great Lakes and the Baltic. Although the ecosystem-management and international political challenges are comparable, scientific research and institutional responses differ. The Baltic approach has been more effective because of more "holistic" thinking by the participants, a result of closer cultural and political ties.

Our concluding chapter invites readers to draw and debate conclusions about the case studies and their implications for understanding the interplay of science and politics in international environmental policy. We suggest several possible common themes from the case studies to illustrate some possible avenues for further discussion. Individually and collectively, the case studies demonstrate the weaknesses in current thinking about the relationship between science and politics in international environmental issues. We show that current models of the interplay between science and politics are too limited, and we suggest three alternate ways of thinking about and researching this relationship. We believe that the findings in the case studies and their analysis, together with the proposed models we present in this final chapter, will stimulate better theory building by both students and scholars alike and eventually may enlighten policy makers.

NOTES

1. Karl Popper, *The Logic of Scientific Discovery*, 14th ed. (London: Routledge, 1992).
2. Thomas Kuhn, *The Structure of Scientific Revolutions* (Chicago: University of Chicago Press, 1962).
3. Paul Feyerabend, *Against Method*, rev. ed. (London: Verso, 1988).
4. David Easton, *The Political System* (New York: Knopf, 1953); Carl Schmitt, *The Concept of the Political* (New Brunswick: Rutgers University Press, 1976); Hannah Arendt, *The Human Condition* (Chicago: University of Chicago Press, 1958).
5. The quote is from John Locke, *Second Treatise of Government*, C. B. Macpherson, ed. (Indianapolis: Hackett, 1980): 19. Locke's ideas provide a foundation for libertarian philosophy.
6. The basic rule paraphrased here is from John Rawls, *A Theory of Justice* (Cambridge, Mass.: The Belknap Press, 1971). For an accessible review of liberalism and other political philosophies, see Will Kymlicka, *Contemporary Political Philosophy: An Introduction* (Oxford: Clarendon Press, 1990).
7. Adam Smith, *The Theory of Moral Sentiments*, D. D. Raphael and A. L. Macfie, eds. (Oxford: Clarendon Press, 1976). Smith saw this work as more fundamental than his work on the wealth of nations, which is considered the foundation of classical economics.
8. Amitai Etzioni, ed., *Rights and the Common Good: The Communitarian Perspective* (New York: St. Martin's Press, 1995).
9. See, for example, Garry D. Brewer and Peter deLeon, *The Foundations of Policy Analysis* (Homewood, Ill.: Dorsey Press, 1983).
10. Richard Elliot Benedick, *Ozone Diplomacy: New Directions in Safeguarding the Planet*, enlarged ed. (Cambridge, Mass.: Harvard University Press, 1998).
11. John W. Kingdon, *Agendas, Alternatives, and Public Policies* (Boston: Little, Brown and Company, 1984).
12. Deborah A. Stone, *Policy Paradox and Political Reason* (Glenview, Ill.: Scott, Foresman and Company, 1988).
13. A similar development can be seen in "science studies," which contend that the believed separation of science from society, that science is "above politics," is a mirage that misdirects Western society. For example, see Bruno Latour, *Pandora's Hope: Essays on the Reality of Science Studies* (Cambridge, Mass.: Harvard University Press, 1999); and *We Have Never Been Modern* (Cambridge, Mass.: Harvard University Press, 1993).

14. Peter M. Haas, "Introduction: Epistemic Communities and International Coordination," *International Organization* 46(1) (Winter 1992): 1–35.

15. Karen T. Litfin, "Framing Science: Precautionary Discourse and the Ozone Treaties," *Millennium: Journal of International Studies* 24(2) (1995): 251–77.

16. Karen T. Litfin, *Ozone Discourses: Science and Politics in Global Environmental Cooperation* (New York: Columbia University Press, 1994): 8.

17. Litfin, "Framing Science."

18. Thus, observed changes in scientific or political discourse are assumed to emanate from new ideas.

19. Simon Shackley and Brian Wynne, "Global Climate Change: Mutual Construction of an Emergent Science-Policy Domain," *Science and Public Policy* 22(4) (August 1995): 218–30.

20. In contrast to explanations in which A is believed to cause B, a counterfactual says that if A had not occurred, B would not have happened.

21. This theory is based on the work of Michel Foucault, a French historian whose ideas borrow from individual and social psychology. How does the structure of discourses constrain individual ideas? What is the psychology of change in discursive structures? How can new ideas be thought? For an introduction to Foucault's writings see Graham Burchell, Colin Gordon, and Peter Miller, eds., *The Foucault Effect: Studies in Governmentality* (Chicago: The University of Chicago Press, 1991).

I

REGIONAL ISSUES

Pollution between countries caused the earliest international environmental issues. In 1938 the Trail Smelter Arbitration established that a country is responsible for environmental harms to foreign countries from activities within its territory. Examples of these types of events in this volume are the chapters on acid rain in Ontario and the presumed drift of dioxin from the northern United States to Arctic Canada. In this part the chapter by Brusca and Bryner looks at the science and politics of the Sonoran Desert, intersected by the international border between the United States and Mexico. However, as here, many apparently bilateral issues are often part of a regional problem. Protection of the Sonoran Desert ecosystem is embroiled not only with the politics and culture of each country, but also with the politics of the border, the North American Free Trade Agreement (NAFTA), and other regional regimes. While the issue is scientifically local, a desert ecosystem, in political terms, it is international and regional.

Intuitively, we may expect that when two countries are jointly responsible for a single ecosystem without significant economic benefit, they may cooperate on the scientific research to understand it and regulate access to preserve it. But as is so often the case in environmental issues, the science and politics of the Sonoran Desert have run afoul of economic and cultural differences between the participants—abiding and largely unrelated historical issues, beliefs and expectations, and institutional rigidities. Chapter 2 relates the science and politics that led to the formation of a huge biosphere reserve in Mexico that the United States subsequently matched. Brusca and Bryner detail the widespread damage to ecosystems as diverse as deserts and fisheries from multiple causes, including population increase, industrialization, and crime. They then track how the Mexican government, not known for its environmental activism and its concern for peasant welfare, responded to grassroots activism when prompted by several NGOs. The United States reluctantly protected its share of the cross-border desert ecosystem, but continues to fail to supply the Colorado River water needed to return the ecosystem to any semblance of its former glory and that it is obligated to provide.

The case generates many questions. How did changes in scientific methods and practices create a perception out of multiple local environmental challenges that a whole river basin ecosystem should be protected? Was accumulating scientific evidence of the importance of the Colorado River the critical factor that created the perception that a whole ecosystem should be protected? What do we learn about the environmental efficacy of democracy from the different effects on reserve formation of Mexico's centralized one-party political system and the United States's decentralized federal system and culture of legal activism? How did NAFTA and the "environmental side agreement" influence Mexican policy? What role did local organizations, grassroots groups, nongovernmental organizations, and others play in the formation of the reserves? Given the importance of economic activity in the region, how did the goals of preserving ecosystem health and economic activity interact? One question that this case cannot answer, but that remains critically important, is: How successfully will the Mexican biosphere reserve restore ecosystem health if states in the U.S. Southwest do not renegotiate the division of the Colorado River flow? Another question from this case should be kept in mind when reading the other cases: how could the grassroots activism (stimulated by U.S. and Mexican NGOs) seen in this case have helped effective policy making?

The second case in this part is Brown's review of the convoluted science and politics of BSE and BH. At first BSE was a scientific curiosity that seemed only to exist in the UK. As it became apparent that the disease could cross species boundaries, it quickly became a major food-security issue. Finally, it was a political crisis as the UK government was accused of a life-threatening cover-up and the European Commission of failing to appropriately research and monitor the problem. Other countries in Europe denied that the problem could arise within their borders, but many EU member countries have now reluctantly admitted to some cases and the rate of incidence in these and other countries is probably higher than national governments are yet prepared to admit. BSE became an EU issue through the export of British feed to other countries, an artifact of open borders within the EU, the Common Agricultural Policy (CAP), and other supranational institutions. In effect, labor-related regulation in one country threatened the health of people in other countries.

Scientifically, BSE is a challenge to accepted knowledge in veterinary science and its research methods. Did scientists act as quickly as they should have, or were they lulled into inaction by assumptions about the development and transmission of animal diseases? Or were they prevented by bureaucratic politics from performing essential research? How should science approach a novel issue that has potential for substantial economic damage and harm to humans? Should it seek to fully understand a novel phenomenon or address only its direct and indirect effects on humans? Did scientists promptly and fully communicate their findings to the right audience, and if they did not, were they gagged or did they choose silence?

BSE also laid bare the failings of political funding for animal- and food-related research, the dangers of noninterventionist free-market politics, and the myth of civil service protection of a "public interest." If the precautionary principle is appropriate for BH (whose effects on human health also were poorly understood), should it not have been applied immediately to BSE? What would have been the economic effect of acting with precaution on BSE? Was the response to BSE aided or hindered by the institutions of the EU (that was founded to increase cooperation between its members and pool resources on important issues)?

This case suggests many other questions. Are food-security issues more emotive and politicized than other policy areas? If so, why? How much did each of the following factors affect policy outcomes in BSE and BH: scientific uncertainty, divergent issue framing, the power of agrarian interests, bureaucratic factors, political ideological factors, cultural factors, legal principles, multilevel governance structures, human foibles? What is the appropriate role of scientists and scientific information in public policy making? Was the BSE case a particularly rare and extreme issue that placed UK and EU governance structures under unusual stress? Is it an outlier that should be ignored when considering environmental policy making and the relations between science and politics? The EU responded to its policy failures with some organizational reform. Is it likely that these reforms will adequately guarantee food security in the EU, and if not, why not? Based on their record in other food-security issues like *E coli*, BH, and salmonella, how would the United States and other governments have reacted to a BSE-like crisis?

2

A Case Study of Two Mexican Biosphere Reserves

The Upper Gulf of California and Colorado River Delta and the El Pinacate and Gran Desierto de Altar Biosphere Reserves

Richard C. Brusca and Gary C. Bryner

The Sonoran Desert straddles the Mexico–United States border and encompasses southern Arizona and most of Baja California and the Gulf of California (Sea of Cortez). The Sonoran Desert forms a diverse landscape of terrestrial, freshwater, and marine ecosystems. It is the most tropical of North America's four great deserts and includes desert scrub and grassland, riparian habitats, marine and coastal ecosystems, and patches of tropical deciduous forest that penetrate from the south. This desert covers 120,000 square miles (330,000 sq km), not counting the 109,961-square-mile (283,000 sq km) Gulf of California. The northern gulf is a unique body of water in many ways. High nutrient levels, shallow waters, and strong tidal mixing combine to make it one of the most productive marine regions in the world.[1]

The region's drylands and maritime habitats are tightly linked both ecologically and economically. Originating in the Rocky Mountains, the Colorado River's 1,700 mile (2,800 km) journey to the Gulf of California is most powerfully felt in the Sonoran Desert, where the river winds like a lifeline, matriarch to the ecological and human history of the American Southwest, draining a basin of 244,000 square miles and supplying water to 30 million people, irrigating 3.7 million acres (1.5 million hectares) of farmland, and linking the desert and gulf. The ecological integrity of the Sonoran Desert and the Gulf of California are threatened by a number of factors, including the reliability of Colorado River water, groundwater overdraft, habitat conversion, pressure from fisheries, unsustainable agriculture and ranching, the introduction of exotic species, and narcotrafficking and related military activities. About 60 percent of the native vegetation of the Sonoran Desert has been converted or destroyed, for example, and nearly all of its rivers have been diverted or dried over the past century.[2]

In response to these threats, nearly three million hectares on both sides of the Arizona-Sonora border have been designated as national parks and monuments, national wildlife refuges, and biosphere reserves. Two biosphere reserves, the Upper Gulf of California and Colorado River Delta Biosphere Reserve and the El Pinacate and Gran Desierto de Altar Biosphere Reserve, which were established in Mexico in 1993, provide protection for over half of this land area, 1.66 million hectares. This chapter discusses the creation and operation of these reserves and the interaction of scientific research and politics in establishing them.

Several questions are at the heart of an assessment of the evolution of science and policy and their interaction in this case. How did knowledge about the threats to the ecosystem evolve, and how did it lead to political action? How has the Mexico–United States border location of the reserves affected their creation and management? How has domestic politics affected international commitments? Given the importance of economic activity in the region, how did the goals of preserving ecosystem health and economic activity interact? What role did local organizations, grassroots groups, nongovernmental organizations, and others play in the formation of the reserves? How successful has the creation of the reserves been in helping to restore and maintain the ecological and societal well-being of the region?

SCIENTIFIC ISSUES SURROUNDING BIOSPHERE RESERVES

The delta of the Colorado River was once one of the richest areas of biodiversity in North America, supporting a marshland and estuary system of 1.9 million acres.[3] Early authors describe an abundance of condors, pronghorns, and bobcats in the delta.[4] Today, almost all of the rich riparian systems (and the large mammals) of the Lower Colorado River and its delta are gone, and the delta has been transformed from a lush subtropical wilderness to a moonscape of sun-baked mud. The twenty dams and thousands of miles of canals, levies, and dikes have converted the river into a highly controlled plumbing system in which virtually every drop is managed. Before the filling of the Hoover Dam (creating Lake Mead), the delta experienced a perennial discharge from the Colorado River. By the time Glen Canyon Dam was completed in 1962, regular input of Colorado River water to the delta and upper gulf had completely ceased. For twenty years after completion of Glen Canyon Dam, as Lake Powell filled, virtually no water from the river reached the sea. In 1968, flow readings at the southernmost measuring station on the river were discontinued, since there was nothing to measure.

Due to the greatly reduced freshwater flow into the delta, the powerful tides of the Gulf of California began to overwhelm the river channel. Today, during spring tides, seawater creates an estuarine basin thirty to thirty-five miles up river, killing most of the freshwater flora and fauna that used to live along the lowermost river corridor.[5] The spectacular El Niño of 1983–1984 brought several years of heavy precipitation that could not be contained by the twenty dams along the Colorado River. During these years, approximately sixteen million acre-feet of water flooded the delta, and the Colorado River once again reached the Gulf of California.[6] Although this flood was a disaster for the urban and agricultural economy of the Mexicali Valley, it was highly beneficial for the native vegetation and fauna, the return of which was remarkable. As a result of these floods, four freshwater/brackish wetland regions were recreated and continue to persist in the delta.

Even though little water now crosses the border into Mexico, the sixty-mile stretch of the river's channel that runs from Morelos Dam to the Gulf of California maintains five times more native wetlands and natural riparian habitat than does the entire Lower Colorado River in the United States. The only true Sonoran Desert wetland left north of the border is the Bill Williams River at Lake Havasu, and the Colorado River Delta is now the last small refuge for Lower Colorado River riparian gallery forest habitat.

Water has always been the ultimate limiting factor in the Sonoran Desert. The region of the biosphere reserves addressed in this chapter lies in the Lower Colorado River Valley region, which is the hottest and driest subdivision of the Sonoran Desert. This region experiences summer highs that may exceed 120°F, with soil surface temperatures up to 180°F. Annual rainfall in this region is less than three inches, with up to three years passing with no rainfall at all in some areas. The rich riparian forests that once crisscrossed the Southwest, including the state of Sonora, are now rare, due to urban growth, agriculture, ranching, surface water diversion, and overpumping of groundwater. The development of powerful mechanized pumps in the 1920s led to massive groundwater overdraft in agricultural areas throughout the Sonoran Desert. Today the overdraft in this region averages 1.25 million acre-feet per year.[7] The Costa de Hermosillo (Sonora) irrigation district peaked at 887 pump-powered wells supplying groundwater to more than 100,000 hectares of irrigated land, exceeding recharge rates by 250 percent. Due to plummeting water tables and saltwater intrusion, millions of acres of farmland have been permanently abandoned throughout the coastal plains of Sonora and southwestern Arizona.

Despite these reductions in freshwater input, the Gulf of California fauna is highly diverse, comprising 5,968 named and described macrofaunal species (i.e., animals larger than a few millimeters in size). Due to the presence of many undescribed invertebrate species, especially in poorly studied taxa (e.g., sponges, tunicates, copepods, planktonic species), this is estimated to be less than 50 percent of the actual animal diversity of the gulf.[8] The northern gulf (from the Colorado River Delta to, and including, the Midriff Islands of Angel de la Guarda and Tiburon) is home to nearly half of this marine life (2,258 macroinvertebrates and 544 vertebrates), including 367 fishes, 7 marine reptiles, 146 aquatic birds, and 24 marine mammals. At least 142 of these northern gulf species are endemic solely to this region.[9]

Islands and coastal wetlands of the gulf are critical habitats for water birds. At least 213 species of birds have been recorded from the delta region alone; 102 of these are waterfowl, and 30 species are known to breed on the delta. At least seventy-four species use the delta as a migratory stopover or winter ground.[10] Included among these is the endangered southwestern willow flycatcher. The two sand islands at the mouth of the delta (Islas Montague and Pelicano) alone provide habitat for forty-two species of shore birds numbering between one and two hundred thousand individuals, and at least eleven bird species breed on Isla Montague. Of the land birds, most are neotropical migrants that use the delta as a stopover. About 163,000 wintering shorebirds use the extensive mudflats of the delta annually.[11] The western population of North America's white pelican, which has been on the decline for many decades, relies on the delta as a migratory stopover. The endangered totoaba and the currently overfished gulf corvina both occur in the delta region, as does the endangered gulf miniature porpoise, or vaquita.

The terrestrial ecosystems of the Sonoran Desert also sustain an extraordinarily high biotic diversity. Five thousand species of vascular plants are reported from the state of Sonora alone—20 percent of Mexico's total flora, in an area comprising less than 10 percent of the country.[12] Other biotic groups are equally rich in the Sonoran Desert: 150 butterfly species, 1,200 moths, 17 hummingbirds, 160 mammals, and 500 birds (roughly half the known number of birds present in the continental United States or in all of Mexico). At least 96 species of reptiles are endemic to the Sonoran Desert, and there are 552 recorded endemic plants in Baja California alone. The deciduous riparian gallery forest of the Sonoran Desert may have the highest breeding-bird densities on the continent, harboring 304 to 847 breeding pairs per 40 hectares.[13]

The spectacular San Pedro riparian corridor, perhaps the healthiest remaining riparian area in the Sonoran Desert, is home to an estimated 400 species of birds, 83 mammal species, and 47 reptile and amphibian species. The "sky islands" of southeastern Arizona and adjacent Sonora are recognized by the International Union for the Conservation of Nature (IUCN)

Box 2.1 Macrofaunal (Animal) Diversity in the Gulf of California

MARINE SPECIES DIVERSITY IN THE GULF OF CALIFORNIA

Invertebrates (excluding copepods and ostracods): 4,853 species (767 endemic to the gulf)
Vertebrates: 1,115 species
  Fishes: 891 species (87 endemic to the gulf)
  Aquatic birds: 181 species (2 essentially endemic to the gulf)
  Marine mammals: 36 species (2 endemic to the gulf)
  Marine reptiles: 7 species (none endemic to the gulf)
Total: 5,968 species (858, or 14.4 percent, endemic to the gulf)

MARINE SPECIES DIVERSITY IN THE NORTHERN GULF OF CALIFORNIA

Invertebrates (excluding copepods and ostracods): 2,258 species (128 endemic to the northern gulf)
Vertebrates: 544 species
  Fishes: 367 species (13 endemic to the northern gulf)
  Aquatic birds: 146 species (none endemic to the northern gulf)
  Marine mammals: 24 species (1 endemic to the northern gulf)
  Marine reptiles: 7 species (none endemic to the northern gulf)
Total: 2,802 species (142, or 5.1 percent, endemic to the northern gulf)

Source: Brusca et al., "Macrofaunal Biodiversity in the Gulf of California (Sea of Cortez)"; Findley et al., "Macrofauna del Golfo de California [Macrofauna of the Gulf of California]."

as one of the great centers of plant diversity north of the tropics. Mexico itself is one of the twelve megadiversity countries (as defined by Conservation International), harboring 10 percent of the planet's biological diversity on only 1.3 percent of the world's land area. The Lower Colorado River supports 24 federally listed (U.S.) endangered/threatened fish species and 67 "at risk" fish species.[14]

Political battles are fought in this region over water and wetlands use, fishing, farming, cattle grazing, and loss of biodiversity. Overfishing in the gulf threatens marine life and has contributed to the destruction of offshore habitats. Most fisheries in the gulf are either exploited to or nearly to their maximum capacity, or they are overexploited. In populated coastal regions, intertidal communities have been decimated by the removal of animals by tourists and visitors and by locals who collect for the "curio trade." Over-

Box 2.2 The Upper Gulf/Delta and Pinacate Biosphere Reserves

The Upper Gulf of California/Colorado River Delta and the El Pinacate/Gran Desierto de Altar Biosphere Reserves lie within the Lower Colorado River Valley subdivision of the Sonoran Desert—the largest, hottest, and driest area of all the Sonoran Desert.

THE UPPER GULF OF CALIFORNIA/COLORADO DELTA BIOSPHERE RESERVE

The reserve is inhabited by 1,048 species of invertebrates (excluding copepods and ostracods), 43 of which are endemic to the reserve, and 379 species of vertebrates (none endemic to the reserve). The reserve also contains 15 species of marine mammals, 230 species of marine fishes, 230 species of fish, 131 species of aquatic birds, and 5 species of marine reptiles. These figures, however, are probably not a fully reliable indication of the gulf's biodiversity. Many species inhabiting the Gulf of California are poorly documented, and the biodiversity of the reserve remains virtually unexplored.

THE PINACATE/GRAN DESIERTO DE ALTAR BIOSPHERE RESERVE

In the Pinacate reserve, sand soils predominate, including the Gran Desierto, the largest dune field in the New World, covering 500,000 ha from the Algodones Dune Fields northwest of Yuma, Arizona, to the great dune fields of El Pinacate and Bahía Adair. The reserve is also home to 200,000 ha of spectacular volcanic formations, with impressive lava flows, one of the greatest concentrations in the world of giant Maar-type craters, and close to 400 cinder cones. Despite the harsh environment, the Pinacate reserve possesses a high biodiversity: 506 species of vascular plants, 38 species of mammals, 237 birds, 44 reptiles, 4 amphibians, and 2 native freshwater fishes have been catalogued from the reserve.

Source: N. Lancaster, R. Greeley and P. R. Christensen, "Dunes of the Gran Desierto Sand-Sea, Sonora, Mexico," *Earth Surface Processes and Landforms* 12 (1987): 277–88; R. S. U. Smith, "Sand Dunes in the North American Deserts," in G. L. Bender, ed., *Reference Handbook of the Deserts of North America* (Westport, Conn.: Greenwood Press, 1982): 481–526.

grazing by cattle has been a serious problem in much of Sonora and Arizona, resulting in degraded riparian areas, increased soil erosion, and the spread of exotic grasses. Ranching experienced a substantial growth in the Pinacate biosphere reserve region until the 1990s. However, the poor quality of this area for ranching is apparent. Nearly all of the permanently or temporarily inhabited farms in the Pinacate had at least a few cattle. The stock range of the Pinacate reserve is low quality, requiring twenty to thirty hectares per animal. When forage is scarce, native plants are consumed, endangering the indigenous flora—overgrazing is evident in some areas. The recent introduction of European and African exotic grasses that out-compete native plants and spread fire is a major threat to the entire region.[15]

Extreme groundwater overpumping is also occurring in the Sonoran Desert. In the Tucson and Phoenix Basins today, groundwater pumpage averages about four times the natural recharge rate.[16] Overall, the overdraft in the Sonoran Desert region averages 1.25 million acre-feet per year.[17] Due to plummeting water tables and salinization, millions of acres of farmland have been permanently abandoned throughout Sonora, the Colorado River Delta, and southwestern Arizona. Overpumping leads to declines in riparian zone water levels, which eliminates key species such as rushes, willows, and cottonwood, allowing mesquite and salt cedar to expand and creating a "desertification" of riparian areas.[18] Regional extirpation has been documented for 36 of the 82 breeding-bird species that formerly used riparian woodlands in the Sonoran Desert.[19] Conservation International estimates that as much as 60 percent of the Sonoran Desert is no longer covered with native vegetation, but is dominated by more than 380 alien species introduced by humans and their livestock. Secondary effects of groundwater overdraft include soil compaction, land subsidence, and fissuring. In some parts of Arizona and Sonora, land has subsided by one meter, causing fissures that extend for more than half a mile. In some areas, subsidence and fissuring are so extensive that roads have been closed for fear that vehicles would disappear into crevices.[20]

THE SOCIAL, POLITICAL, AND ECONOMIC CONTEXT

Most of the Colorado River Delta lies in the state of Baja California. The delta region includes the Mexicali Valley, most of the Colorado River mainstream from Morelos Dam to the gulf, the Río Hardy and its floodplain, the Sierra Cucapá Range, and the Laguna Salada Basin. This region also includes the two border cities of Mexicali (with nearly 1 million people) and San Luis Colorado (160,000 people), as well as many small towns and communal land arrangements called *ejidos*. In the heart of the delta is the Cucapá Indian communal land, which includes both the Sierra Cucapá and the Laguna Salada floodplain.

In 1904, the "Colorado River Land Company" was formed by six Americans. The company came to own some of the best agricultural land on the delta, rich alluvial soils derived from the river's seasonal flood history. The Mexican government gave the company permission to exploit virtually all of the resources of the delta and the Mexicali Valley. The company brought in illegal immigrants from China to work the land. In 1937, an insurrection movement born in the Mexicali Valley resulted in the expulsion of the company and the distribution of land and water rights among the *ejidos*. By the end of 1937, 44 *ejidos* had been formed in the Mexicali Valley, and 144,000 hectares were allocated to the new communities.[21]

During the years of occupation by Americans and Mexicans, the native Cucapá people that stayed on the Mexican side of the border were never considered in any land or water decisions, despite the fact that they were the original owners of the Mexicali Valley. As the Cucapá (or "River People") lost their lands (and the reliable Colorado River summer floods), they started to segregate and nearly vanished from the delta. Today there are only two to three hundred Cucapá living in the area; most of them live in the border cities, but a few of them (about eighty-five) are still living on the river banks, especially in the settlement of El Mayor.

The El Pinacate and Gran Desierto de Altar Biosphere Reserve has only a few small scattered settlements and between one and two hundred permanent residents. Population density in the reserve is very low (about 0.01–0.02 person/square kilometer). The lack of local employment forces residents to search for work in the nearby cities of Puerto Peñasco, Sonoyta, and San Luis Río Colorado. Basic services are generally lacking, as is access to necessary goods such as food and clothing. The majority of the inhabitants of the reserve are men, whose families choose to live in one of the cities mentioned above. Settlements are concentrated in the northeast region of the buffer zone, near the town of Sonoyta, primarily along two major highways.

Today, the Southwest is dominated by people who have lived in the region for less than a decade. The population of the Sonoran Desert doubled between 1970 and 1990, to nearly 7 million.[22] By the late 1970s, the deltas of all Sonoran rivers had been almost entirely converted to agriculture, and over a million hectares of mesquite, cottonwood, and willow riparian forest and coastal thorn scrub had been destroyed.[23] The progressive salinization of aquifers and increasing cost of water extraction have caused a decrease in land devoted to crop production, and today only about half the cleared land is still used. The rest lies bare with little vegetative cover.

Approximately 23 million people live in the Lower Colorado Basin today and are largely dependent upon water from the Colorado River. By 2020 it is estimated that more than 38 million people will be living in the Lower

Colorado Basin (table 2.1). This is the fastest growth and most massive land conversion in North America's history. There are no signs that this growth is tapering off, although dwindling water supplies may soon begin to slow it down.

U.S. urban communities are heavily dependent on Colorado River water. However, most (about 80 percent) of the Colorado River's water reaching Mexico is used for agriculture (table 2.2). The Mexicali Valley, at the northern end of the modern delta, is one of the most important agricultural regions in Mexico. It maintains 3,000 kilometers of irrigation canals and 17 agricultural drains discharging into the delta. However, emerging cities, especially along the border, are demanding more water for their growing urban needs, and it is probable that the percentage of Colorado River water that goes to agriculture will diminish significantly in coming years.

The Gulf of California is one of the most productive marine ecosystems in the world, and it has supplied the Mexican, U.S., and Asian markets with abundant fish resources throughout the twentieth century. Mexico is the world's sixth largest fish producer. Since the early 1900s, three coastal fishing communities have become established in the northern gulf: San Felipe in Baja California and El Golfo de Santa Clara and Puerto Peñasco in Sonora. By the 1960s, Mexico's offshore shrimp trawling fleet had become the nation's most important fishing sector, politically and economically.[24] Until the late 1980s, shrimp provided the main source of income for the fishing sector in

Table 2.1 Population Projections in the Lower Colorado River Basin

Area	*Year 1990*	*Year 2020*	*Percent Increase*
Arizona	3,665,000	6,980,000	90
Southern California	16,757,000	26,318,000	57
Southern Nevada	800,000	1,630,000	104
Mexico (using Colorado River water)	1,700,000	3,240,000	91
Lower Colorado River	22,922,000	38,168,000	67

Source: Morrison et al., *The Sustainable Use of Water in the Lower Colorado River Basin.*

Table 2.2 Water Use by Sector for the Lower Basin in the United States and Mexico (%)

Use	*Arizona*	*Southern California*	*Southern Nevada*	*Mexico*	*Regional Total*
Urban	24	44	85	7	33
Agriculture/livestock	76	48	15	93	64
Environmental/other	0	8	0.2	0	4

Source: Morrison et al., *The Sustainable Use of Water in the Lower Colorado River Basin.*

these communities and was the catalyst for their growth. In the past two decades, there have been dramatic changes in the targeted species, and as marine-resource exploitation has grown, the number of species harvested has steadily increased.[25] Today the waters of the Gulf of California supply 40 percent of Mexico's total fisheries, and the northern gulf itself provides 15 percent of the national fisheries. Small-scale ("artisanal") fisheries in the gulf contribute more than 25 percent of the national fisheries production of Mexico, and this percentage is growing rapidly. Approximately 150,000 jobs are generated indirectly from small-scale fishery activities. Recent studies estimate that as many as 30,000 fishers (industrial and small-scale) work in the Gulf of California.[26]

In addition to finfish and shrimp, San Felipe and Puerto Peñasco also fish for blue crab and various mollusks (e.g., black and pink murex, clams, scallops, octopus), and fishers in El Golfo de Santa Clara exploit the large clam beds of the delta.

Maquiladoras ("offshore" manufacturing plants, south of the border) have thrived in cities like Tijuana, Mexicali, and San Luis Río Colorado, all of which are dependent on Colorado River water. *Maquiladoras* have priority over water intended for urban uses, and with their high profits, these industries can afford to pay for the water.[27] Some industries have even bought water from treatment plants in the United States, while others are trying to buy agricultural water rights from the Mexicali Valley.[28] These events mark the beginning of a new period, where agricultural activities are being replaced by industrial activities on the border.

In the 1990s, drug trafficking increased considerably within the upper gulf/delta and Pinacate reserves. The physical and climatic characteristics of the region are ideal for this activity—isolated, inhospitable, and rugged, with uninhabited expanses of lava flows and sand dunes. As part of Mexico's war on drugs, the Ministry of Defense (via the army) undertakes patrol activities to combat the narcotraffic. The Ministry of Environment and Natural Resources (SEMARNAT), through the Directorate of the Pinacate Biosphere Reserve, collaborates with the Mexican army. However, this cooperation is largely in the form of mitigating the impacts that are generated on the natural resources during military operations. Military groups working in the reserves are commonly not sensitive to the fragility of the land, causing off-road vehicle damage to the environment.

THE EVOLUTION OF SCIENTIFIC KNOWLEDGE

Although the flora and fauna of the northern gulf is fairly well known, the biota of the biosphere reserve and delta region are largely unexplored. The coastline in the uppermost gulf is not easily accessed, nor are the expansive mudflats of the delta itself, and fundamental biological exploration has yet to be made in this region. There has never been a comprehensive survey of the flora or fauna of the Colorado delta. Despite the lack of systematic data, the research available has identified numerous threats to the Sonoran Desert and gulf ecosystem in the area of these two biosphere reserves.

Water Quality

Much of the available data on the health of the Colorado River Delta focuses on water-quality problems. Most of the freshwater supply to the reestablished wetlands of the delta derives from agricultural runoff, although some Colorado River surplus water and municipal raw sewage also drains into the river's channel. There are at least 17 agricultural drains in the Mexicali Valley that flow directly into the Colorado River/Hardy River channel. These agricultural waters also carry 70,000 tons of fertilizers each year, and 400,000 liters of insecticides each year across the delta.[29] The salinity of the drains is so high that its water is not tolerated by native riparian species. Hence, large stretches of delta wetland are dominated by salt cedar (a salt-tolerant invasive tree introduced from Asia in the early 1800s to stabilize stream banks in Arizona and California) that has low value for wildlife because few bird species appear to adapt to its use.[30]

Different sources of water to the delta carry a variety of contaminants that are toxic to wildlife and humans. Pesticides are of special concern since the majority of water is derived from agricultural runoffs, but there is no program in place that systematically monitors for a spectrum of pesticides on the delta. In the 1970s, levels of dichlorodiphenyltrichoroethane (DDE), the most persistent metabolite of dichlorodiphenyldichloroethylene (DDT), an insecticide that once was widely used but is now banned almost everywhere, in clams from the Mexicali Valley were as high as eleven parts per million (ppm) (wet weight). DDT was banned from agricultural use in Mexico in 1978, and a subsequent study in the mid-1980s, showed that concentrations of DDE in clams collected from the Mexicali Valley were much lower, averaging less than 0.2 ppm. More recently, in 1998, levels of DDE in clams from the Colorado riverbed of the delta, and agricultural drains, also tested low (<0.2 ppm). However, concentrations of DDE were much higher in birds of the Mexicali Valley, ranging from 0.04 to 11 ppm. Compared with other agricultural valleys in northwestern Mexico, the Mexicali Valley sustains higher levels of DDE.[31]

Other trace elements of concern that are carried to the delta are boron, arsenic, and selenium.[32] Selenium is a naturally occurring element that concentrates in the Colorado River due to high evaporation rates, especially in the reservoirs. When selenium reaches the lower section of the river and the delta, it can concentrate to levels that are toxic to wildlife. Bioaccumulation to toxic levels causes high rates of embryonic mortality and deformity.[33] High levels of selenium have been reported from water, sediment, and fish tissues throughout the Lower Colorado River.[34] Selenium levels 1.8 to 14.2 times higher than the U.S. Environmental Protection Agency's (EPA) criterion of 5 μg/L for the protection of freshwater aquatic life (8 μg/L in Mexico) have been reported from delta waters and sediments.[35] Recent studies have shown selenium to be low in invertebrate and fish tissues, whereas higher concentrations are found in bird tissues (e.g., double-crested cormorants, where values were near the threshold at which reproductive effects might occur).[36] The periodic high flows of Colorado River water since the 1980s have had the effect of flushing selenium through the riparian system, and selenium values seem to stay low as long as this periodic flushing occurs.

These studies indicate that DDT (DDE) and selenium are present in low concentrations in the delta's riparian system, but concentrate in greater levels higher in the food chain, especially if periodic adequate flushing from the Colorado River does not take place. Contaminants in general, and organochlorine pesticides and selenium in particular, constitute a permanent threat to the delta's wetlands. Although no standardized monitoring program has been implemented in the delta yet, there is awareness of the threat among the residents of the region, especially the Cucapá who regularly consume locally caught fish and waterfowl. In addition, radionuclides apparently are now leaching into the Colorado River from uranium tailings in Utah, but no information exists regarding the transport of these contaminants to the delta.

Reduction of freshwater input from the Colorado River has impacted numerous species of riparian and marine life. Reduction of the brackish estuarine habitat has, in combination with overfishing, driven the totoaba to near extinction. Absence of freshwater may drive the endemic Palmer's Saltgrass, which needs periodic freshwater flooding to germinate, to extinction. Loss of marshland eliminates key habitat used by many species of shellfish and finfish as nursery-ground during their preadult lives.

Fisheries

In 1941, John Steinbeck and Ed Ricketts wrote passionately about the destruction of the Sea of Cortez by fishing trawlers.[37] Little has transpired since then to address their concerns, and fishing pressure in the gulf remains extreme. In the northern gulf, the historical development of this heavy exploitation is enmeshed in a complex political ecology. Beginning in the 1950s, shrimp fishing grew exponentially in the gulf. Until the late 1980s, shrimp provided the main source of income for the fishing sector in these communities and was the catalyst for their growth. In addition to causing an increase and industrialization of the large-scale, offshore fishing fleet (especially trawl fishing), shrimp fishing also stimulated the growth of the small-scale fishing sector that uses *pangas* (small skiffs powered by outboard motors) as fishing vessels.

Government policies have consistently encouraged the expansion of both the industrial and small-scale fishing sectors. Commercial boats overexploit the offshore waters, and *panga* fishers take shrimp from coastal lagoons and *esteros* before they have even reached reproductive maturity. The number of large shrimp boats in the gulf grew to over 1,200 by 1999, despite warnings as early as the 1970s of a possible crisis resulting from overexploitation.[38] In the late 1980s, catch per unit of effort (CPUE) began to fall dramatically, making shrimp fishing less and less profitable.[39] Production diminished to the point of near financial collapse for the shrimp industry. In Puerto Peñasco alone, the active trawling fleet was reduced from 220 to 60 boats by 1993, when the upper gulf/delta biosphere reserve was established.[40]

The region's fisheries often operate under practically open-access conditions, existing fishing regulations are commonly not enforced, federal subsidies support overcapacity in industrial fleets, the biology of commercial species is poorly known (or unknown), and monitoring programs measuring the ecological impact of Mexico's fishing operations are almost nonexistent. The American Fisheries Society's official list of marine fishes at risk of extinction notes (an underestimated) six species from the Gulf of California (four endemic); all are large serranids and sciaenids, sensitive to overharvesting because of late maturity and formation of localized spawning aggregations.[41] The sciaenids require estuarine habitats in the rapidly diminishing Colorado River Delta for spawning and nursery grounds. The society also lists the entire gulf (especially its northern part) as one of five geographic hot spots in North America, where numerous fish species are at risk; certainly, the same could be said for the invertebrates of this region.

In the Sea of Cortez, reduction of freshwater inflow, chemical pollution from agriculture and urban areas, and coastal habitat destruction have combined with overfishing, use of nonselective fishing gear, and lack of reliable scientific data to drive such highly visible species as the endemic giant croaker-like totoaba and the vaquita porpoise to near extinction, to cause local extirpation of five species of sea turtles, and to substantially reduce the gulf's important commercial finfish and shrimp populations. Many once abundant, but less visible, species, such as the endangered giant brown sea cucumber and many shark species, are now practically gone from the gulf. Cucumbers and sharks have vanished at the hands of local fishers who take them primarily for the Asian market. The black murex snail fishery in Puerto Peñasco alone landed 600 tons in 1993, but by 1999 the catch had fallen to just 90 tons.[42] Though nationally insignificant, this murex fishery supports a local economy, and this species is also a key predator in the northern gulf rocky shore ecosystem.

Industrial shrimp trawling exacts a harsh toll on the gulf's marine environment. Around a thousand shrimp trawlers annually rake an area of sea floor equivalent to twice the total size of the gulf (in the shallow northern gulf, it has been estimated that every place on the sea floor is dragged with a shrimp net an average of four times per year). This constant bottom trawling damages fragile benthic (bottom) habitats and captures 10 to 30 kilograms of bycatch for every kilogram of shrimp caught.[43] Catch per unit of effort has been declining for decades, while fuel and export subsidies artificially sustain the overcapacity of industrial fishing fleets. Limited scientific and anecdotal information suggests that sweeping changes in benthic community structure have taken place over the past fifty years of these disturbances.[44] Loss of maritime habitats due to coastal development has reduced the rich *esteros* and mangrove communities of the gulf that served as critical spawning and nursery grounds for shrimp and other invertebrate and fish species. Loss of these wetlands also reduces stopover sites for migratory birds and for reptiles such as sea turtles and giant chuckwalla.[45]

Large artisanal fleets operating in nearly open-access conditions also contribute to overharvesting. In the three main communities of the northern gulf, the large commercial shrimp boat fleet has recently fallen to 175 boats, but the small-scale (*panga*) fishing fleet now exceeds 1,300 boats, which exploit over 70 species of fishes, mollusks, and crustaceans. A recent survey by Conservation International Mexico estimated that there are between 9,000 and 18,000 *pangas* active at any one time throughout the gulf. Approximately 40 percent of this harvest is exported to U.S. and Asian markets, primarily in California, Korea, China, and Japan.[46] Not only are the targeted resources diverse, but the fishing methods and the gear employed are also highly variable, including using gillnetting and long-lining, diving with air-delivery compressors, and employing traps. In the northern gulf, gillnet fishing from *pangas,* targeting *chano* and other finfishes, also incidentally captures the critically endangered and endemic vaquita porpoise and sea turtles. Poaching and incidental catch of sea turtles is a problem throughout western Mexico, although turtle-excluder devices are now mandatory (though commonly not employed) for industrial fishing vessels. Sea turtles have been essentially extirpated from the northern gulf. Today, hundreds of artisanal shark fishers roam freely throughout the gulf, and enforcement mechanisms, clear delimitation of take zones, and harvest limits for shark still need to be developed.

One of the greatest threats introduced by narcotraffic is the increase in small fishing boats (*pangas*) in the northern gulf. The number of *pangas* engaged in fishing in El Golfo de Santa Clara in 1996 was 254; today estimates are 400 to 500 *pangas.*[47] Large commercial boats have been virtually eliminated in El Golfo. Many of the small boats are obtained by local fishers by way of the narcotraffic through the region, because drug traffickers often abandon *pangas* when they have completed their drug runs along the coast. Some fishers may even be paid with a *panga* to lead traffickers through unknown terrain or to transport a drug shipment.

Many of the fisheries issues in the northern gulf have coalesced around the endemic vaquita porpoise. In 1985, scientists working in the region discovered some of the first complete and fresh specimens of the vaquita, or Gulf of California miniature porpoise. This small porpoise was first discovered and described by scientists in 1958 based upon a single skull specimen. The vaquita is endemic to the northern gulf and extremely rare. Its "rediscovery" immediately caught the attention of the environmental community. The vaquita was being caught incidentally in trawls and gillnets, especially illegal nets set for totoaba and legal nets set for shrimp, *chano*, sierra, sharks, and rays.[48] The demise of the charismatic vaquita, coupled with dramatic declines in shrimp and totoaba catch, focused national (and international) attention on issues of overexploitation in the northern gulf.

With the most recent comprehensive estimate of vaquita abundance at just 567 individuals (95 percent confidence interval equals 177 to 1,073), and mortality at an estimated 39 to 84 deaths per year,[49] the vaquita is the most endangered marine cetacean in the world. Based on mortality compiled for El Golfo de Santa Clara, it is estimated that 13 vaquitas are killed annually in shrimp nets, 17 by the *chano* fishery, 7 in nets set for sharks and rays, 2 in nets set for sierra, and few (or perhaps none) in the corvina fishery, for a total minimum annual mortality of 39 vaquitas. Though the primary cause of mortality prior to and in the 1980s was the now illegal ten- and twelve-inch mesh nets set for totoaba, it appears that the shrimping and *chano* fisheries using smaller-mesh-size gillnets pose the most significant threat to this porpoise. The high market value of shrimp increases the challenges for vaquita conservation.

A lack of biological information on vaquita and fisheries and the almost total lack of control over fisheries activities have hindered progress toward protecting the vaquita and toward more sustainable fisheries management in the reserve and northern gulf. Recent studies on small-scale fisheries have provided important insights and suggestions for management of the gulf.[50] In 1996, the IUCN placed the vaquita on its list of critically endangered animals and the Mexican government established an International Committee for the Recovery of the Vaquita (CIRVA). Using current information on vaquita abundance and mortality and fishing activities and zones, CIRVA has made the following recommendations for its recuperation:

- Reduce vaquita bycatch to zero as soon as possible.
- Extend the southern boundary of the upper gulf/delta biosphere reserve

to include the entire range of the vaquita and exclude gillnets and trawlers from the enlarged reserve.

The need to actively involve the fishing sector in the development of regulations and management systems is recognized by the management team of the biosphere reserve, as well as by NGOs working in the region. Many members of the shrimp fishery are interested in modernizing and downsizing the fleet in preparation for an anticipated free-trade pact with the EU. Since an EU free-trade pact would most likely require that diesel and export subsidies be reduced or eliminated, the older and more inefficient boats would become unprofitable anyway. It has also been suggested that funds from multilateral donors such as the Global Environment Fund be used to buy out the older part of the shrimp fleet. The estimated cost to purchase 400 boats and fishing licenses is about $60 million.[51]

The blame for fishery management inefficiencies and failures is not easily addressed and is certainly not unique to Mexico. Throughout the world, the inherent management complexity of marine fisheries commonly has a reverse effect, resulting in increased utilization and misuse of marine resources. In 1995, the Food and Agriculture Organization (FAO) reported that 69 percent of the world's fisheries are fully or overexploited or depleted and in need of urgent conservation and management measures. In Mexico, management of small-scale fisheries has proven to be a great challenge, and it has led to increasing concern as fishery managers and government agencies have worked to integrate the social and economic importance of this sector into management practices. For historical and logistical reasons, small-scale fisheries have been largely neglected in the formulation of national fisheries policies. Small-scale fisheries are particularly important in developing countries where they provide employment opportunities and the bulk of the domestic day-to-day food for most coastal communities.

Other Ecosystem Threats

Cattle production in the Southwest has played a major role in the transformation of grasslands to scrublands, the spread of exotic plants, and the degradation of fragile desert and riparian areas. Many researchers consider the cumulative impacts of cattle grazing to be irreversible. Statistics from Mexico's Comisión Téchnica Consultiva de Coeficientes de Agostadero (COTECOCA) confirm that two to five times the recommended stocking rates occur in the state of Sonora.[52] For more than fifty years, grazing issues have led to arguments among cattlemen, conservation biologists, and ecologists. The antibovine faction argues that overgrazing leads to habitat destruction, terracing and erosion on slopes, loss of riparian communities, and increased threat of Old World grass invasions. In Sonora, the widespread introduction of exotic grasses for grazing has been on a large scale due to government support. Currently, these grasses occur in very low numbers in the biosphere reserves, but they are spreading rapidly toward the region from both the north and south.

At least 384 naturalized non-native plants now occur in the Sonoran Desert region.[53] In addition to plants, at least 5 to 10 mammals, 2 amphibians, 50 to 60 fishes, and several reptiles have been introduced. There are no estimates of how many invertebrates have been introduced to this region. The intentional introduction of numerous species of African grasses (especially Lehmann love grass and buffel grass) in this region has profoundly changed vegetation structure, fire frequencies, and migratory wildlife corridors.[54] Plant species diversity is ten times lower in buffel grass communities, compared with native desert scrub communities. Conversion of desert scrub vegetation to buffel grass pasture also can cause a fourfold reduction in aboveground standing-crop biomass.[55] Lehmann love grass now covers more than 400,000 acres of Arizona. Buffel grass is now the dominant herbaceous perennial in much of the Sonoran Desert, covering at least 2.3 million acres of cultivated pasture and growing wild (feral) in most of the rest of the region.[56] These grasses have already spread over an estimated 600,000 hectares in the state of Sonora, and they are now intruding into Baja California. Ranchers now harvest buffel grass seed stock by hand along Sonora's main highways, where it grows wild throughout the state. At present, buffel grass areas occur in the Pinacate reserve (and in Baja California) in low numbers, but the chance of its spreading, either naturally or by planned introductions, is a growing threat to native communities.

Salt cedar ("tamarisk") was introduced to North America for stream-bank erosion control in the 1850s and has since spread to over 4,000 hectares of riparian habitat. It was naturalized in the Colorado River watershed by the 1920s. A mature salt cedar consumes as much as 800 liters of water per day, 10 to 20 times the amount used by the native species that it has replaced.[57] Bird diversity can plummet when salt cedar replaces native riparian woodlands. One comparison found 154 bird species for every 40 hectares of native vegetation, compared to 4 species per 40 hectares in salt cedar–dominated areas.[58] The islands of the Gulf of California have also been hit hard by the introduction of rats, goats, rabbits, cats, and various invertebrate pests (e.g., cockroaches, ants, crickets), which have decimated native bird and plant communities. It is estimated that 20 percent of the mammals endemic to the gulf islands and 12 percent of the island birds are now extinct.

THE EVOLUTION OF POLICIES TO PROTECT THE RESERVES

Many binational working groups are seeking ways to establish legal assurances for the quantity and quality of Colorado River water needed to sustain the delta's wetlands. However, a host of institutional impediments challenges the implementation of a conservation plan. The present system of water appropriation for the Colorado River is the result of a complicated history of conflict and negotiation over U.S. water uses, and between Mexico and the United States. River flows are controlled and regulated by a complex legal framework of treaties, interstate compacts, state and federal laws, regulations, Supreme Court and U.S. Department of the Interior decisions, and contracts collectively known as the Law of the River. Environmental considerations have only recently been considered in water management decisions.[59] The complexity and specificity of institutions governing the management of the Colorado River, combined with institutional inertia and political disagreements, have frustrated efforts to implement binational conservation strategies.

Colorado River Water Law

Western U.S. water law has played a critical role in the declining fortunes of the Sonoran Desert and northern gulf ecosystems. Completion of the Hoover Dam in 1935 began the "development" of the Colorado River. Arguments over who would own the water began as soon as the Hoover Dam was proposed, and before the dam was even completed, the Upper and Lower Colorado River Basin States Water Compact had been drafted and signed. Embedded in the Law of the River are two key treaties and countless laws, acts, minutes, and court decisions. The 1922 Colorado River compact allocated 7.5 million acre-feet per year (1 maf = 1,230,000 m^3, or 326,000 gallons) to the Lower Basin states (California, Nevada, Arizona), 7.5 maf/yr. to the Upper Basin states (Utah, Wyoming, Colorado, New Mexico), and 1.5 maf/yr. to Mexico. However, California has always taken more than its allotted 4.4 maf/yr., and by 2001 was using about 5.2 maf/yr. from the river. This has been allowed because of river "surpluses" declared by the U.S. Department of the Interior (the agency that controls Colorado River diversions) over the past decade because of wetter-than-normal weather and because the other states were not using all of their allocations. Utah, for example, currently draws about half its entitlement of 1.7 maf/yr. In the late 1990s, the Department of the Interior negotiated with California an agreement to limit that state's use of Colorado River water to the original allocation. In January 2003, when the state failed to comply with the first step in implementing the accord (a plan to transfer water from Imperial Valley farmland to the city of San Diego), the federal agency cut the flow of river water to California by more than 260 billion gallons.[60]

The 1944 United States–Mexico Water Treaty guaranteed Mexico 1.5 maf/yr. from the Colorado River (plus another two hundred thousand acre-feet in wet years, and a reduction of two hundred thousand acre-feet during drought years), but was silent on water quality. A 1973 amendment to the treaty (Minute 242) guarantees Mexico relatively pure water (an obligation the United States has, arguably, never lived up to). Both the water compact and the treaty were based on river-flow data from 1910 to 1920, which we now know was one of the wettest decades in the history of the Colorado River, during which a mean flow of about 20 maf/yr. occurred. Averaged out over many decades, the mean flow of the river is closer to 13.2 maf/yr. As a result, the "paper rights" to Colorado River water actually grant more than the amount that is normally available.

Until very recently all legal instruments dealing with the Colorado River focused on urban and domestic issues and farming and ranching, with no concern for environmental issues. In an attempt to partially correct this omission, in December 2000, Minute Number 306—the "Conceptual Framework for the United States–Mexico studies for Future Recommendations Concerning the Riparian and Estuarine Ecology of the Limitrophe Section of the Colorado River and Its Associated Delta"—was signed.

Regulating Fisheries

After the passage of Article 27 of the Mexican Constitution in 1933, the government claimed all rights to commercial exploitation of resources within inshore waters. This same year, the Ley General de Cooperativas (General Law of Cooperatives) was passed, and with it began an era of fishery cooperatives. The government bought out private companies operating in coastal inshore waters and established some of Mexico's first cooperatives. It also integrated local cooperatives and packing plants into a government-owned system to produce fishery resources, primarily shrimp, for export purposes. Cooperatives not only offered access to credit for equipment and guaranteed fishing permits and access to fishing grounds, but also reduced the uncertainties of fishing. The cooperatives, in theory, are based on a share system in which all participants share a portion of the profit, even if the profit is not large. These advantages enabled cooperatives to quickly recruit many rural fishers.

During President Echeverría's administration (1970–1976), government policies encouraged the expansion of the *panga*, or "artisanal," fishing sector, and small-scale fishing cooperatives were formed.[61] *Panga* fishers were the principal exploiters of the giant gulf corvina-like totoaba. The large spawning aggregations of totoaba in the delta region made it easy prey for *panga* fishers, and by the late 1960s its populations had been decimated. In 1975, the totoaba became Mexico's first endangered (listed) marine fish species, and its fishery was banned. *Panga* fishers also took to the coastal lagoons and *esteros* to fish for shrimp, a practice that has long been banned in the United States because these are critical habitats that serve as "grow out" nurseries for shrimp and many finfish species. In these shallow nursery areas, Mexican fishers capture juvenile shrimp before they have yet had a chance to reproduce. As of 2001, fewer than three thousand *panga* permits existed in Sonora, but an estimated 9,000 fishers were operating along this coast on any given day.

During the 1970s, the selling price of shrimp increased significantly and a large portion of the population of San Felipe and Puerto Peñasco began fishing for shrimp. There was a large influx of people from inland Mexico to these communities—a new "gold rush" was taking place, but this time it was "pink gold" or "oro rosado," as the large commercial shrimps (*Penaeus spp.*) came to be known. The Gulf of California was experiencing a booming harvest of its fishery resources, as the government moved rapidly to maximize production by giving credits and consolidating fishing infrastructure.[62] The number of large shrimp boats in the gulf nearly doubled from 1970 to 1999 (from 700 to 1,200). With this boom also came the necessary establishment of institutions such as the Secretaría de Pesca (PESCA) to enforce newly defined regulations. Trawler and *panga* fleets grew rapidly in the 1970s, despite warnings as early as the 1970s of a possible crisis resulting from overexploitation.[63] Although total shrimp production continued to climb, catch per unit of effort began to fall, making shrimp fishing less and less profitable. Although the country was benefiting from increased overall production, each new *panga* added to the fleet meant increased competition among fishers. Conflicts arising from competition began to grow as resources were divided among more and more users.[64]

By the end of the 1980s and beginning of the 1990s, shrimp catches were falling precipitously in the northern gulf.[65] Production diminished to the point of near financial collapse for the shrimp industry. Banks took possession of boats, cooperatives closed, and local residents began searching for other work. Despite increased fishing efforts, the total catch fell 33 percent between 1995 and 1999.[66] In Puerto Peñasco alone, the active trawling fleet was reduced from 220 to 100 boats.[67] Besides the consequences to local economies, this near collapse of the northern gulf shrimp fishery provoked more diversification of the fishing activities of small-scale fishers. Species that were sporadic target catches or that had never been commercialized, such as *chano* (gulf croaker), suddenly became actively fished.[68]

Creating the Biosphere Reserves

By the late 1980s, the northern gulf was experiencing an economic and ecological crisis.[69] A series of multiconstituent workshops was convened, beginning in 1990, to develop an action plan for conservation in the upper gulf/delta region. In February 1993, Wetlands for the Americas called for recognition of the Colorado River Delta as an international reserve in the

Programa Red Hemisférica de Aves Playeras. In March 1993, a proposal to declare the upper gulf and Colorado River Delta a biosphere reserve was presented to the Mexican government by a group of NGOs and government agency officials. PESCA fought against the reserve idea, fearing its economic impact on northern gulf fisheries.

In recognition of the extraordinary biological and cultural importance of the upper gulf and Colorado River Delta ecosystems, in addition to increased international pressure to protect the endemic vaquita porpoise and totoaba, the Mexican government declared this region a 934,756 hectare (2,336,890 acre) biosphere reserve in June 1993; 164,779 hectares (411,948 acres) of this reserve is in the *Zona Núcleo* (core zone). Except for traditional practices by the Cucapá people living in the delta and clam harvesting by local residents, the reserve was to prohibit all commercial fisheries within the core zone and to increase regulations for most fisheries within the buffer zone (*Zona de Amortiguamiento*). In 1995, the reserve was accepted into the United Nations Economic, Scientific, and Cultural Organizations's (UNESCO) Man and the Biosphere (MAB) system of worldwide biosphere reserves, and today it is combined with the Pinacate biosphere reserve as one unit in the MAB Program. The upper gulf/delta biosphere reserve has an official "sister reserve" on the Colorado River in the United States, the Imperial National Wildlife Refuge (U.S. Forest Service). This partnership has proven highly useful in coordinating efforts to protect wildlife, especially migratory waterfowl.

The El Pinacate/Gran Desierto Biosphere Reserve was declared on June 10, 1993, and accepted into the UNESCO-MAB Program in 1995. When the Pinacate biosphere reserve was created, there were three main economic activities in the area: agriculture, ranching, and surface mining. These activities were considered incompatible with the conservation objectives of the reserve. In this dry region of the Sonoran Desert, agriculture is an activity with high economic and environmental cost and low return, and it has resulted in the bankruptcy of many *ejidos*. For this reason, market forces alone are gradually driving agriculture from the area. Currently, agricultural activities within the reserve are at a very low level.

Ranching was regulated immediately after the declaration of the reserve, because it was creating direct environmental degradation, destruction of archaeological sites, and contamination of natural water sources. Since the reserve's establishment, no additional cattle have been allowed into the area. Mining in the reserve was also considered a particularly harmful activity, due to its considerable impact on the landscape. Since the creation of the reserve, all mining activity has been stopped in the core zone, and this activity is now permitted only at two sites in the buffer zone.

Prior to establishment of the Pinacate reserve, there were three other low-level economic activities: extraction of wood, hunting, and tourism. Hunting and wood extraction were prohibited immediately upon declaration of the reserve, and these are prohibited in both the core and buffer zones. The *ejidatarios* may use wood only for personal purposes, such as cooking. Hunting is permitted only for the local Tohono O'odham people for religious purposes.

Taken together, these activities had been important for the inhabitants of the Pinacate biosphere reserve, and this regulation has driven a search for new productive alternatives for local inhabitants. However, despite these efforts, the enforcement of current laws, and the poor conditions for agriculture within the reserve, the rural communities persist in engaging in low-level irrigated farming and open-range ranching. Some *ejidos* have requested authorization to exploit new sand and gravel quarries for sale to the construction industry.

In May 1997 the Technical Advisory Council of the Pinacate reserve was created with representatives from governmental, social, and academic sectors, as well as representatives from the *ejidos* within the reserve. Through deliberations of the council, negotiations with *ejido* residents have been reinvigorated. The council recommended a study to determine the socioeconomic status of the communities within the reserve and to define new economic activities compatible with the conservation of natural resources. The Colegio de la Frontera Norte (COLEF) was contracted to produce a feasibility study of sustainable projects for Pinacate communities. Among the possibilities suggested, high-priority ideas were those related to tourist activities and the construction of nurseries for the production and marketing of native plants. Projects related to agriculture (such as cultivation of date palms) were not recommended.

There are now several projects related to ecotourism planned within the Pinacate reserve, and negotiations are underway to obtain support for their development. In addition, with support from the Centro de Investigaciones en Biotecnología of the Universidad Autónoma del Estado de Morelos (CIBUAEA), a proposal is being developed for native cacti research and nurseries. Finally, with the support of national and international organizations, programs are being developed for environmental education within the reserve communities, primarily focused on children and youth. However, all of these efforts are in their infancy and need increased funding and partnership support (especially from NGOs) to move forward more rapidly.

Recognizing the limitations of the upper gulf reserve, in the spring of 2001 numerous institutions joined forces to develop a more effective strategic plan for the recuperation of the vaquita porpoise. The plan lays out a strategy for establishing a new protected area that would overlap the existing reserve, but extend beyond it to encompass the majority of the vaquita's known range. The plan also urges enforcement of new Mexican laws established for protected areas (in November 2000) that make it illegal to destroy the sea floor and to use fishing methods having high incidental capture of nontarget species. With these legal tools in place, elimination of trawling and gillnetting within the reserve becomes an issue of implementation and enforcement, both of which are most effectively done by working directly with fishers.

Reserve managers are incorporating some of the new fishing restrictions into their management plan, including the elimination of trawling activities and banning gillnets with a mesh size under six inches. But today the greatest threat to the vaquita appears to be the *panga* shrimp fisheries that use nets with a two-inch mesh size. This activity will be difficult to eliminate, and alternative ways to fish for shrimp are not clear-cut. If alternatives could be found, certification of these products under the Marine Stewardship Council's certification program could help provide its implementers with an added value for their efforts to save the endangered vaquita. Such certification programs are underway further south in the gulf, with the blue crab fishery (Callinectes spp.). Some experts say banning gillnets immediately is the only chance for saving the vaquita from extinction, whereas others say these nets cannot be eliminated until fishers have alternatives. With a population size smaller than 600 individuals, however, time is running out for the vaquita. Because saving the vaquita is linked to the development of sustainable fisheries and general protection for marine habitats of the region, the vaquita provides a powerful symbol and focus for protecting the entire northern gulf ecosystem.

Assessing the Mexican Biosphere Reserves

From one perspective, the creation of the biosphere reserves in the gulf region seems to be a success story. Although scientific information on the health of habitats and species in the upper gulf was incomplete, evidence of widespread loss of biodiversity was sufficient to provoke the Mexican government to take action to create the biosphere reserves. At least as important in motivating action as concerns about endangered species and habitats were fears that the fisheries would not recover and a critical source of livelihood would be lost. Scientific arguments about the value of biodiversity and the importance of preserving it interacted in complex ways with economic arguments about protecting industries and ensuring sustainable harvesting of resources. The vaquita played a key role in turning the issue from largely a local concern to an international issue. There is now in place a network of organizations dedicated to developing ecotourism, encouraging nature education, protecting wetlands, ensuring sustainable resource use, and protecting threatened and endangered species.

Efforts by NGOs, local stakeholders, and grassroots groups were crucial to the creation of the biosphere reserves and played a key role in formulating proposals, educating local groups, and lobbying the government to take action. Local stakeholders sharing a commitment and identity, a "sense of place," and a common understanding of the local ecosystem are trying to find ways to use the regional sociopolitical structure to take positive steps toward sustainable economic ventures. Local efforts have centered on seeking sustainable employment for regional inhabitants, ecotourism, sustainable aquaculture (e.g., oyster ranching) and halophyte farming; local initiatives to participate with government to reduce overfishing and fishing that threatens the extinction of key species; and NGO and community efforts to educate local stakeholders in sustainable economic practices.

Important citizen movements are coalescing around the wetlands of the upper gulf/delta reserve. In one of the gulf communities, Puerto Peñasco, where development is literally threatening to consume the entire coast, citizens involved in aquaculture activities and tourism projects are joining together to state their interests and concerns and to learn more about how to protect the ecosystem functions that they depend on. Petitions to government have produced responses. When Puerto Peñasco fishers requested a shortened shrimp season in the 1980s, the government acted. Such efforts are strengthened when they come from unified groups such as those currently emerging in support of the biosphere reserves. With the scientific information provided by the Center for the Study of Deserts and Oceans (CEDO), commercial divers at Puerto Peñasco recently requested that the government ban the black murex fishery for one year, to allow for new recruitment. The government responded by establishing a year moratorium on the take of this species. Effective and strong communication of the issues to government decision makers is one of the major challenges facing the isolated communities of the reserves. In this regard, local NGOs can be of considerable assistance, helping give communities communication tools and a louder voice with Mexico City.

However, lack of clear development guidelines and conflicting interests within the government often result in approval of developments that are not in the long-term interests of the local communities or the biosphere reserves. Recently, for example, representatives of the Federation of Cooperatives of Small-Scale Fishermen from Puerto Peñasco wrote a letter of concern about a tourism megaproject located within the upper gulf/delta reserve that has the potential to destroy one of their prime fishing grounds. Though their complaint was registered, it has not been resolved, and the commercial tourism interests of the area remain strong. Though the upper gulf/delta reserve was established to help in the management of fisheries in the northern gulf, the reserve staff literally has no power to make or enforce decisions with regard to these fisheries.

One of the greatest challenges for managers of Mexican biosphere reserves is to find a balance between the use of the area's natural and cultural resources by local residents and visitors and the conservation of these resources. One of the most suitable and popular sustainable uses for nature reserves is ecotourism. The advantages of ecotourism over most other activities are obvious to reserve managers and visitors, but sometimes not to the local inhabitants, who are typically reluctant to switch from one well-known activity to another completely new activity. Ecotourism is a new concept in the Pinacate/delta region. Although visitation to these areas has increased since declaration of the biosphere reserves in 1993, the activities conducted by most visitors do not fit the concept of ecotourism and still tend to degrade the resources, although usually on a smaller scale than traditional uses.

Small communities and *ejidos*, such as those in the biosphere reserves, typically lack formal structures and organizations that can facilitate communication and action. An important part of the work of the reserves and supporting organizations in recent years has been assisting in the establishment of such structures. As mandated by the management program, the upper gulf/delta reserve established a Citizens' Technical Advisory Council in 1998. This council is composed of representatives of the different stakeholder groups in the reserve, including nonprofit organizations, scientists, fishers from each community, the tourism sector, and the Cucapá people.

From another perspective, the future of the reserves is partly dependent on increasing the flow of water into the area, and the prospects for that are uncertain at best. Strong political commitment from the United States in securing adequate water to restore the gulf ecosystem has yet to occur. In 1997, the U.S. Department of the Interior and the Mexican Ministry of the Environment and Natural Resources signed a letter of intent recognizing the need for binational cooperation in protecting the reserves and other critical areas along the United States–Mexico border. However, since 2001, there have been no signs of further cooperative conservation efforts from the U.S. administration, and the George W. Bush administration's strong propensity to act unilaterally in international affairs makes it less likely that the United States will work with Mexico to secure sufficient water to protect the reserves. The mismatch between costs and benefits makes cooperation difficult. Since the benefits of action will largely occur in Mexico, while the burdens of diverting water for the delta, rather than consuming it, lie largely in the United States, there are strong political forces arrayed against sending more water south.

One option is to use U.S. environmental law to help secure the reserve's future. Environmental groups have filed lawsuits aimed at protecting Colorado River and delta threatened and endangered species.[70] The North American Agreement on Environmental Cooperation (NAAEC), often refereed to as the Environmental Side Agreement to NAFTA, authorizes citizens, NGOs, and others to bring action to enforce domestic environmental laws and regulations.[71] In December 1999, Defenders of Wildlife and fourteen other Mexican and U.S. environmental groups notified the U.S. government of their intent to sue the Department of the Interior's Bureau of Reclamation for operating and maintaining dams, reservoirs, and other facilities along the Colorado River in ways that failed to supply sufficient freshwater to the gulf to protect endangered species, particularly the totoaba, vaquita, desert pupfish, Yuma clapper rail, and southwestern willow flycatcher. Plaintiffs argue that the U.S. Endangered Species Act (ESA) authorizes the bureau to allocate water in the Colorado River to protect threatened and endangered species.[72]

As of the writing of this chapter, the case has not been decided, but federal courts have upheld the authority of federal agencies to reallocate water under contract to meet the purposes of the ESA.[73] The Supreme Court has repeatedly held that U.S. laws do not apply extraterritorially unless Congress declares their intent to be otherwise. Plaintiffs argue that they are challenging the application of U.S. laws by a federal agency within the United States that have an impact in another country and that Congress intended that the ESA apply to protecting species outside the United States.[74] A study by Environmental Defense concluded that perennial flows of at least 32,000 acre-feet and pulse flows of 260,000 acre-feet at 3,500 to 7,000 cubic feet per second, on average, every four years would be required to protect the habitat of the threatened and endangered species.[75]

A nonprofit group has been formed in Baja California, Asociación Ecológica de Usuarios del Río Hardy y Río Colorado (AEURHYC, the Ecological Association of Users of the Hardy and Colorado Rivers), represented by the stakeholders in this area: farmers, tourism businesses, the Cucapá people, and Mexican and American tourists. This Mexican grassroots movement is support by several similar organizations in the United States that are also lobbying for more Colorado River water to be delivered to the delta. Scientists and conservation groups from Mexico and the United States have begun to pool efforts to understand the complex issues associated with management of the Colorado River water and the delta region. Numerous meetings have taken place, which involve a wide range of both citizen and government participation alongside scientists, with the ultimate goal of providing high-quality water resources for the restoration of the northern gulf/Colorado River Delta ecosystems.

NOTES

The authors greatly appreciate the reviews of this work by Tom Van Devender and Mark Dimmitt. Appreciation is also extended to Christine Norodom and the Man and the Biosphere Program of UNESCO for funding the development of this case study. The original compilation of data for this paper came from two workshops and contributions by J. Campoy-Favela, C. Castillo-Sánchez, R. Cudney-Bueno, L. T.

Findley, J. García-Hernández, E. Glenn, I. Granillo, M. E. Hendrickx, J. Murrieta, C. Nagel, M. Román, P. J. Turk-Boyer, and the senior author. The authors are grateful to these collaborators for their assistance.

1. D. E. Brown, ed., *Biotic Communities: Southwestern United States and Northwestern Mexico* (Salt Lake City: University of Utah Press, 1994); S. J. Phillips and P. W. Comus, *A Natural History of the Sonoran Desert* (Tucson: Arizona-Sonora Desert Museum Press, 2000); R. S. Felger, *Flora of the Gran Desierto and Río Colorado of Northwestern Mexico* (Tucson: University of Arizona Press, 2000); R. S. Felger, M. B. Johnson, and M. F. Wilson, *The Trees of Sonora* (New York: Oxford University Press, 2001); and R. C. Brusca, *Common Intertidal Invertebrates of the Gulf of California, Revised and Expanded Edition* (Tucson: University of Arizona Press, 1980).

2. R. F. Dasmann, "Biotic Provinces of the World." International Union for the Conservation of Nature and Natural Resources," *Occ. Pap.* 9 (Gland, Switzerland, 1974); L. R. Dice, *The Biotic Provinces of North America* (Ann Arbor: University of Michigan Press, 1943); G. P. Nabhan and A. R. Holdsworth, *State of the Desert Biome: Uniqueness, Biodiversity, Threats and the Adequacy of Protection in the Sonoran Bioregion*, 2nd ed. (Tucson: Arizona-Sonora Desert Museum Press, 1999).

3. R. S. Felger, "Vegetation and Flora of the Gran Desierto, Sonora, Mexico," *Desert Plants* 2 (1980): 87–114.

4. D. T. MacDougal, "Delta and Desert Vegetation," *Bot. Gazette* 38 (1904): 44–63; D. T. MacDougal, "Botanical Explorations in the Southwest," *J. New York Bot. Garden* 553 (1904): 89–108; D. T. MacDougal, "Botanical Explorations in Arizona, Sonora, California and Baja California," *J. New York Bot. Garden* 666 (1905): 91–102; D. T. MacDougal, "The Desert Basins of the Colorado Delta," *Bull. Amer. Geog. Soc.* 3912 (1907): 705–29; R. C. Murphy, "Natural History Observations from the Mexican Portion of the Colorado Desert," *Proc. Linn. Soc. New York* 28–9 (1917): 43–101; A. Leopold, *A Sand County Almanac with Other Essays on Conservation from Round River* (New York: Oxford University Press, 1966); and E. W. Funcke, "Hunting Antelope for Museum Specimens," *Field and Stream* (March 1919): 834–6.

5. J. M. Payne, F. A. Reid, and E. Carrera-Gonzalez, *Feasibility Study for the Possible Enhancement of the Colorado Delta Wetlands, Baja California Norte, Mexico* (Sacramento, Calif.: Ducks Unlimited, Inc., 1992).

6. E. P. Glenn, C. Lee, R. Felger, and S. Zengel, "Effects of Water Management on the Wetlands of the Colorado River Delta, Mexico," *Conservation Biology* 104 (1996): 1175–86.

7. J. I. Morrison, S. L. Postel, and P. H. Gleick, *The Sustainable Use of Water in the Lower Colorado River Basin* (Oakland, Calif.: Pacific Institute, Global Water Policy Project, UNEP, Turner Foundation, 1996).

8. R. Brusca, personal observation.

9. R. C. Brusca, L. T. Findley, P. A. Hastings, M. E. Hendrickx, J. Torre Cosio, and A. M. van der Heiden, "Macrofaunal Biodiversity in the Gulf of California (Sea of Cortez)," in J.-L. E. Cartron, G. Ceballos, and R. Felger, eds., *Biodiversity, Ecosystems, and Conservation in Northern Mexico* (New York: Oxford University Press, in press); L. T. Findley, M. E. Hendrickx, R. C. Brusca, A. M. van der Heiden, P. A.

Hastings, and J. Torre, "Macrofauna del Golfo de California [Macrofauna of the Gulf of California]," CD-ROM Ver. 1.0. Macrofauna Golfo Project. Conservation International, Washington, D.C. [Spanish and English editions], in press.

10. E. Mellink, E. Palacios, and S. Gonzalez, "Non-breeding Water Birds of the Delta of the Río Colorado, Mexico," *J. Field Ornithology* 681 (1997): 113–23; G. Ruiz-Campos and M. Rodríguez-Meraz, "Composición Taxonómica y Ecología de la Avifauna de los Ríos El Mayor y Hardy, y Áreas Adyacentes en el Valle de Mexicali, México," *Anales Inst. Biol., Ser. Zool.* 68 (1997): 291–315; M. Cervantes, M. J. Roman, and E. Mellink, "AICA: NO-17 Delta del Río Colorado," in H. Benitez, C. Arizmendi, and L. Marquez, *Base de Datos de las AICAS* (Mexico City: AICAS, CIPAMEX, CONABIO, FMCN, y CCA, 1999), available at www.conabio.gob.mx, accessed 6 March 2000; G. W. Kramer and R. Mogoya, "The Pacific Coast of Mexico," in M. Smith, R. L. Pederson, and R. M. Kaminski, eds., *Habitat Management for Migrating and Wintering Waterfowl in North America* (Lubbock: Texas Tech University Press, 1989): 507–28; J. Campoy and M. J. Roman, *Avifauna de la Reserva de la Biosfera del Alto Golfo de California y Delta del Río Colorado Reporte de Investigacion* (Mexicali: INE/SEMARNAP—IMADES, 1999).

11. B. W. Massey and E. Palacios, "Avifauna of the Wetlands of Baja California, Mexico: Current Status," *Stud. Avian Biol.* 15 (1994): 45–57.

12. Felger et al., *The Trees of Sonora.*

13. S. W. Carothers, R. R. Johnson, and S. W. Aitchison, "Population Structure and Social Organization in Southwestern Riparian Birds," *Amer. Zool.* 14 (1974): 97–108; R. R. Johnson, L. Haitht, and J. M. Simpson, "Endangered Species Versus Endangered Habitats: A Concept," in R. R. Johnson and D. A. Jones, eds., *Importance, Preservation and Management of Riparian Habitat: A Symposium* (USDA Forest Service General Technical Report RM-43, 1987), 68–79; R. D. MacArthur and J. W. MacArthur, "On Bird Species Diversity," *Ecol.* 42 (1961): 594–8.

14. Morrison et al., *Sustainable Use.*

15. T. Van Devender, R. S. Felger, and A. Búrquez, "Exotic Plants in the Sonoran Desert Region, Arizona and Sonora," in M. Kelly, E. Wagner, and P. Warner, eds., *Proceedings of the California Exotic Plant Pest Council Symposium,* vol. 3, October 2–4,1997, Concord, California, 10–15.

16. Personal communication, City of Tucson, 1999.

17. Morrison et al., *Sustainable Use.*

18. J. C. Stromberg, R. Tiller, and B. Richter, "Effects of Groundwater Decline on Riparian Vegetation of Semi-arid Regions: The San Pedro, Arizona," *Ecol. Applications* 6 (1996): 113–33.

19. L. L. Jackson, J. P. McAuliffe, and B. A. Roundy, "Desert Restoration: Revegetation Trials on Abandoned Farmland," *Restoration and Management Notes* 9 (1991): 71–80.

20. Nabhan and Holdsworth, *State of the Desert Biome.*

21. F. Contreras-Mora, *El Movimiento Agrario en el Territorio Norte de la Baja California* (Aguascalientes: Instituto de Investigaciones Históricas del Estado de Baja California, 1987).

22. Nabhan and Holdsworth, *State of the Desert Biome.*

23. A. Búrquez and A. Martínez-Yrízar, "Conservation and Landscape Transformation in Sonora, Mexico," *J. Southwest* 39 (1997): 371–98.

24. J. B. Greenberg, "Territorialization, Globalization, and Dependent Capitalism in the Political Ecology of Fisheries in the Upper Gulf of California," in A. Biersack and J. B. Greenberg, eds., *Culture, History, Power, Nature: Ecologies for the New Millennium* (Tucson: University of Arizona, in press).

25. R. Cudney-Bueno, "Management and Conservation of Benthic Resources Harvested by Small-Scale Hookah Divers in the Northern Gulf of California, Mexico: the Black Murex Snail Fishery," M.S. thesis, University of Arizona, 2000.

26. Conservation International, personal communication to Brusca.

27. D. Calbreath, "Maquiladora Industry in Baja California," *Union-Tribune,* San Diego, 1998.

28. I. Coronado, "Conflicto por el Agua en la Region Fronteriza," *Borderlines* 76 (1999): 2–4.

29. Dirección General de Ecología, 1995.

30. V. K. Rosenberg, R. D. Ohmart, W. C. Hunter, and W. B. Anderson, *Birds of the Lower Colorado River Valley* (Tucson: University Arizona Press, 1991).

31. M. A. Mora and D. W. Anderson, "Seasonal and Geographical Variation of Organochlorine Residues in Birds from Northwest Mexico," *Arch. Environ. Contam. Toxicol.* 21 (1991): 541–8.

32. S. Alvarez Borrego, J. A. Gaxiola Castro, J. M. Acosta Ruíz, and R. A. Schartzlose, "Nutrients en el Golfo de California, *Ciencias Marinas,* 5(2) (1978), 53–71.

33. H. M. Ohlendorf, D. J. Hoffman, M. K. Saiki, and T. W. Aldrich, "Embryonic Mortality and Abnormalities of Aquatic Birds: Apparent Impacts of Selenium from Irrigation Drainwater," *Sci. Total Environ.* 52 (1986): 49–63.

34. D. Welsh and O. E. Maughan, "Concentrations of Selenium in Biota, Sediments and Water at Cibola National Wildlife Refuge," *Archives Enivron. Contam. Toxicol.* 26 (1994): 452–8; K. A. King, D. L. Baker, W. G. Kepner, and C. T. Martinez, *Contaminants in Sediment and Fish from Natural Wildlife Refuges on the Colorado River, Arizona* (Phoenix, Ariz.: U.S. Fish and Wildlife Service, 1993); D. B. Radtke, W. G. Kepner, and R. J. Effertz, *Reconnaissance Investigation of Water, Bottom Sediment, and Biota Associated with Irrigation Drainage in the Lower Colorado River Valley, Arizona, California and Nevada, 1986–87,* (Tucson, Ariz.: U.S. Geological Survey, 1988).

35. C. Valdés-Casillas, E. P. Glenn, O. Hinojosa-Huerta, Y. Carillo-Guerrero, J. García-Hernández, F. Zamora-Arroyo, M. Muñoz-Viveros, M. Briggs, C. Lee, E. Chavarría-Correa, J. Riley, D. Baumgartner, and C. Condon, *Wetland Management and Restoration in the Colorado River Delta: The First Steps,* special publication of CECARENA-ITESM Campus Guaymas and NAWCC México, 1998.

36. M. A. Mora and D. W. Anderson, "Selenium, Boron, and Heavy Metals in Birds from the Mexicali Valley, Baja California," *Bull. Environ. Contam. Toxicol.* 21 (1995): 541–548; Valdés-Casillas et al., *Wetland Management*; J. García-Hernández, E. P. Glenn, J. Artiola, and D. J. Baumgartner, "Bioaccumulation of Selenium (Se) in the Ciénega de Santa Clara Wetland, Sonora, Mexico," *Ecotox. Enviromental Safety* 46 (2000): 298–304; K. A. King, A. L. Velasco, J. Garcia-Hernandez, B. J. Zaun, J. Record, and J. Wesley, *Contaminants in Potential Prey of the Yuma Clapper Rail: Arizona and California, USA, and Sonora and Baja, Mexico, 1998–1999* (Phoenix, Ariz.: U.S. Fish and Wildlife Service, 2000).

37. J. Steinbeck and E. F. Ricketts, *Sea of Cortez* (New York: Viking Press, 1941).
38. E. Snyder-Conn and R. C. Brusca, "Shrimp Population Dynamics and Fishery Impact in the Northern Gulf of California," *Ciencias Marinas* 1(3) (1977): 54–67.
39. D. Hoyos, "Auto Veda de la Pesquería de Camarón? Caída Drástica Estimula la Acción," *Noticias del CEDO* 3 (1991): 2.
40. R. Cudney-Bueno and P. J. Turk-Boyer, "Pescando entre Mareas de Alto Golfo de California. Una Guía sobre la Pesca Artesanal, Su Gente y Sus Propuestas de Manejo," *CEDO Tech. Ser.* no. 1, 1998.
41. J. A. Musick, M. M. Harbin, S. A. Berkeley, G. H. Burgess, A. M. Eklund, L. T. Findley, R. G. Gilmore, J. T. Golden, D. S. Ha, G. R. Huntsman, J. C. McGovern, S. J. Parker, S. G. Poss, E. Sala, T. W. Schmidt, G. R. Sedberry, H. Weeks, and S. G. Wright, "Marine, Estuarine and Diadromous Fish Stocks at Risk of Extinction in North America (Exclusive of Pacific Salmonids)," *Fisheries* 25(11) (2000): 6–30.
42. Cudney-Bueno, "Management and Conservation."
43. J. M. Garcia-Caudillo, "El Uso de los Excluidores de Peces en la Pesca Comercial de Camarón: Situación Actual y Perspectivas," *Pesca y Conservación* 3(7) (1999): 5; J. Pérez-Mellado and L. T. Findley, "Evaluación de la Ictiofauna Acompañante del Camarón Capturado en las Costas de Sonora y Norte de Sinaloa, México," in A. Yañéz-Arancibia, ed., *Recursos Potentiales de México: La Pesca Acompañante del Camarón*, Programa Universitario de Alimentos, Instituto de Ciencias del Mar y Limnología, Instituto Nacional de la Pesca (Rio Hondo, Mexico, D.F.: Universidad Nacional Autónoma de México, 1985): 201–54; R. Brusca, interviews with shrimp fishers.
44. J. M. Nava-Romo, "Impactos a Corto y Largo Plazo en la Diversidad y Otras Características Ecológicas de la Comunidad Béntico-Demersal Capturada por la Pesquería de Camarón en el Norte del Alto Golfo de California, México," unpublished M.S. thesis, Instituto Tecnológico y de Estudios Superiores de Monterrey-Campus Guaymas, Sonora, México, 1994.
45. Búrquez and Martínez-Yrízar, "Conservation and Landscape Transformation."
46. Cudney-Bueno, "Management and Conservation."
47. Personal communication from José Campoy-Favela, director of the upper gulf/delta biosphere reserve, to R. Brusca.
48. O. Vidal, "Population Biology and Incidental Mortality of the Vaquita, Phocoena Sinus," *Report of the International Whaling Commission*, no. 16 (Special Issue, 1995): 247–72.
49. C. D'Agrosa, O. Vidal, and W. C. Graham, "Mortality of the Vaquita (Phocoena Sinus) in Gillnet Fisheries during 1993–1994," in A. Bjorge and G. Donovan, eds., *Biology of Phocoenids*, Special Issue 16 (1995): 283–91; Vidal, "Population Biology"; O. Vidal, R. L. Brownell Jr., and L. T. Findley, "Vaquita, *Phocoena sinus* Norris and McFarland, 1958" in S. H. Ridgway and R. Harrison, eds., *The Second Book of Dolphins and the Porpoises,* Vol 6., *Handbook of Marine Mammals* (San Diego: Academic Press, 1999): 357–378.
50. Intercultural Center for the Study of Deserts and Oceans, *Pescando entre Mareas del Alto Golfo: Una Guía sobre la Pesca Artesanal, Su Gente y Sus Propuestas de Manejo* (Puerto Penasco Sonora, Mexico: CEDO Intercultural, undated); Cudney-Bueno and Turk-Boyer, *Pescando entre Mareas de alto Golfo de California.*

51. Packard Foundation, *A Strategic Focus for the Mexico Program* (Los Altos, Calif.: David and Lucile Packard Foundation, Conservation Program, 2000).

52. Búrquez and Martínez-Yrízar, "Conservation and Landscape Transformation."

53. Felger, *Flora of the Gran Desierto*; Nabhan and Holdsworth, *State of the Desert Biome*.

54. A. Y. Cooperrider and D. S. Wilcove, *Defending the Desert: Conserving Biodiversity on BLM Lands in the Southwest* (Washington, D.C.: Environmental Defense Fund, 1995); Johnson et al., *Endangered Species Versus Endangered Habitats*; M. P. McClaran and T. R. Van Devender, eds., *The Desert Grassland* (Tucson: University Arizona Press, 1995).

55. McClaran and Van Devender, *The Desert Grassland*; Nabhan and Holdsworth, *State of the Desert Biome*.

56. McClaran and Van Devender, *The Desert Grassland*.

57. Cooperrider et al., *Defending the Desert*.

58. S. R. Anderson, "Potential for Aquifer Compaction, Land Subsidence, and Earth Fissures in the Avra Valley, Pima and Pinal Counties, Arizona," *U.S. Geol. Surv. Hydrol. Invest. Atlas HA-718* (Washington, D.C.: U.S. Government Printing Office, 1989); S. Johnson, "Alien Plants Drain Western Waters," *Nature Conserv. News* 36(5) (1986): 24–5.

59. J. Bolin, "Of Razorbacks and Reservoirs: The Endangered Species Act's Protection of Endangered Colorado River Basin Fish," *Pace Environmental Law Review* 11 (1993): 35–87.

60. Shaun McKinnon, "California Water Flow Is Slashed," *The Arizona Republic*, 5 January 2003, at www.arizonarepublic.com, accessed 6 January 2003; William Booth and Kimberly Edds, "California's Supply of Surplus Water Shut Off," *The Washington Post*, 6 January 2003, at www.washingtonpost.com/ac2/wp-dyn/A14658–2003Jan5?, accessed 15 July 2003.

61. J. B. Greenberg and C. Vélez-Ibáñez, "Community Dynamics in a Time of Change: An Ethnographic Overview of the Upper Gulf," in T. R. McGuire and J. B. Greenberg, eds., *Maritime Community and Biosphere Reserve: Crisis and Response in the Upper Gulf of California*, Occasional Paper No. 2, Bureau of Applied Research in Anthropology (Tucson: University of Arizona, 1993).

62. J. B. Greenberg, "Territorialization, Globalization, and Dependent Capitalism."

63. E. Snyder-Conn and R. C. Brusca, "Shrimp Population Dynamics and Fishery Impact in the Northern Gulf of California," *Ciencias Marinas* 1(3) (1977): 54–67.

64. Cudney-Bueno, "Management and Conservation."

65. Hoyos, "Auto Veda de la Pesquería de Camarón."

66. Ignacio Ibarra, "Shrimped out in Sonora," *Arizona Daily Star*, 9 March 2001, A1.

67. Cudney-Bueno and Turk-Boyer, *Pescando entre Mareas de alto Golfo de California*.

68. Cudney-Bueno, "Management and Conservation."

69. McGuire and Greenberg, *Maritime Community and Biosphere Reserve*.

70. Peter Nichols, "Water for the Colorado River Delta," unpublished paper, University of Colorado School of Law, 2000.

71. NAAEC, Article 1(g).

72. Letters from Defenders of Wildlife et al., to Bruce Babbitt, Secretary, United States Department of the Interior, et al., "Notice of Violation of the Endangered Species Act Relating to Lower Colorado River Activities," 14 December 1999, at www.defenders.org, accessed 15 June 2001.

73. See, for example, *O'Neill v. United States*, 50 F3d. 677 (9th Cir. 1995) (contractual arrangements, including those where the government is a party are subject to the provisions of subsequent legislation), and *Natural Resources Defense Council v. Houston*, 146 F.3d 1118 (9th Cir. 1998) (the Bureau of Reclamation has discretion to reduce the amount of water in a contract in order to comply with the ESA).

74. For a discussion of the Supreme Court's Foley/Aramco doctrine, see David Hunter, James Saltzman, and Durwood Zaelke, *International Environmental Law* (New York: Foundation Press, 1998): 1429.

75. Environmental Defense Fund, *A Delta Once More: Restoring Riparian and Wetland Habitat in the Colorado River Delta* (New York: EDF, 1999).

3

Scientists and Scientific Uncertainty in EU Policy Processes

Bovine Spongiform Encephalopathy and Bovine Hormones

M. Leann Brown

Modern policy problems can be so novel, complex, interrelated, and technical that policy makers may be unaware of the dilemmas until scientists identify them, may lack understanding of the nature of problems, including their intrinsic cause-and-effect relationships, and may be unable to craft effective policy solutions. Policy makers increasingly must address highly controversial issues, such as those associated with food quality, for which there is inadequate scientific knowledge and no scientific consensus. Legislation must be undertaken in the dark, without a definitive scientific basis, often in a confrontational political context. This study investigates the roles of scientists and the effects of scientific uncertainty in EU policy making in response to a new disease in cows, bovine spongiform encephalopathy (BSE, or "mad cow" disease) that has been linked to Creutzfeldt-Jakob Disease (CJD), a lethal human illness.[1] Comparison with the approximately contemporaneous issue of whether hormones should be administered to enhance beef and dairy production—bovine growth hormones (BGH) to fatten cattle and increase milk production—highlights other factors that affect EU response to food-quality issues. BSE and the use of bovine hormones to fatten cattle are novel examples of food-security issues, a fundamental concern of modern governments, in which scientists have been integrally involved in the policy processes; scientific uncertainty remains high, and political pressures have been intense. However, the scientific and political processes involved and the EU policies that emerged differ in important ways.

The food-security issues of this study provide an arresting reminder that the consequences of humankind's intervention into the natural environment are difficult to predict and may yield both benefits and costs in human health and welfare. These cases raise several important questions: How do high levels of scientific uncertainty and a lack of scientific consensus affect political decision making? What roles do scientists play in policy making? How do high levels of scientific uncertainty affect those roles?

The next section discusses the scientific issues raised by BSE and and the use of bovine hormones, including the empirical and methodological challenges scientists encountered trying to understand the phenomena, how divergent issue "framing" affected these processes,[2] and how scientific uncertainty and a lack of scientific consensus affected scientists' ability to provide effective input into the policy processes. Challenges such as the potential cross-species migration of disease, the unknowns associated with a new disease, acquiring valid and reliable evidence linking BSE to CJD, and assessing the risk to humans and animals in the face of incomplete evidence confronted BSE scientists. Private- and public-sector scientists often differed on whether BSE should be framed as an animal health and welfare issue, a human health issue, or both, and on the extent to which economic, social, and political concerns should figure in their policy recommendations.

SCIENTIFIC KNOWLEDGE AND UNCERTAINTY

The BSE epidemic and use of artificial hormones to enhance beef and dairy production are novel, highly complex, human and veterinary medical challenges fraught with high levels of scientific uncertainty. As a result, no consensus exists among EU and global scientists regarding the various cause-and-effect relationships, appropriate research design and methods, the levels of risk involved, and the most effective policy solutions.

BSE and Science

Animal remains have been used as cattle feed in Britain since the 1930s, but during the 1970s and 1980s, solvents thought to be a health risk to rendering workers were banned and cost-saving lower temperatures were used to process the feed. Some scientists believe these new techniques allowed a resilient strain of a sheep disease, scrapie, to enter the cattle feed and reemerge as a new disease, BSE. Other scientists do not accept the scrapie theory. In March 1991, Iain Pattison, a retired British scrapie researcher, wrote in the

Veterinary Record that the scrapie theory "could have been tested experimentally if known sheep scrapie had been fed to normal cattle." The head of the British Ministry of Agriculture, Food, and Fisheries' (MAFF) epidemiology team, John Wilesmith of the Central Veterinary Laboratory in Surrey, dismissed Pattison's proposal in the same journal two weeks later: "It would only confirm what epidemiological study has already shown."[3] Other potential sources of the disease being investigated include the injection of hormones derived from the pituitary glands of slaughtered cows to improve breeding, vaccines, a reaction to microbes or organophosphate treatments, or a mutation of the prion protein gene.[4]

A veterinarian in Surrey, England, first brought BSE to the attention of authorities in 1985, but scientists only discovered that BSE had been transmitted across species to cats and pigs in 1990.[5] Table 3.1 illustrates the ever-widening scope of the epidemic. In 1990, there was no suggestion that the disease could be transmitted through food to humans. By 1995, BSE had been identified in cattle in three other EU countries.

In March 1996, a joint memo from the Health and Agriculture Ministers to Prime Minister John Major confirmed that scientists were prepared to assume a link between BSE in cattle and CJD in humans. As many as 500,000 BSE-contaminated cattle carcasses are thought to have been used in cattle feed and entered the human food chain.

At the onset on the BSE crisis, no consensus obtained as to whether the disease posed a danger to humans, and uncertainty persisted around such issues as the incubation period for the disease, the procedures necessary to deactivate the disease, and how rapidly and far the disease had spread beyond Britain. In 1988, roughly two years after BSE was first identified,

Table 3.1 Number of Reported Cases of BSE in the EU (as of May 22, 2001)

EU Member State	*1995*	*1996*	*1997*	*1998*	*1999*	*2000*	*2001*
Austria	0	0	0	0	0	0	0
Belgium	0	0	1	6	3	9	12
Denmark	0	0	0	0	0	1	1
France	3	12	6	18	31	162	37
Germany	0	0	0	0	0	7	50
Ireland	16	73	80	83	96	149	0
Italy	0	0	0	0	0	0	1
Luxembourg	0	0	1	0	0	0	0
Netherlands	0	0	2	2	3	0	7
Portugal	14	29	30	106	159	114	34
Spain	0	0	0	0	0	2	42
UK	14301	8013	4309	3178	2254	1352	196

Source: www.cnn.com/interactive/world/europe/0010/cjd/content.html (accessed October 12, 2002).

epidemiologists in Britain's MAFF announced that the source was probably scrapie, a familiar sheep disease considered harmless to humans. Critics charge that the assumed link between scrapie and BSE resulted in insufficient research into BSE's possible connection to CJD in humans. The European Parliament's (EP) 1997 BSE Inquiry Report (hereafter the EP's 1997 Report) is unequivocal on this point:

> The necessary research effort was not carried out, nor were the research fields properly defined so as to obtain information rapidly and determine the risk to humans; indeed, obstacles were put in the path of scientists adopting more critical attitudes to the inadequacy of the precautions being taken.[6]

The discovery of BSE in cats and pigs in 1990 signaled a variance from scrapie in its ability to transcend the species barrier and suggested a potential threat to humans.

Before 1990, the lack of scientific information and understanding allowed UK and EU policy makers to assume no harm and act on interests other than public health. Between 1989 and March 22, 1996, British officials and the EU's Scientific Veterinary Committee (SVC) maintained that the risk of transmission of BSE to humans, should it exist, was remote. The following quotation confirms UK officials' framing of BSE as a public relations problem rather than a human-health crisis. In September 1990, Britain's chief veterinary officer stated at a meeting of the EU Standing Veterinary Committee:

> Given that there is no risk to human beings, what we are now proposing is once again based on an extremely cautious approach, because there is public concern. People are worried about this new disease and, to tell the truth, it is more an issue of consumer confidence than consumer protection. . . . But we say that there is no risk, or indeed any proof of such a risk.[7]

The British government finally acknowledged the serious and imminent risk of the disease spreading to humans in March 1996.

Until July 1994, no consensus among experts existed regarding the procedures required to deactivate BSE agents in meat-and-bone meal manufacturing processes. In January 1993, the EP adopted a resolution calling for amending EU regulations to require 134°C for twenty minutes at three bars pressure for processing the feed. The Commission rejected the proposal on the grounds that these parameters were not acceptable to the scientific community, did not guarantee safety, and might convey a false impression of security. When the EU Council (the Council) finally legislated on this issue (Decision 96/449/EEC, which entered into force April 1997), it imposed similar regulations (133°C for twenty minutes at three bars of pressure), which the EP cited as evidence of the Commission's failure to adopt a precautionary policy in the face of scientific uncertainty.[8]

Another area of uncertainty centered on the extent to which BSE had spread to other western European states. Many animals with unusual, undiagnosed symptoms have likely been slaughtered and ended up on supermarket shelves. Between 1989 and 1997, Ireland, Portugal, and nonmember Switzerland reported a total of five hundred cases of BSE caused by British feed. Austria, Belgium, Finland, Luxembourg, Spain, and Sweden reported none. Denmark, Germany, and Italy reported one, five, and two cases, respectively, whereas France reported twenty-seven cases (see table 3.1). The head of BSE research at the Dutch Institute for Animal Science and Health believed that these figures were too low to be true. Based on the number of cows infected in Britain and the quantity exported to western Europe, at least 1,688 should have become ill. In 1997, only France and Germany reported to the Commission that they had in place surveillance procedures for the disease.[9]

Bovine Hormones

Although its causes and processes are not completely understood, BSE is likely a consequence of changes in industrial practices designed, in part, to protect workers and lower production costs. In contrast, the problem of bovine hormones was created by scientists like other issues such as the use of other biotechnologies, global warming, and nuclear power. Cattle naturally produce the protein hormone bovine somatotropin (BST), which affects distribution of feed to bodily functions like growth and lactation. In the early 1980s, the gene governing the synthesis of BST was isolated and cloned by the U.S. corporation Genentech, and since 1982, it has been possible to produce large quantities of what is called recombinant BST (rBST) through genetically engineered bacteria. Administering hormones may increase yield by as much as 15 and 10 percent in beef and dairy cattle, respectively; and the Canadian and U.S. beef industries commonly administer several kinds of hormones to enhance production. Genetically engineered bovine hormones have been on scientific agendas for only two decades.

Due to the novelty and the complexity of these substances, there are no generally accepted and adequately validated methods for identifying and monitoring endocrine disruptors (EDs), which alter the functioning of organs and glands that secrete hormones. The sheer numbers of potential EDs and the complexity of their molecular effects and control mechanisms hamper assessment. BGH is a protein composed of approximately 190 amino acid residues released from the anterior pituitary gland as four molecular variants. Preformed BGH, stored in pituitary somatotroph cells, may be released in response to several stimuli, including BGH releasing factor and somatostatin from the hypothalamus, blood concentrations of glucagon, insulin-like growth factors and estrogen, and psychological stimuli such as stress and sleep. BGH action on its target tissue is mediated by a variety of factors such as concentrations of hormones and metabolites in the blood, the types and level of blood plasma binding proteins, the tissue distribution and concentration of BGH receptors, and transmembrane signaling mechanisms.[10] In 1999, of the more than 560 suspected EDs identified by European Commission[11] experts, literature pertaining to only 116, mostly high-production-volume chemicals and some metals, had been analyzed.[12]

In addition to the high levels of scientific uncertainty deriving from the novelty and complexity of these substances, much of the debate concerning bovine hormones derives from value differences among EU and other scientists (shaped or determined by their disciplines, institutional roles, and national historical and cultural influences) on a host of issues, including scientific innovation, human health, animal welfare, environmental protection, and economic growth. How scientists prioritize these values affects how they frame the issues and the advice they provide political decision makers. For example, evidence suggests that some EU scientists are more concerned about animal welfare than North American scientists. A report of the EU's Scientific Committee on Animal Health and Animal Welfare delineates these differences:

> The fact that farm animals are reared for commercial purposes should not cause us to forget that they are living and sensitive creatures which need to regulate their lives and avoid suffering. The concept of welfare has to be defined in such a way that it can be scientifically assessed and the term can be used in legislation and in discussion amongst animal users and the public.[13]

Concerning the U.S. Food and Drug Administration (FDA) assessment of the effects of rBST on dairy cattle, the EU scientific committee commented, "it appears that no animal welfare concerns were considered at all."[14] Most studies indicate no adverse animal health and welfare effects from administering rBST, and veterinary advisory boards in the EU and United States approved the use of rBST. However, a few studies suggest a negative impact on the health and welfare of cows associated with higher milk production (e.g., higher incidence of mastitis, or inflammation of the mammary glands), foot and leg problems, and swelling around the sites of the biweekly injections.[15] Some EU scientific advisers regard these latter studies as justification for banning the use of bovine hormones.

Although most scientists believe that beef derived from cattle given several growth hormones represents no threat to human health, the EU Scientific Committee on Veterinary Measures Relating to Public Health concluded that:

> In the case of Estradiol-17 *B* here is a substantial body of recent evidence suggesting that it has to be considered as a complete carcinogen, as it exerts both

tumor initiating and tumor promoting effects. The data available does not allow a quantitative estimate of the risk.[16]

Similarly, although most scientists believe that the use of hormones to promote milk production does not pose a threat to human health, others disagree. The higher concentration of insulin-like growth factor-l (IGF-l) concerns some.[17] There is also concern that the necessity of treating animals for mastitis associated with higher milk production may compound existing problems of antibiotic resistance in human consumers of the milk.

The case might be made that the novelty, complexity, scientific uncertainty, and divergent framing associated with these cases precluded scientists' bringing consensus and authoritativeness to EU policy processes. However, scientific input is a necessary component of policy making. Later we will discuss how research findings and advice from scientists from multiple disciplines is funneled into EU policy processes through government bureaucracies and scientific advisory committees. The next segment investigates various economic, political, and social factors that shaped and interacted with the state of scientific understanding and scientists' roles to shape EU policy. As you read, you might consider the following questions: Did UK and EU policy makers inappropriately privilege economic interests relative to public-health concerns? How did such factors as sectoral and bureaucratic clout, political ideology, and communication affect policy processes? Were governmental actors and processes overwhelmed by the BSE issue, such that they became chaotic or irrational? How did historical cultural factors influence how scientists, policy makers, and the general public conceptualized these issues?

ECONOMIC, POLITICAL, AND SOCIAL INFLUENCES

As both BSE and bovine-hormone use relate to food and reached the political agenda in the late 1990s, they evolved within the same economic, political, and social context, and some of the same actors—especially farmers, food scientists, and agricultural ministries—appear in both cases.

Economic Influences

Both BSE and bovine hormones created mixed incentives for the principal economic participants. Individual farmers and the agrarian sector as a whole seek to increase production and profits and reduce costs. Feed regulations to prevent the spread of BSE would directly hurt the feed producers' profits and, by increasing feed costs, would also reduce farmers' profits. Farmers also wanted to limit the cost of BSE in destruction of herds and reduced domestic sales and exports, but destroying herds would help restore confidence among domestic and export consumers that British beef was safe. Similarly, the use of hormones in beef and dairy production would increase farm profitability, but regulations restricting the importation of hormone-enhanced products also would protect farmers from import competition.

Member-state governments and the EU have their own economic and political objectives, including limiting economic costs associated with disease and reduced exports, meeting consumer demands for a safe food supply, and fulfilling multilateral trade obligations. These food crises undermined EU free-trade objectives and its public legitimacy and embroiled the organization in disputes with its trading partners. The EU's CAP had produced burgeoning surpluses of agricultural goods, and, at any given time, approximately 50 percent of the EU budget is required to sustain CAP programs. Thus, the EU has no economic incentive to increase agricultural production and strives to limit import competition. For a host of historical, cultural, and institutional development reasons, agricultural interests are among the most powerful in member-state governments and politics and clearly are the most powerful among EU advocacy groups.

In October 2000, it was reported that the BSE crisis had cost the UK £4 billion and that some of its EU partners still refused to import British beef. The cost of dealing with BSE grew as the epidemic spread. In 2000, France's largest farmers' union called for the slaughter of millions of cattle at a cost of approximately FF20 billion (US$2.61 billion). The cost of the crisis also spread to additional sectors, including other foodstuffs, retailers, and restaurants.

In contrast with BSE, which was predominantly a British and EU public-health, economic, and political problem, the EU also had to take into account international economic legal considerations in the bovine-hormone case. On January 1, 1995 the World Trade Organization (WTO) Agreement on Sanitary and Phytosanitary Measures, which allows signatories to limit imports on sanitary-protection grounds only on the basis of scientific risk assessment, entered into force. Canada and the United States charged that the EU's banning use of six beef hormones[18] contravened WTO provisions. The WTO found in favor of the plaintiffs and subsequently its appellate body found that the EU had provided:

> general studies which do indeed show the existence of a general risk of cancer; but they do not focus on and do not address the particular kind of risk here at stake—the carcinogenic or genotoxic potential of the residues of those hormones found in meat derived from cattle to which the hormones had been administered for growth promotion purposes. . . . [T]hose general studies are in

other words relevant but do not appear to be sufficiently specific to the case at hand.[19]

In response to these findings and more specifically to the WTO's finding that general studies are an inadequate basis for the import bans, the Commission initiated seventeen new scientific studies and requested its Scientific Committee on Veterinary Measures relating to Public Health (SCVPH) to conduct hormone- and residue-specific risk assessment for the six hormones in question on the basis of existing data and to review EU legislation. The Commission instructed that in addition to scientific data, SCVPH risk assessment should take into account the difficulties of control, inspection, and enforcement of the requirements of good veterinary practice, in addition to three particular EU interests: maintaining a high level of consumer-health protection; establishing objective, transparent, and reliable procedures for evaluation of risk (which alludes to EU desires to promote process and organizational legitimacy); and avoiding WTO members' erecting reciprocal barriers against EU exports.[20]

Political Influences

When the European Economic Community was established in 1957, its primary objectives were the liberalization of intraregional trade and the establishment of a customs union and a common agriculture policy. Over time, the European Community has evolved into a fifteen-member European Union (growing to twenty-five members on May 1, 2004) that also institutionalizes cooperation in home and justice affairs and foreign policy and security matters. The EU is commonly conceptualized as a "hybrid" intergovernmental and supranational entity, wherein the member states retain veto power over issues like treaty making and military security, but have ceded sovereignty over most economic matters, including agriculture and trade, to the regional body. Prerogatives and responsibilities associated with protecting human health and the environment are shared between the member states and the EU. In 1992, the "subsidiarity principle" was incorporated in the Treaty on European Union (the Maastricht Treaty), which prescribed a division of labor between the member states and the EU wherein the union undertakes actions only if the objectives of the actions cannot be sufficiently achieved by the member states and the proposed actions could be better achieved by the EU due to the scale of the policy effects.[21]

When they were first confronted with these novel and scientifically unclear issues, the first inclination of member-state governments and the EU was to frame these issues as agricultural and trade, rather than public-health, problems. Thus, British and EU responses to the issues were shaped by the political clout of the agrarian sector, the nature of scientific advice entering decision-making processes, and the bureaucratic and ideological context of decision making. Scientific experts contribute to policy processes from inside and outside of government. However, government experts' ability to influence policy is often a function of bureaucratic context, politics, ideology, and other institutional forces in play at the time. Such factors as bureaucratic clout and resources, the level of competition and conflict within and between bureaucracies, the quality of communication among bureaucracies, and the level of political and ideological support or antipathy for the experts' recommendations influence their success in shaping policy options. The BSE case provides robust evidence that governmental experts' influence was directly proportionate to the power of their bureaucracy. In both Britain and the EU, agricultural bureaucracies demonstrated significantly more political clout than those charged with public health. In Britain, rivalry between MAFF and the Department of Health was so severe that at one point John Gummer, then agriculture secretary, implored then Health Secretary Kenneth Clark not to respond to questions in the British parliament on food safety to avoid revealing policy differences.[22] At the outset, MAFF experts' opinions on the BSE crisis, rather than those of the Department of Health, dominated British government efforts to deal with the disease, and MAFF experts provided leadership on the relevant expert committees at the EU level.

A lack of resources and conservative governments' impetus to deregulate also hamstrung experts' ability to deal with the problem. For example, the British former chief medical officer claimed that in the midst of the BSE crisis, the pressure to streamline government spending was "almost continuous and took away much of my energy and intellectual resources." Conservative governments' antipathy for governmental regulations also figured in bureaucratic experts' ability to implement policy. British environmental-health officers complained that they were not given the resources to enforce regulations passed in the early 1990s, including the ban on "specified bovine offals." This provides an important reminder that scientific knowledge may be intentionally or inadvertently ignored or misapplied in all stages of the policy process, in this case implementation. The chairman of the British government's senior scientific advisory body on the BSE crisis reported that "the situation would have been transformed if the regulations had been applied rigorously."[23]

An additional bureaucratic factor affecting the influence of scientific experts relates to communication and coordination between and among the various sources of scientific advice and with political decision makers. For example, there were no overt mechanisms of communication between the EU SVC and the Standing Veterinary Committee to guarantee that the scientific opinions of the SVC were accurately presented to the Standing Veterinary Committee, although this is the Commission's responsibility.[24] Each organizational subunit has its own interests and agendas, and these, as well as quality of communication between them, affect how scientific information is used in decision making.

High levels of scientific uncertainty combined with politicization may profoundly affect policy processes. The BSE case demonstrates that when confronted with novel, complicated, and scientifically uncertain issues with the potential for devastating public-health consequences and public agitation, policy makers may exhibit denial and avoidance behaviors, such as refusing to adequately fund research and investigations into the problem, with the intended or unintended consequence of perpetuating scientific uncertainty. Officials may also transfer the issue to scientific committees to delay decisions or shift responsibility for dealing with the problem to actors and venues less susceptible to public scrutiny. For example, Richard Southwood, chairman of the British Working Group on BSE established in 1989, later told EP officials that his government had not asked the working group whether exporting meat-and-bone meal might put herds in other countries at risk. This possibility, however, was discussed on numerous occasions in the EU SVC and was raised by the European Renderers' Association at several meetings in 1990 and in at least one association correspondence to the SVC in February 1990.[25]

Veterinary checks for BSE were suspended in the UK between 1990 and 1994. The EP's 1997 Report describes the situation in this way:

> Research can be carried out when some stimulus exists . . . where there is a problem and when there is sufficient raw material to carry out such research, in this case the organs of diseased animals. Although both these conditions were present in the UK during the years when the disease was at its height, no scientific findings were made. This is a further area where responsibility can be seen to lie with the UK.[26]

The case can be made that British political actors sought to obscure the nature and extent of the problem by perpetuating the scientific uncertainty surrounding the issue. For example, the UK "put pressure on the Commission not to include anything related to BSE in its general inspections of slaughterhouses, as periodically carried out between 1990 and 1994 in the context of their adaptation to the internal market."[27]

Social and Cultural Influences

Cultural factors exert contextual and direct influence on EU food policies. Intense and deep-seated emotions and myths are associated with farm life and food in all societies, a good example of which is stereotypically ascribed to the French. In the last century, some western Europeans suffered starvation as a consequence of war and are thus more sensitive to food-security issues than are most North Americans. Two additional cultural factors most immediately relevant to these cases are western Europeans' mistrust of scientific innovation and technology as compared with North Americans and (as previously discussed) their sensitivity to animal-welfare issues. During the twentieth century, scientific experimentation undertaken by the Nazi regime and the thalidomide disaster[28] provided tragic evidence that scientists and science can be malevolent. Recent controversies surrounding biotechnology provide ample evidence of how these cultural concerns can easily erupt into the political arena. These cultural differences have resulted in the EU's embracing of the so-called precautionary principle, which subsumes a cluster of concepts, including preventative action, proportionality of response, and duty of care.

The precautionary principle, derived from German sociolegal tradition, gained recognition in the 1980s, along with the rapid development of EU law. Although there is no single agreed-upon definition, the principle has been ensconced in various international agreements, including Principle 15 of the 1992 Rio Declaration on Environment and Development and several important marine protection conventions. The International Court of Justice (ICJ) has ruled that the precautionary principle represents customary law and should be taken into account by legislators internationally.

Industry is concerned that this principle is dangerous in its lack of a common definition and in the growing number of calls for more radical interpretations of the principle—for example, demands for reduction or cessation of activities without any evidence of their detrimental effects on the environment or human health. Several EU industrial organizations have called for a reasonable interpretation of the principle (e.g., to incorporate cost-benefit analysis and scientific-risk analysis) that also recognizes societal concerns.[29] The United States apparently rejects such application of the principle. These cases suggest that the use of the precautionary principle may be case-specific and influenced by political and economic interests. When this divergent issue framing that derives from cultural and experiential differences is taken into account, it is not surprising that EU and North American policy makers advocate different policies. The following section will explore the sources of primary research on these food issues and how that information is translated into various policy alternatives via scientific advisory committees.

THE DEVELOPMENT OF KNOWLEDGE

As the BSE and bovine-hormone issues claimed the attention of the scientific and policy communities, primary research on the scientific puzzles discussed in the section titled "BSE and Science" continued apace within EU and non-EU states (including Canada, Japan, and the United States), other intergovernmental organizations, such as the World Health Organization (WHO), and transnational corporations. The findings of "pure research" were channeled into EU decision processes via various bureaucratic bodies, including scientific advisory committees.

Member-State and EU Research

As would be anticipated, EU member states and the organization itself commissioned and funded primary research programs to address these food-security issues. The EP's 1997 Report stated that to date Britain and the EU had devoted £60 million and ECU 3.75 million, respectively, to BSE research.[30] In an interesting twist, it appears that in 1994 the (British) head of the Commission's Directorate for Employment, Industrial Relations, and Social Affairs (DG V) used one research program to deter initiating research into the link between BSE and CJD. When the (German) head of DG V's public-health division advised his superior to respond to pressure from the German Council delegation and immediately launch such an investigation, the director rejected the advice, stating that no such investigation could be started pending the results of a planned research program into unusual diseases.[31] Later, the official would claim that DG V lacked the resources to carry out its responsibilities. The EP's 1997 Report assessed DG V's research role: "It thwarted research into links between CJD and BSE at a time when the risks to humans were already largely foreseeable." In its final assessment, the EP asserts that the necessary research was not carried out with regard to BSE, the research undertaken did not properly define the areas of investigation to obtain information rapidly and ascertain the risk to human health, and the Commission supported activities to complement British research, rather than funding wide-ranging EU research. The EP further charges that obstacles were placed in the path of scientists adopting more critical attitudes to the inadequacy of the precautions being taken. In February 2001, the EU announced a new €971 million package (nearly $1 billion) of anti-BSE measures.

With regard to bovine hormones and other EDs, the EU was more inclined to assume research leadership and proactively engage existing scientific uncertainties. The EU's Fourth European Framework Programme for Research, Technology, and Development funded research on environmental aspects of EDs and began to investigate links between endocrine disruption and human health. One project addressed the possibility that exposure to EDs may result in decreases in human sperm counts and increases in testicular cancer. Another developed a bioassay to detect EDs in the environment. Achievements of the Fourth European Framework Programme were outlined in the report "Indocrine Disrupters, How to Address the Challenge?" issued as the proceedings of a Joint Conference of the European Commission, its Directorate General for Environment, Nuclear Safety, and Civil Protection (DG XI), and the Austrian Council Presidency in November 1998.

Under the Fifth Framework Programme, a new project funded under the "Environment and Health" Quality of Life Programme sought to identify dietary and environmental factors responsible for male urogenital malformations and low sperm counts. Additional research was funded between 1999 and 2001 in both the Quality of Life and the Environment programs.[32] In 2000, the Parliament expressed concern in the decline in funding for ED in the Fifth Framework Programme relative to previous programs. It called on the Commission and member states to ensure that sufficient resources for independent research were allocated in the Sixth Framework Programme. The Parliament further recommended that interdisciplinary and cross-border research projects be coordinated, involving independent participants of different scientific backgrounds. Issues of transparency and public education also concerned the Parliament. It requested that member states and the Commission make information on EDs widely available on the basis of the right-to-know principle, opining that "such information should also seek to accurately communicate all the uncertainties to the public and the actions that are being taken to address the uncertainties and the problem in general."[33]

Intergovernmental Organizations

Intergovernmental health organizations played consultative roles in managing the BSE crisis. For example, the 24–25 June 1996 BSE inquiry convened by the EP heard testimony and took evidence from several external experts including the general director of the Office International des Epizootes (OIE) headquartered in Paris, the director of the WHO's Division of Emerging and Other Communicable Disease Surveillance and Control, and a member of the technical committee of the European Renderers Association in Rotterdam. The general director noted that the OIE, with 143 member countries, is the only world organization devoted to animal health. OIE assets include a permanent staff of 30, an estimated 100 expert volunteers, and 104 reference laboratories. The organization's objectives include global surveillance of 104 animal diseases (including BSE) and harmonization of regulations relating to trade in animals and animal products through the OIE International Health Code. The EU and WHO participate in the OIE's fact-finding meetings.[34]

The WHO initiated regular monitoring of BSE's potential risk to human health in 1986. It convened four expert meetings in 1991, 1993, 1994 (with the OIE), and 1995. Following the report of ten cases of the new variant of CJD (vCJD) in the UK in March 1996, the WHO convened two consultations to review the public-health issues related to BSE and the new variant of CJD.[35] In December 2000, the WHO announced fresh initiatives to address global concerns over BSE, including convening a global conference of experts and officials.

Corporate Research

As might be anticipated, most research relating to BSE and vCJD was undertaken by EU actors and the international health community, but the bovine-hormone case well illustrates the salient role of transnational corporations in global ED research. As was noted, in the early 1980s U.S. corporations developed hormones to enhance beef and dairy production, undertook field studies to convince the U.S. FDA of their safety, and were required by the FDA to monitor the consequences of their usage. For example, Monsanto was required to establish a postapproval monitoring program for Posilac (PAMP), the commercial version of BST. Monsanto research findings, available to EU decision makers and the public via published corporate reports and the FDA's Freedom of Information guidelines, are specifically cited in EU scientific advisory committees reports. The EU's Scientific Committee on Animal Health and Animal Welfare 1999 Report on Animal Welfare Aspects of the Use of Bovine Somatotropin also cited a 1998 report by scientists from Canada's Health Protection Branch. The Health Canada report said that antibiotic resistance in farm-borne human pathogens linked with the increased risk of mastitis associated with the use of BST had not yet been properly investigated, although it has obvious human-health implications. Although Canadian authorities permit use of hormones to enhance beef production, in January 1999 they denied approval for use of rBST in Canada due to "a sufficient and unacceptable threat to the safety of dairy cows." This conclusion was supported by a scientific report from a committee of veterinary experts headed by an internationally recognized veterinary epidemiologist, which found increased risks of mastitis, infertility, and lameness. The EU also cited data from Canadian scientists in its submission regarding the effects of bovine hormones to the Joint FAO/WHO Expert Committee on Food Additives (JECFA) meeting in February 1998. The JECFA expressed the view that mastitis induced by bovine hormones was not within the committee's purview; however, it was willing to consider the risks associated with possible increased use of antibiotics. In this document, the JECFA specifically referenced PAMP data.[36]

Scientific Advisory Committees

In addition to supporting "pure research" to generate scientific understanding, governments at all levels receive systematic, ongoing scientific input into policy processes from both external and internal scientific experts. This input customarily is formalized as the findings of scientific advisory committees to provide risk assessment and guidance as to the scientific merits of alternative policy choices (i.e., risk management). Governments need to be seen to be taking advice, but it matters how, why, and from whom they accept it. The degree to which scientific advice and government policy are regarded as legitimate rests on the legitimacy of the processes. Political culture and constitutional context determine the legitimate ways to receive information and advice.[37] From the perspective of the EU Commission, scientific committees may serve as an instrument for generating scientific and policy convergence and consensus, or hammering out compromises, in addition to providing informational and legitimizing functions. Expert committees are most influential when they have significant resources, are constituted to provide both scientific expertise and interest representation, have a clearly delineated mission, and operate within a consensual decision-making culture. Some analysts criticize the growing importance of expert committees in EU decision making on the grounds that they may lack transparency, act as corporatist cartels, and distort the institutional balance. In many cases, expert committees meet behind closed doors, do not publish their minutes and opinions, and inform only the Council and Commission of their conclusions.[38]

As was previously noted, several EU advisory committees have provided input regarding these food-security issues, including the Scientific Committee for Food (SCF), the Standing Committee for Foodstuffs (StCF), and the SVC. The SCF was established by a Commission decision in April 1974 and revised in July 1995.[39] The EC commissioner for consumer affairs has appointed its members since 1997. Early in its existence, the SCF worked mainly in food toxicology and the majority of appointees were toxicologists. As other issues appeared on its agenda, its composition broadened to include nutritionists, microbiologists, and biotechnologists. Although most of the experts come from within the EU, occasionally some particularly qualified non-EU expert may be invited to contribute information. It is often noted that the role of scientists is to provide risk assessment and the responsibility of politicians is to provide risk management. The primary task of the SCF is the interpretation of data and its translation into criteria that may be used by EU legislators. The committee gives its opinion at the request of the Commission, but also may draw the attention of the Commission to any food-safety problem on its own initiative.[40]

Debate centers on the extent to which SCF serves the political interests of the Commission. The Commission is not required to solicit input from the SCF and may decline to do so in cases where it anticipates a ruling incongruent with its preferences. By 1998, the Commission had received the SCF's assessment of thousands of substances and only in one case, the safety of gelatin derived from bovine bones, did the Commission propose use of a substance that had received an unfavorable SCF assessment. Jurgen Neyer discusses the objectivity of the SCF:

> The SCF is acknowledged among nearly all delegates for the high quality of its work and its unique expertise. That said, member states are well aware that the SCF as an institution is not as objective as it is supposed to be: sometimes it is used by the Commission to give additional evidence only to those policies which are in line with the Commissions' interests. Moreover, the BSE . . . case has highlighted the fact that even scientific institutions can easily be captured by certain interest groups and be used for political purposes by the Commission.[41]

Despite this charge, the SCF is the EU body most directly charged with scientific assessment with regard to food safety.

In 1969, the Council of Ministers established the StCF to institutionalize intraorganizational cooperation with regard to implementation of food legislation. The Commission's Directorate General for Industry (DGIII) chaired the StCF, and its members represent the member states congruent with its role to support (and control) the Commission with regard to implementation of foodstuffs regulations. Legislation referring to its role notes the necessity for consulting the most recently available scientific evidence, and its mandate focuses exclusively on the regulatory need to meet safety requirements on a case-by-case basis.

In practice, the Commission organized the StCF sessions, set the agenda, and controlled the flow of information to committee members. StCF delegates heavily criticized the Commission's using SCF findings as an instrument to further economic harmonization. The Commission used its agenda-setting power to introduce multiple topics in a single session, which facilitated "package deals"; it also placed high demands on the technical expertise of member-state delegations, which they were not always able to meet. The Commission was also known to impose a tight time schedule such that national delegations had little time to consult with domestic constituents, formulate their positions, and respond to Commission proposals. Despite these attempts to control, votes in the StCF were unanimous less than 50 percent of the time. Of twenty-five votes analyzed by Neyer, 24 percent were disputed, but resulted in positive votes; however, 8 percent of Commission proposals were rejected.[42]

With specific regard to BSE, concern about the lack of objectivity of the BSE subgroup of the EU SVC was generated by the fact that it was almost invariably chaired by a British national and its membership dominated by British experts and MAFF officials.[43] As might have been anticipated, this derived initially from the fact that Britain had more experience with BSE than other member states, but the persisting imbalance raised issues of bias.

The head of the UK's Central Veterinary Laboratory, subsequently adviser to MAFF, acted as *rapporteur* on BSE at full meetings of the SVC. Some SVC members suggested that he withheld information.[44] The EP's 1997 Report charges:

> The preponderance of UK scientists and officials, therefore, meant that the SVC tended to reflect current thinking at the British Ministry of Agriculture, Fisheries and Food. . . . It made a partial and biased reading of the advice and warnings of the scientists. The views of certain scientists who could be considered more critical were not taken into account.[45]

The EP opined that the Commission placed too much weight on the role of the SVC given its lack of reliable data and domination by the British. Commission officials from the EC Directorate General for Agriculture (DG VI) repeatedly told Parliamentary investigators that they could not go beyond the recommendations of the SVC, because to do so would risk losing member states' support on the Standing Veterinary Committee.

Confidence in impartiality of expert bodies is enhanced by the perception that credible divergent opinions have been given a hearing. There is evidence that the Commission did not encourage the expression of views of scientists whose views differed from those of majority of members SVC or the BSE subgroup on the basis of recent research findings. The SVC's opinions are adopted by consensus, and divergent opinions were not recorded. Further, there appears to have been some overt effort to silence minority opinions. For example on September 21, 1994, the director general of DG VI called upon the German Ministry of Health to ensure that German scientists whose assessment of the risks posed by BSE to humans diverged from the consensus opinion did not express their opinion publicly. This letter read in part:

> I find it quite unacceptable that officials of a national government should seek to undermine Community law in this way, particularly on such a sensitive subject. The persons concerned have had their opportunity in the Community committees to debate their opinions. These have been rejected by the vast majority of EU experts. I would ask you, therefore, to ensure that this debate is not continued, particularly in an international forum.[46]

After its traumatic experience with the BSE crisis, in November 1997 the Commission completely restructured its scientific advisory system to include eight new scientific committees and a scientific steering committee. Although the scientific committees have an advisory role, many of their opinions are the basis for Commission legislative proposals. The scientific committees are consulted whenever EU law requires it, and they are increasingly consulted on a voluntary basis. They also may act as whistleblowers, directing the Commission's attention to new risks for consumer health.[47]

The next section provides a chronological summary of EU actions to address these food challenges. As you read, you might remain alert to the primary actors and institutions involved in policy making and the roles scientific information and scientists play in those processes.

THE EU GRAPPLES WITH BSE AND BOVINE HORMONES

Food-security issues pose significant political challenges to the EU and member-state governments. They are always potentially emotive and politicized because they relate to every citizen's basic needs. At the outset, Britain and the EU accepted the recommendations of agricultural scientific experts and bureaucrats and acted to support agricultural and trade interests. However, when human-health risks became more apparent and public alarm mounted, governments were anxious to ameliorate panic and demonstrate that everything humanly possible was being done to understand and address the problems. There is good evidence that a desire to avoid panic shaped the way the British government responded to the advice of its scientific advisory committees. Public confidence in the EU and member governments has steadily eroded as the number of food issues has grown over the past two decades; these crises have threatened the political survival of member governments. Many regard the BSE crisis as the most damaging political and legitimacy crisis sustained by the EU since the mid-1960s. And, in Britain, the BSE crisis contributed to the electoral defeat of John Major's government.[48]

BSE

In 1979, a British Royal Commission under Lord Zuckerman, set up to examine potential pathogenic transmissions in meat-rendering processes, first warned against feeding animal remains to ruminants.[49] In 1985, a veterinarian in Surrey identified a new disease in cattle, BSE. British government ministers were apprised of the disease two years later; they were told that meat-and-bone meal was the only viable hypothesis to explain the emergence of BSE. In June 1988, the UK passed the Bovine Spongiform Encephalopathy Order banning the use of certain types of meal, and a month later the UK announced a slaughter policy for animals with BSE symptoms. In 1989, export of British cattle born before July 1988 and the offspring of affected animals was banned. In 1990, scientists discovered BSE-like symptoms in cats and pigs. Despite these developments, British Agriculture Minister John Gummer publicly proclaimed beef "completely safe" for human consumption. On television, he was portrayed encouraging Cordelia, his four-year-old daughter, to bite into a beefburger. That same year, Britain set up a National CJD Surveillance Unit in Edinburgh to monitor CJD cases and investigate a possible link with BSE. The BSE epidemic reached its peak in Britain in 1992–1993, with about 100,000 confirmed cases of the disease. Nineteen-year-old Stephen Churchill, the first known victim of vCJD, died on May 21, 1995. As previously noted, British scientists and government officials were prepared to conclude a linkage between BSE in cattle and CJD in humans in 1996, more than a decade after the first appearance of BSE in the Southeast of the country.

Initially, the EU was prepared to allow Britain to assume responsibility for addressing the disease. Although the EU also banned export of British cattle born before July 1988 and the offspring of affected animals in July 1989, as was noted, EU activities to address BSE were suspended between June 1990 and May 1994, except for regulation on embryos, in a blatant example of avoidance in the face of the crisis. An extraordinary meeting of the EU Council to deal with BSE was convened 6–7 June 1990. During these meetings, an assistant director general for agriculture (DG VI) was excluded from the room when he presented a proposal drawn up by veterinary services permitting exports of UK beef in deboned form only. The Council noted that slaughterhouse inspections would be undertaken in the UK, but subsequently EU veterinary inspectors were not invited to publish their findings or present them to the EU Council or SVC.[50] When questioned later by EP investigators, the Commission admitted that it had received political pressure from the British government not to include BSE checks in the general slaughterhouse inspections carried out between 1990 and 1994. The EP's 1997 Report reproved the Council for not actively seeking to know about BSE and for hiding behind scientific committees: "There was also an attitude of 'benign neglect' of the issue (a willingness to let a British problem be dealt with by the British) on the part of the Commission and, through the veterinary committees, by the other Member States."[51] The evidence suggests that EU willingness to allow the British to address the BSE issue did not derive from legal requirements of the subsidiarity principle. Rather it was more a function of the UK's proximity to the issue and the fact that food-quality issues have historically been predominantly the province of member states. Even when BSE became a regional trade and health concern, the EU was satisfied to allow the British to provide leadership in providing analysis and policy prescriptions, given their experience with the crisis, the persisting scientific uncertainties, and the political volatility of the issue. The case may be made that the EU simply wished that the "mad cow" issue would go away.

In March 1996, after UK health and agricultural ministers confirmed scientists' opinion that there was a link between BSE in cattle and CJD in humans, the EU banned exports of British beef worldwide. Despite this, agri-industry-sponsored scientific research often continued to influence EU decision making or was used to justify political decisions with regard to BSE. For example, a proposal to lift the embargo on gelatin presented to the EU SVC by the head of the Commission's DG VI in April 1996 was blocked by opposition from a majority of the committee delegations. Between 11–15 April 1996, the Scientific Committee on Cosmetology,[52] the SCF, and the European Medicine Evaluation Agency all expressed opposition to ending the embargo on gelatin. Disregarding these recommendations, the Commission, citing the provisional 1994 Inveresk Report, which had been commissioned by the Association of Gelatin Manufacturers of Europe, persuaded the SVC that gelatin derived via certain procedures are low risk and obtained a report favoring lifting the embargo. No one had expressed confidence in the Inveresk Report: the research cited pertained to scrapie, rather than BSE; the veterinary services of DG VI were aware of the provisional and incomplete nature of the report; and a representative of the Association of Gelatin Manufacturers said of the Inveresk Report, "the results are not clear enough and we had to extend the study."[53] The EP's 1997 Report charges that the agriculture commissioner or the director of DG VI decided to reduce the scope of the embargo before the several scientific committees delivered their opinions and two weeks before the Inveresk Report was submitted to the SVC. This demonstrates that policy makers may cite as a basis or employ incomplete or inapplicable scientific evidence to persuade or justify policy preferences without actual scientific input. In testimony before the EP in 1996, the president of the European Public Health Alliance lamented, "BSE/CJD was not the first problem to result from economic logic being considered above and in isolation from other criteria."[54] In late August 1999, the EU lifted its export ban on British beef, but several EU members resolutely refuse to import these products. By 2000, vCJD had killed approximately eighty people in Britain. A baby girl born to a mother with vCJD was also found to have contracted the disease. The political fallout spread across Europe with the epidemic. In 2001, Germany's health and agriculture ministers were forced to resign in the face of heavy criticism over their handling of BSE.

Bovine Hormones

As with the BSE case, the initial EU policy response to the creation of bovine-hormone technologies appears to be a consequence of economic motivations deriving from an already burgeoning supply of agricultural goods generated by the CAP and a desire to limit agricultural imports. In 1981 and 1985, the EU promulgated directives prohibiting the use of hormones for fattening cattle by January 1, 1988.[55] In 1989, Directive 96/22/EC broadened the scope of legislation to prohibit domestic producers' administering to farm animals substances having a thyrostatic, oestrogenic, or gestagenic action for growth promotion and prohibited meat imports of animals treated with growth hormones from third world countries.

The U.S. FDA approved the use of a milk-production-enhancing BST in 1993, which Monsanto introduced into U.S. markets as Posilac in February 1994. Immediately, two applications for marketing authorization for this product came before the EU Commission. The Commission turned for expert advice to its Committee for Veterinary Medicinal Products, which recommended that trial herds be investigated for health and welfare effects in two-year postmarketing studies. At the end of 1994, the Council decided instead to extend its moratorium on the marketing and use of BST until 1998 and called for preparation of a report on its use by independent scientists by that year.[56] Thus, a veterinary advisory committee provided input into EU decision making regarding BST, but was not the determining factor in shaping the legislation.

As was noted, in the late 1990s, in response to WTO findings that the EU import ban violated trade agreements, the Commission undertook to defend itself by initiating multiple inquiries drawing upon information from in-house and member-state scientific communities, as well as Canadian, United States, WHO, and corporation research. Divergence in EU values, goals, and framing resulted in EU advisors interpreting the emerging global pool of knowledge, characterized by high levels of scientific uncertainty, in different ways. This uncertainty allowed politicians to privilege other objectives and advice such as focusing on the dangers of genetically modified products and concern about the welfare of farm animals. The 1999 Report of the Scientific Committee on Animal Health and Animal Welfare encapsulates the cluster of issues that framed EU policy makers' and scientists' thinking about the use of BST in dairy cattle: ethical views of and moral obligations to animals, animal welfare, the political debate, and the quality and changing nature of scientific evidence. In the face of scientific uncertainty, the WTO (and the United States) maintain that free-trade obligations should take precedence and the ban on beef imports should be lifted, in contrast with the EU's resort to the precautionary principle heuristic. At the end of the day, the EU has chosen to continue its ban on imports of hormone-enhanced beef products and to pay the penalties imposed by the WTO. Further, it has broadened its concerns about the potential negative effects of bovine hormones to include other EDs and is funding primary research and considering legislation to deal with the more than 500 of these substances.

The bovine-hormone and BSE cases are only two among several recent "food scares" experienced in EU countries, and the organization is deliberating the crafting of comprehensive and coherent food-safety legislation. In 1997 the Commission issued a green paper on the general principles of EU foodstuffs legislation, which proposed a framework directive and the revision and consolidation of directives on food labeling, among other things. The EU also authorized the establishment of a new European Food Safety Authority with functions similar to the U.S. FDA that began operations in 2002.

CONCLUSION

This chapter has encouraged thinking about how novel food-security issues, characterized by intricacy and high levels of scientific uncertainty, are addressed by the complex multilevel governing structure that is the EU. Gaps in scientific understanding persisted throughout, including such important issues as the cause-and-effect relationships in the natural phenomena, appropriate research designs, the risk to human beings, and the most effective policy solutions. The BSE and bovine-hormone cases suggest that these uncertainties precluded experts' sharing "causal beliefs" and formulating a scientific consensus, thereby robbing them of an "authoritative claim to policy-relevant knowledge."[57] Thus, scientists in and outside of the policy processes interpreted and framed the scientific data in different ways, and political decision makers privileged other values and criteria, such as economic and political interests over scientific advice.

EU member states and the regional organization itself fund "pure research" on these issues. And, experts from private corporations, sectoral associations, universities, governmental bureaucracies, and intergovernmental organizations are active participants in the policy process. Scientists wear multiple hats, and are part and parcel of the maze of governmental bureaucracies. The complex multilevel nature of EU decision making, as well as institutional factors such as ideology, bureaucratic rivalries, and communication difficulties, affected how scientific advice was used. Scientists, member states, and the EU also operated within the broader context of cultural values and public opinion, as well as a global arena where actors such as the WTO and the United States framed the issues according to different value priorities. The scientific experts of this study did not just rationalize and reinforce existing political preferences. However, it must be concluded that while scientists were an integral part of the political process and competition, scientific uncertainty and other factors deterred their exerting independent determining influence over policy decisions.

CHRONOLOGY

BSE

1979 A British Royal Commission under Lord Zuckerman warns of potential problems arising from changes in meat-rendering processes.

1985	A veterinarian in Surrey, England, identifies a new disease in cattle, BSE.
1987	UK government ministers are first told about the new disease; meat and bone meal are identified as the "only viable hypothesis for cause of BSE."
1988	The UK passes a law, the Bovine Spongiform Encephalopathy Order, banning the use of certain types of meal and announces a slaughter policy for animals with BSE symptoms.
1989	The EU bans export of British cattle born before July 1988 and the offspring of affected animals.
1990	Scientists discover BSE-like symptoms in cats and pigs. British agriculture minister John Gummer claims beef is "completely safe" and appears on television encouraging his daughter, Cordelia, aged four, to bite into a beefburger. The UK sets up a National CJD Surveillance Unit in Edinburgh to monitor CJD cases and investigate a possible link to BSE.
1992–1993	BSE reaches its peak in the UK with 100,000 confirmed cases.
1995	The first known victim of vCJD, nineteen-year-old Stephen Churchill, dies on May 21. BSE is identified in three additional EU member states.
1996	Scientists conclude a linkage between BSE in cattle and CJD in humans. The EU imposes a worldwide ban on all British beef exports. The ban on gelatin derived from cattle is ended. The EP initiates an inquiry into EU handling of BSE.
1997	Scientific studies on mice provide convincing evidence for a link between vCJD and BSE; the UK imposes the "beef-on-the-bone" ban.
1998	The EU Commission lifts the ban on British beef, but France continues to enforce the embargo.
2000	A baby girl born to a mother with vCJD is also found to have contracted the disease; vCJD has killed eighty people in Britain and infected at least five more, one in France. Italy bans the import of adult cows and beef on the bone from France. The first case of mad cow disease is detected in Spain. Germany confirms the first two cases of mad cow disease in cows born on German soil.
2001	Germany's health and agriculture ministers resign after heavy criticism over their handling of mad cow disease.

Bovine Hormones

1981, 1985	The EU issues directives prohibiting the use of hormones for fattening cattle.

1982	rBST is genetically engineered.
1989	The EU prohibits imports of meat from animals treated with growth hormones.
1993	The U.S. FDA approves the use of rBST.
1993	The EU Committee for Veterinary Medicinal Products' response to two applications for marketing authorization for rBST recommends allowing trial herds and two-year postmarketing studies.
1994	The EU Council bans the use and sale of rBST until December 13, 1999.
1995	The WTO Agreement on Sanitary and Phytosanitary Measures enters into force and provides that sanitary measures should be based on scientific risk assessment, but "not only risk ascertainable in a laboratory operating under strictly controlled conditions, but also risk in human societies as they actually exists, in other words the actual potential for adverse effects on human health in the real world where people live and work and die."
1999	The WTO Appellate Body finds that the EU ban on the use of six beef hormones contravened WTO provisions.

NOTES

1. BSE symptoms include tremors, weight loss, and loss of coordination. Scientists believe that BSE is connected to a new variant of CJD (vCJD) in humans. By May 2001, there were at least one hundred confirmed cases of vCJD in Britain and at least five more in other countries.

2. "Frames" are structures of meaning whereby scientists and policy actors attempt to make sense of and assign importance to issues. Information and experience are interpreted and evaluated within the context of these frames, which are generally resistant to change. James G. March and Barbara Levitt, "Organizational Learning," in *The Pursuit of Organizational Intelligence*, ed. James G. March (Malden, Mass.: Blackwell Publishers, 1999): 80–1.

3. Peter Aldhous, "Scrapie Theory Fed BSE Complacency," *New Scientist*, 150, #2025, 13 April 1996, 4.

4. James Meikle, "Fertility Jab Linked to BSE Epidemic," *The Guardian*, 9 August 1999, 8; European Parliament (EP), Committee on the Environment, Public Health, and Consumer Protection and the Committee on Agriculture and Rural Development of the European Parliament, "Bovine Spongiform Encephalopathy (BSE)—(Creutzfeldt-Jakob Disease) CJD: Our Health at Risk?" Info Memo: Hearing no. 17 (Brussels, 24–25 June 1996): 8.

5. Robin Oakley, "Analysis: Hope and Blame in Britain," 26 October 2000, at http://europe.cnn.com/2000/WORLD/europe/10/26/bse.oakley/index.html, accessed 10 November 2000.

6. EP BSE Inquiry Report A4–0020/97/A (7 February 1997): 9; available at www.mad-cow.org/final_EU.html, accessed 15 July 2003.

7. EP BSE Inquiry Report, 8.

8. Uncertainties also persist regarding the transmission of BSE among cattle: maternal transmission; time-dependent and age-dependent exposure to infected feed; variability in infectiveness over the incubation period; age-dependent susceptibility; and the influence of host genetics on the course of the disease. See EP, EP BSE Inquiry Report, 30 and 1996: 7. Uncertainties regarding new variants of CJD include its incubation period and the eventual number of deaths that will result from the disease. Professor John Collinge of St. Mary's Hospital, London, a member of the government's BSE advisory committee, estimates the incubation time for CJD to be a matter of decades. Studies of related diseases, such as kuru, suggest incubation periods of up to forty years. See Roger Highfield, "Warning of Major CJD Epidemic," *The Telegraph,* #1519, 23 July 1999, at www.telegraph.co.uk, accessed 18 July 2003.

9. Debora MacKenzie, "Secrets and Lies in Europe," *New Scientist* 154, #2080, 3 May 1997, 14.

10. Scientific Committee on Animal Health and Animal Welfare, Report on Animal Welfare Aspects of the Use of Bovine Somatotrophin (10 March 1999): 17; available at http://europa.eu.int/comm_food/fs/sc/scah/out21_en.pdf, accessed 18 July 2003.

11. The European Commission (commonly "the Commission") provides executive functions for the EU; it is responsible for drafting legislation for consideration by the Council of Ministers (known colloquially as the European Council or "the Council") and the EP and implementing the decisions of the Council.

12. EP, Opinion of the Committee on Industry, External Trade, Research and Energy for the Committee on the Environment, Public Health and Consumer Policy on Public Health: Effects of Endocrine Disrupters on Human and Animal Health (Communication) COM (1999) 706-C5–0107/2000–2000/2071 (COS): 15.

13. Scientific Committee on Animal Health and Animal Welfare, Report on Animal Welfare, 6.

14. Scientific Committee on Animal Health and Animal Welfare, Report on Animal Welfare, 30.

15. Jos Bijman, "Recombinant Bovine Somatotropin in Europe and the USA," *Biotechnology and Development Monitor* 27 (1996): 2–5.

16. The correct name for the drug commonly known as Estradiol-17 *B* is "Estradiol-17 beta-dehydrogenase." European Commission (XXIV/B3/SC4), Opinion of the Scientific Committee on Veterinary Measures Relating to Public Health, Assessment of Potential Risks to Human Health from Hormone Residues in Bovine Meat and Meat Products (30 April 1999): 73.

17. IGF-l stimulates cell division in infants' intestines and promotes intestinal development. Increased quantities of IGF-l in the milk of cows treated with rBST could affect the intestines of human consumers. Many unknowns also persist regarding how elevated levels of insulin-like growth factors affect the welfare of cows, calves in utero, and calves that consume the milk. See Scientific Committee on Animal Health and Welfare, Report on Animal Welfare, 28.

18. Estradiol-17 beta-dehydrogenase, progesterone, testosterone, zeranol, trenbolone acetate, and melengestrol acetate.

19. Commission of the European Communities, Communication from the Commission to the Council and the European Parliament, WTO Decisions Regarding the EC Hormones Ban, COM (1999) 31 final, Brussels, Belgium (10 February 1999a): 1.
20. Commission of the European Communities, Communication from the Commission to the Council and the European Parliament: 2–3.
21. See Council of the European Communities/Commission of the European Communities. Treaty on European Union (Luxembourg: Office for Official Publications of the European Communities, 1992): 13–14.
22. Meikle, "Fertility Jab Linked to BSE Epidemic."
23. Meikle, "Fertility Jab Linked to BSE Epidemic."
24. The SVC is comprised of experts appointed by the Commission from a list of names submitted by the member states. They are nominated in principle on the basis of their professional qualifications, and there is no criterion for nationality balance in its membership. EP BSE Inquiry, 23, available at www.mad-cow.org/final_EU.html, accessed 18 July 2003.
25. EP BSE Inquiry, Annex 1.
26. EP BSE Inquiry, paragraph 1c.
27. EP BSE Inquiry, 8, 46, Annex 32.
28. Thalidomide, a drug introduced to the market in West Germany in 1957, was prescribed as a sedative and widely used by women to alleviate symptoms associated with morning sickness. When taken in the first trimester of pregnancy, thalidomide resulted in horrific birth defects in children around the world.
29. The EC Committee of the American Chamber of Commerce in Belgium, *EU Environment Guide, 1995* (Brussels, 1994): 70.
30. EP BSE Inquiry, 9, 17.
31. EP BSE Inquiry, 22.
32. EP, Opinion of the Committee on Industry, 14.
33. EP, Committee on the Environment Public Health and Consumer Policy, Report on the Commission Communication to the Council and the European Parliament on a Community Strategy for Endocrine Disrupters—A Range of Substances Suspected of Interfering with the Hormone Systems of Humans and Wildlife, COM (1999) 706-C5–0107/2000–2000/2071 (COS), final A5–1097/2000 (2000): 7, 9.
34. EP, "Our Health at Risk?" 7, 9.
35. EP, "Our Health at Risk?" 2–4, 8–9.
36. Scientific Committee on Animal Health and Animal Welfare, Report on Animal Welfare, 30–1.
37. Richard Topf, "Conclusion, Advice to Governments—Some Theoretical and Practical Issues," in *Advising West European Governments; Inquiries, Expertise and Public Policy,* eds. B. Guy Peters and Anthony Barker (Pittsburgh, Penn.: University of Pittsburgh Press, 1993): 190.
38. Peter Bursens, Jan Beyers, and Bart Keeremans, "The Environment Policy Review Group and the General Consultative Forum," in *EU Committees As Influential Policymakers,* ed. M. P. C. M. Van Schendelen (Aldershot, UK: Ashgate, 1998): 7–13, 25–42.
39. *Official Journal of the European Union* L 167, vol. 38, 18 July 1995, 22 and L 167, vol. 17, 20 May 1974, 1.

40. Paul Gray, "The Scientific Committee for Food," in *EU Committees As Influential Policymakers,* 69–72.

41. Jurgen Neyer, "The Standing Committee for Foodstuffs: Arguing and Bargaining in Comitology," in *EU Committees As Influential Policymakers,* 158.

42. Neyer, "The Standing Committee for Foodstuffs," 150–5.

43. EP BSE Inquiry, Annex 3.

44. EP BSE Inquiry, 23.

45. EP BSE Inquiry, 9.

46. EP BSE Inquiry, 24.

47. Commission of the European Communities, DGXXIV (Health and Consumer Protection) of the European Commission, Midterm Review of the Commission's Scientific Advisory System (Brussels, 21 May 1999c): 1.

48. Robin Oakley, "Analysis: Hope and Blame in Britain."

49. EP BSE Inquiry, 6.

50. EP BSE Inquiry, 19.

51. EP BSE Inquiry, 8.

52. This issue fell within the mandate of the Scientific Committee on Cosmetology because many cosmetics, such as lipstick and lotions, have among their ingredients gelatin, a tasteless protein produced by animal tissues.

53. EP BSE Inquiry, 31.

54. EP BSE Inquiry, 17.

55. *Official Journal of the European Union*, L 222 vol. 21, 7 August 1981; L 275 vol. 21, 26 September 1981; L 352 vol. 25, 31 December 1985.

56. Council Decision 94/936/EC.

57. Peter Haas, "Introduction: Epistemic Communities and International Policy Co-ordination," *International Organization* 46(1) (1992): 3.

II

GLOBAL ISSUES

Among global issues, none is potentially more dangerous and demands bigger and more complex science than climate change. Both chapters in this part address aspects of this issue. Taken together, these chapters offer a new look at climate change that raises questions about the science and politics of climate change and their respective influences on international policy.

In chapter 4 a recognized expert in the political perspectives of the issue reviews the issues and progress in the science and politics of climate change. Soroos raises several important questions about climate science and politics. Chapter 5 looks more closely at the interaction of science and politics by tracing the political response to each of the major scientific developments. The final section of chapter 5 assesses the usefulness of the usual suspects in search of a theoretical explanation for the relationship between science and politics in climate change. It finds that none is of more than marginal value.

Climate change is an interestingly complicated issue that is much discussed in the media. Nearly every academic discipline has something to say about climate change. Natural scientists, atmospheric physicists, chemists, biologists, oceanographers, computer modelers all participate in climate-research programs. They are joined by economists, political scientists, geographers, and demographers, who assess the social causes and human consequences of a changing climate. Ideally, the social scientists must predict future social norms for family size, political and economic systems, and lifestyles.

Climate forecasts are being made despite multiple uncertainties. To estimate future anthropogenic greenhouse gas (GHG) emissions, scientists rely on intelligent extrapolations of historical trends in population growth, technology, energy use, and economic production and consumption in a rapidly changing world. Natural scientists have an incomplete understanding of most of the underlying natural processes, including the uptake of GHGs by oceans and terrestrial sinks and the sensitivity of the climate to atmospheric GHG concentrations. But future climates will impact fertility, agricultural production, and energy use and potentially emigration and war, to say nothing of the environmental systems on which human societies are founded.

The issue is unique in another way: every one of the six million people on Earth contributes daily to anthropogenic climate change. Energy is the key to modern life in the industrial developed countries: every aspect of production, distribution, and consumption consumes fossil fuel energy. In poorer countries, subsistence farmers raise rice, landless peasants clear forests for grazing or fuel wood, wealthy landowners burn forests for plantations or log them, and factories use older, less environmentally efficient technologies. And everyone's weather will change in ways not yet fully understood, but few countries want to make the policy changes necessary to prevent, or even mitigate, those changes. National interests are difficult to calculate—unless the future is ignored. Issues of fairness or ethical treatment are lost among safe legal language that protects the lifestyles of the rich.

These two chapters raise many questions. Has the central importance of general circulation models (GCMs) improved policy efficiency or diminished the reputation of science and its value in policy making? Should scientists researching policy-relevant issues be held to the same scientific standards as scientists researching so-called pure science? Are all future climate scenarios equally valid because scientific uncertainty is endemic to the problem, or should the scientists that create them indicate which ones they believe are most probable? Or does that inappropriately make scientific reports policy prescriptive? Are scientists responsible for communicating their findings to policy makers and appropriately interpreting them, or is interpretation through values best left to the political world? Should natural scientists make the strongest possible interpretation of available data (supporting the precautionary principle) or should they weigh only the data that can be empirically justified?

Does international policy only reflect the will of the powerful and ignore scientific findings? What would fairness mean in climate change? If policy is made by the powerful, what are the uses of science in issues like climate change? What is the balance of influence on international policy between domestic politics, the beliefs of national leaders, national interests, and scientific information and interpretation? Does this balance change over time and, if so, why? Do boundary institutions like the Intergovernmental Panel on Climate Change (IPCC) protect science from political influence, mute the influence of scientists, or make science more relevant to policy? By rejecting the Kyoto agreement, is the United States morally responsible for the harm that other countries may suffer from a changing climate? Should ethics be considered in national policy making on environmental issues?

4

Science and International Climate Change Policy

Marvin S. Soroos

The prospect that human beings are modifying the Earth's climate system has become one of the preeminent global environmental issues of the early twenty-first century. In recent decades, scientists of many nationalities and diverse disciplines have sought to determine the extent to which human activities may have contributed to rising global temperatures over the past century. Others are engaged in forecasting the amount of warming that is likely to occur over the next century, as well as the impacts that such changes will have on the global climate, environmental systems, and human communities. More scientific effort and resources have arguably been expended on studying the human impact on climate than on any other environmental problem. Nevertheless, considerable uncertainty remains about how sensitive the global climate system is to human activities and the consequences that rising temperatures will have for natural and human systems. Is enough known about the dangers associated with human-induced global warming—including the possibility of sudden, large, nonlinear, and irreversible changes—to warrant preventive actions that may have significant economic repercussions?

Mounting scientific evidence of human impacts on global climate prompted a series of international meetings in the late 1980s to discuss a response to the problem. Formal negotiations that began in 1991 have led to the adoption of two major treaties, the Framework Convention on Climate Change (FCCC) of 1992 and the recently finalized Kyoto Protocol, which commits the developed countries to reducing their emissions of GHGs. But will these agreements, even if they are fully implemented, contribute significantly toward stabilizing the Earth's climate?

SCIENTIFIC ISSUES

In contrast with the other planets in the solar system, Earth has an environment with a moderate climate that has permitted a wide diversity of species to evolve and flourish. The planet's favorable environment is largely attributable to the chemical composition of its atmosphere, which contains gases that not only moderate the flow of energy coming from the sun, but also capture a substantial portion of the heat radiated from the Earth. Paradoxically, the atmosphere's two most plentiful gases, nitrogen and oxygen, which account for 99 percent of its volume, have virtually no effect on the Earth's climate. Rather, it is much smaller concentrations of several other gases, in particular carbon dioxide (CO_2), methane, ozone, and nitrous oxides, along with water vapor, that shape the planet's weather. These so-called greenhouse gases, or GHGs, warm the climate by holding in part of the long-wave infrared energy radiated back from the Earth that would otherwise escape to outer space. CO_2, which accounts for only 0.03 percent of the atmosphere's volume, is responsible for about 26 percent of the greenhouse effect. Water vapor, which constitutes between 0 and 4 percent of the atmosphere, is also an important GHG, accounting for about 60 percent of the warming effect of the atmosphere. Venus, with a much greater abundance of GHGs in its atmosphere, has a climate that is much too hot for life forms, while Mars with a smaller volume of GHGs, has weather conditions that are too frigid for life.[1]

Climate is affected to a lesser extent by microscopic liquid and solid particles known as aerosols, such as pollen, dust, soot, and sulfates. Some of these aerosols are present in the atmosphere as a result of natural forces, such as volcanic eruptions and forest fires. Others are generated by human activities, such as the burning of sulfur containing fossil fuels and land-clearing operations. Most of these particles have a cooling effect because they reflect approaching solar radiation back into space before it has a chance to reach and warm the Earth's surface. However, dark-colored soot, which is rich in carbon, absorbs energy and, thus, has a localized warming effect. Aerosols reside in the atmosphere for relatively short periods of time, such as a few days, before they gravitate to Earth or are washed out by precipitation. Thus, their impacts on climate are brief compared with those of the predominant GHGs, such as CO_2, which may remain in the atmosphere for a century.[2]

Concern over global warming arises because myriad human activities have added to the naturally occurring concentrations of GHGs. Over the past two centuries, the fossil fuel–driven industrial revolution has been primarily responsible for an increase in atmospheric concentrations of CO_2 from approximately 280 to 367 ppm in 2000, a rise of more than 30 percent. Current concentrations of CO_2 are believed to be substantially higher than the natural levels that have occurred at any time over the past 420,000 years, and quite possibly during the past 20 million years. Concentrations of methane, a much more potent GHG, have risen 151 percent since 1750 due to emissions from fossil fuels, cattle, rice fields, and landfills. Concentrations of nitrous oxides, which are given off by agricultural soils and cattle feed lots, have risen 17 percent since 1750.[3]

What impacts have these increases in GHG concentrations had thus far on global climates? The task of detecting the human imprint on the global climate is complicated by two basic features of climates. First, climates naturally undergo changes over periods ranging from decades to centuries, millennia, and even longer, as is evident from the swings that have taken place between ice ages and warmer interglacial periods. The lack of an identical planet devoid of human pollutants to serve as a control case greatly complicates the task scientists face in isolating the extent of human-induced climate. Second, climate is determined not only by solar activity and atmospheric processes, but also by complex interactions between the atmosphere and other components of the Earth system, in particular the oceans, land masses, and the biosphere.

Two features of the relationship between the oceans and the atmosphere have received increasing attention in recent decades. The first is the naturally occurring oscillations that have major implications for weather, the most notable being the El Niño/Southern Oscillation, which is centered in the southern Pacific Ocean, but has impacts on weather globally. During El Niño periods, a weakening of westward moving trade winds causes warm waters to build up in the eastern Pacific, triggering heavy rains and severe flooding over coastal regions of the Americas, whereas other regions as far away as Africa and Southeast Asia experience intense droughts. While there is evidence that these events have been natural occurrences from prehistoric times, they appear to have become more frequent, intense, and longer lasting in recent decades, possibly due to human impacts on the atmosphere and global climate.[4] The second is the role that a network of ocean currents plays in transporting warmth from the tropics toward the polar regions. Thus, the Gulf Stream and North Atlantic Current provide northern Europe with a more moderate climate than would be expected for a region so far from the equator.

Temperature records reveal that the Earth's surface has warmed approximately 0.6°C over the past 142 years, during which reasonably reliable weather data has been collected. This rise has not been steadily upward as might have been expected as GHG concentrations gradually rose. The most pronounced periods of warming have occurred from the 1920s through the

1940s and since the late 1970s, while somewhat cooler average temperatures prevailed between these two periods. Since 1976 warming has been occurring three times faster than over the past century, and for 23 consecutive years the average temperature has been above the 1960–1990 annual average.[5] Nine of the 10 warmest years over this period have occurred since 1990, with 1998 and 2001 being the first and second warmest, respectively. Inferences from other indicators of climate suggest that for the Northern Hemisphere, the 1990s may have been the warmest decade of the past thousand years.[6] Given the natural variability of climate, however, scientists have been cautious about attributing part or all of these increases in global average temperatures to human factors, such as GHG emissions or land-use changes.

Scientists have noted a number of other climatic and environmental developments that may be harbingers of a trend toward global warming. Among the more recently reported potential indicators of climate change are the following:

- A decrease of approximately 10 percent in the extent of snow cover since the late 1960s
- A 14 percent reduction in the amount of perennial Arctic Sea ice between 1978 and 1998 and a 40 percent reduction in ice thickness over the last 20 to 40 years
- An increase of 0.31°C in the temperatures of the top 300 meters of the Atlantic, Pacific, and Indian Oceans over the past 50 years
- An advance of 9.8 days in freshwater spring thaw dates at 38 sites in the Northern Hemisphere between 1846 and 1996 and an 8.7-day delay in fall freeze-up dates
- A retreat of mountain glaciers in nonpolar regions during the twentieth century
- An increase of 2 to 4 percent in the frequency of heavy precipitation events during the latter half of the twentieth century
- A rise in sea levels of 0.1 to 0.2 meters during the twentieth century
- A substantial drop in the amount of long-wave radiation escaping from the Earth between 1970 and 1997.[7]

Scientists express caution about attributing these anomalies to human-induced changes in the climate system. Nevertheless, the way in which these varied environmental developments have coincided with the marked rise in global temperatures over the past two decades and a continuing upward trend in concentrations of atmospheric GHGs reinforces concerns that significant changes are taking place in the global climate system that are largely attributable to human activities. What temperature trends can be expected in the future? Concentrations of CO_2 in the atmosphere will rise further as the world's population, which is projected to grow from approximately 6 billion in 2000 to 9.3 billion by 2050, continues to rely heavily on fossil fuels as an energy source.[8] If current trends continue, concentrations of atmospheric CO_2 will be twice preindustrial levels by 2100, which, by some estimates, could cause global average mean temperatures to rise by as much as 5°C over the next century. Such a warming would be substantially greater than the warming observed over the twentieth century and would be without precedent over the last 10,000 years.[9] As will be discussed later in the chapter, questions have been raised in some circles about the scientific data and methods upon which such forecasts are based.

What makes global warming a major threat is not so much the gradual rise in mean global temperatures, as the multiplicity of impacts such a warming would have on other aspects of climate and the environment, as well as the consequences these impacts could have for human communities. These impacts, which will vary considerably by region, are more difficult to anticipate than trends in global average temperatures. While some of the impacts may be beneficial to certain human communities, especially if temperature increases remain moderate, other consequences could be highly disruptive. Let us consider a few of the potential impacts of global warming that have received the most attention.[10]

Impacts on Weather

Global warming is expected to alter not only average temperatures, but also the incidence of unusual or extreme weather events. Droughts are likely to become more severe in arid areas, which will further aggravate shortages of water for agricultural, industrial, and residential uses. Forest and range fires in these areas could become more frequent and difficult to contain. For other regions, however, an energized hydrological cycle will stimulate more precipitation, including an increased incidence of extreme rain events leading to flooding. Tropical storms may become more frequent and of greater intensity. A larger proportion of the planet's precipitation can be expected to fall as rain, rather than snow, altering seasonal river flows that impact the availability of drinking water, irrigation, fishing, navigation, and the generation of hydroelectric power.

Rising Sea Levels

Potentially one of the most disruptive impacts associated with global warming would be an increase in sea levels due to both the expansion of the oceans waters as they warm and an accelerated rate of melting of glaciers in the polar regions. Scientists project a rise in sea levels between 9 and 88 centimeters over the period 1990 to 2100.[11] A collapse of the West Antarctic ice sheet, which is a possibility, but not considered probable for the foreseeable future, could cause a much greater rise in sea levels.[12] Higher sea levels may impact the ecology of coastal zones by inundating wetlands and by increasing the salinity of rivers, bays, and groundwater. Low-lying coastal areas, where many of the world's major cities are located, are threatened by rising waters, tides, and storm surges. Large amounts of productive agricultural land may also be lost to rising waters. Some small island states, such as Kiribati, Tuvalu, and the Maldives, are especially vulnerable to rising seas and storm surges.

Impacts on Flora and Fauna

Terrestrial and marine species of plants and animals are affected in varied and complex ways. Species that are highly adapted to specific climatic conditions are likely to decline, if not become extinct, unless they are able to migrate along with the changing weather conditions. Warming ocean waters are already believed to be a factor in the bleaching of coral reefs around the world. Certain species of trees may die off in regions as temperatures rise or they are exposed to pests that take advantage of changing climates. The survival of some animal species may be jeopardized if climate change triggers a decline in their food sources. The more adaptable species, many of which are regarded as pests, may thrive in a warmer world as they migrate into ecological niches vacated by species that were not able to cope with changes to their environments.

Impacts on Human Health

Under a warmer climate regime, periods of extremely hot and humid weather will become more frequent, posing a threat to older people with heart and respiratory problems, in particular in urban areas of developing countries. Health-endangering levels of ground-level ozone also occur more frequently under warmer conditions. Global warming may also expand the geographical range of diseases, such as malaria, as the insects that transmit them migrate with warmer weather. Alternately, in areas far from the equator, a warming climate may have the benefit of reducing the number of fatalities from exposure to extreme cold.

The Potential for Surprises

In attempting to anticipate the future, there is a tendency to assume that global warming and its impacts will occur as gradual, linear trends. The complexity of the atmospheric dynamics and their relationships to the larger Earth system may give rise to surprises in the form of more abrupt, discontinuous impacts. Such surprises could be triggered by unnoticed trends, the amplifying effects of positive feedback loops, the crossing of critical thresholds, synergies caused by a combination of several changes, and cascading chains of cause-and-effect developments.[13] Concern about these possibilities has been heightened by recent findings from paleoclimatological research reported by the U.S. National Research Council, which suggest that major prehistoric swings in climate have occurred much more abruptly than previously believed.[14]

To illustrate the potential for nonlinear changes, let us consider two possibilities. A positive feedback loop could follow from the shrinking of Arctic snow and ice cover that reflects most of the solar radiation that hits it. The exposed land and water will absorb substantially more of the sun's energy and thus add to a surface warming of these areas, which in turn may cause more snow and ice to melt. A pronounced warming in the Arctic region could also result in a thawing of vast areas of permafrost, which would release large additional quantities of carbon, thereby adding to the atmosphere's burden of GHGs and to the rate of global warming.

Another intriguing possibility is that a pronounced warming in the Arctic area could slow, if not entirely shut down, the northward flowing Atlantic currents. Such a development might occur if accelerated melting of snow and ice in the Arctic released greater amounts of relatively light freshwater into the North Atlantic, where it mixes with the heavier saltwater carried up from the tropics by the surface currents. The diluted and, thus, lighter, saltwater would not sink as readily to become a deep southward-moving current, which is essential to keeping the oceanic conveyer belt moving. Such a slowdown or stoppage of what is known as the thermohaline circulation could cause a dramatic cooling of the weather of the British Isles and northern Europe. While scientists believe that such an occurrence is unlikely over the next few decades, it is a distinct possibility for the more distant future if atmospheric concentrations of GHGs continue to rise.[15]

In recent years, hardly a week has passed without reports of additional climatic anomalies, harbingers of global warming, or new warnings about the possibility of catastrophic future impacts. Have these developments had an impact on public perceptions of the urgency of global climate change and the need to respond to these threats? How do these scientific revelations about climate change play out in the domestic political arenas of the United States and other countries?

POLITICAL AND SOCIAL DIMENSIONS OF THE PROBLEM

Aristotle once observed, "what is common to the greatest number has the least care bestowed upon it. Everybody thinks chiefly of his own, hardly at all of the common interest."[16] More recently, biologist Garrett Hardin, in a seminal article entitled "The Tragedy of the Commons," noted a similar tendency for natural resources that are available for the private gain of multiple actors to be overused and misused.[17] The degradation of commons may result not only from excessive removal of these resources, as in the case of unsustainably harvesting a fishery, but also from overtaxing the capacities of commons to absorb pollutants.

The atmosphere is a commons in the sense of being a convenient medium for the world's population to dispose of a wide variety of airborne wastes.[18] In past eras, the atmosphere was sufficiently vast to harmlessly disperse whatever pollutants were emitted by a much smaller and less industrialized world population. In recent decades, however, human pollutants have approached and, in some cases, exceeded the capacity of the atmosphere to absorb them, as in the case of GHGs that contribute to global warming. The incentives for emitting GHGs are especially strong because these pollutants are generated by activities that are basic to economic development and modern life styles—such as generating electricity, industrial production, transportation, and agriculture. Thus, despite growing evidence that the world's climate system is already being altered by these pollutants, and despite warnings by scientists about accelerated warming in the future, humanity continues those practices that have caused the problem and may be on course toward an impending "global tragedy of the atmosphere."

To avoid such a tragedy, it will be necessary to stabilize concentrations of GHGs in the atmosphere by curbing emissions of these gases or by enhancing natural systems, such as forests that absorb them. The IPCC estimates that stabilizing atmospheric CO_2 at 450 ppm (compared to the preindustrial level 280 ppm) would require reducing global emissions to pre-1990 levels within a few decades and then continuing to reduce them steadily thereafter, until they are a very small fraction of current emissions.[19] Achieving international agreement on how to accomplish such reductions is complicated by several economic and social factors.

Responsibility for GHG Emissions

The developed countries, with only one-fifth of the world's current population, have contributed approximately 74 percent of human additions to CO_2 in the atmosphere since 1950. Their proportion of annual emissions dropped to 60 percent by the 1990s, as development has proceeded elsewhere and the emissions of the countries of the former Soviet bloc plunged dramatically during their transition to market economies. By 2020 more than half of CO_2 emissions could come from the developing countries. While the United States continues to be the largest emitter of CO_2 with about 25 percent of the global total, American emissions may be surpassed by China as early as 2015.[20]

The differential contributions to the climate change problem become more apparent when emissions are compared on a per capita basis (see table 4.1). These divergent levels of emissions of CO_2 have significant implications for devising international strategies to address the climate change problem in that they raise an important question of equity. Do some peoples have a right to use a much greater proportion of the atmosphere's capacity to absorb GHGs than others?

Differential Impacts and Vulnerabilities

Nations and peoples are impacted in very different ways by global warming and its many impacts. In the early stages, climate changes may bestow benefits on some countries, such as an extended growing season in far northern regions. For most countries, however, the effects will on the whole be decidedly negative and increasingly so with the passage of time. Countries also differ considerably in their vulnerability to global warming, which is a function of the magnitude of climatic and environmental impacts in their regions and the capacity of their societies to adapt to them.

On both criteria of vulnerability, developing countries tend to have a significantly greater level of vulnerability than do the developed industrialized ones. While the amount of warming is most pronounced in regions distant from the equator, tropical and subtropical areas will be affected more by heat stress, extended drought, intense storms and rain events, and the spread of diseases and pests. Moreover, the agricultural sector, which is especially sensitive to temperature and precipitation, accounts for a significantly greater

Table 4.1 Per Capita CO_2 Emissions for Selected Countries in 1990 (in metric tons)

Developed	*mmt*	*Countries in Transition*	*mmt*	*Developing*	*mmt*
United States	19.7	USSR	12.8	Mexico	3.8
Canada	15.7	Poland	9.1	China	2.1
Australia	15.5	Bulgaria	8.3	Egypt	1.5
Germany	12.3	Romania	6.7	Brazil	1.3
UK	10.0	Hungary	6.2	Peru	1.0
Netherlands	9.3	Albania	2.1	India	0.8
Japan	8.7			Kenya	0.2
Italy	6.8			Bangladesh	0.14
France	6.2			Cambodia	0.05
		World Average	4.2		

Source: Figures from Robert Engelman, "Imagining a Stabilized Atmosphere: Population and Consumption Interactions in Greenhouse Gas Emissions," *Journal of Environment and Development* 4 (1995): 111–40.

proportion of the economies of the developing countries than of the industrial ones. Many of the third world's major cities are located at sea level, as are important food-growing areas in countries like Bangladesh, Egypt, and China. The plight of developing countries is further complicated by the scarcity of economic resources and access to technologies that would facilitate their adaptation to the changes taking place. The irony of the situation is that the problem of climate change has been caused largely by the disproportionately high emissions of GHGs of the developed countries, while the developing countries, whose emissions are very low by comparison, are the most vulnerable to the impacts of climate change. Do the developed countries owe "new and additional financial resources" to the developing countries to compensate them for the historical and forecast future damage from climate change?

Differing Perceptions and Priorities

The leaders and publics of countries vary considerably in their perceptions of the seriousness of the threat of global warming and the priority that should be given to addressing it. For example, Europeans tend to be more persuaded than Americans of the urgency of the climate change problem as conveyed by the IPCC and, thus, are more inclined to take precautionary steps that would slow the accumulation of GHGs in the atmosphere, even while there are scientific uncertainties about the extent of human impacts on the global climate. Surveys suggest that a majority of Americans are concerned about global warming, but are more willing to tolerate risk—though their risk is estimated to be relatively small because of their geography and wealth—until they are satisfied that the scientific evidence on global warming is more conclusive.[21] The apparent ambivalence of Americans about climate change may be attributed in part to the widely publicized views of a relatively small group of outspoken scientists, who have vigorously questioned whether there is definitive evidence of a human impact on climate.[22] Thus, American policy makers and legislators have been reluctant to make a commitment to limit GHG emissions in the belief that it could be harmful to the nation's economy.

For most developing countries, concerns about global environmental problems have been secondary to their aspirations for economic development and reduced poverty. To achieve these development objectives, they argue that it will be necessary to increase substantially their use of fossil fuels and, accordingly, their emissions of CO_2. Representatives of the developing countries compare their "survival" emissions of GHGs to what they refer to as the "luxury" emissions of the highly industrialized countries. Thus, the developing countries have put off committing to a schedule for limiting their GHG emissions. They have also made it clear that if the developed world wants them to adopt technologies that will limit their GHG emissions, then the developed countries should provide technical assistance and bear the burden of the additional costs that would be entailed in using them.[23]

Domestic Politics As a Complicating Factor

The positions that countries take in international negotiations are often constrained by the interplay of powerful interests in domestic political arenas. Domestic politics has been an especially important factor in shaping the positions the United States has taken in the climate change negotiations. While the executive branch of government is responsible for negotiating international treaties, formally committing the United States to these agreements requires a two-thirds vote of the Senate. Through much of the 1990s, numerous senators from both political parties have expressed strong opposition to committing the United States to international agreements that would require limiting GHG emissions. In July 1997, just months prior to the initial adoption of the Kyoto Protocol, the Senate adopted Resolution 98 by a vote of ninety-five to zero. The resolution advised the Clinton administration not to agree to a treaty requiring emission reductions that would "result in serious harm to the American economy" or would not require a specific schedule of emission reductions or limits by all countries, including developing ones.[24] The opposition of some senators can be explained in part by the lobbying efforts of key industries in their states, such as coal and petroleum producers or automobile manufacturers, whose profits could be adversely affected by policies that would discourage consumption of fossil fuels. Senators also fear a backlash from voters if unpopular policies or taxes are needed to achieve emission-reduction targets.[25]

Nongovernmental groups with diverse perspectives on global climate change have played an important role in shaping the debate on the American response to the problem. On one side of the issue are various advocacy and scientific groups, such as GreenPeace, the Sierra Club, the World Conservation Union, the World Wildlife Federation, the Worldwatch Institute, and the Union of Concerned Scientists, which have sought to inform Americans about the potentially serious consequences of climate change and to mobilize public opinion in favor of strong national and international policies that would substantially reduce the flow of GHGs into the atmosphere. On the other side are groups such as the Cato Institute, the Western Fuels Association, and the Global Climate Coalition (GCC) that question the conclusiveness of the science of climate change and the need to reduce the American reliance on fossil fuels. The GCC, an alliance of trade and consumer associations with a stake in energy policies, sponsored an intensive media campaign in 1997–1998 to arouse public opposition to the Kyoto Protocol. However, the corporate community is by no means united in its response to the global warming issue. In recent years, several companies including Ford, General Motors, Daimler Chrysler, Dupont, Texaco, and Shell have defected from the GCC.[26] The Pew Center on Climate Change has organized a coalition of major companies, including Boeing, IBM, DuPont, Shell, BPAmoco, Alcoa, Intel, Cynergy, Weyerhauser, and Toyota into its Business Environmental Leadership Council, which favors responsible corporate leadership in efforts to address climate change.[27]

The combination of the factors mentioned in this section complicates the challenge of forging an effective international response to the threat of climate change. Will it be possible for a highly diverse group of stakeholders having disparate perspectives and interests to agree on a strategy for stabilizing concentrations of atmospheric GHG gases that would require many of them to make considerable adjustments and sacrifices?

THE SCIENTIFIC RESEARCH PROGRAM

Nineteenth-century European scientists laid the foundation for our contemporary knowledge of the potential human impacts on climate. In 1827 French mathematician and scientist Jean-Baptiste-Joseph Fourier hypothesized that the atmosphere retains heat in much the same way that a greenhouse remains warm on sunny winter days. In the 1860s the English scientist John Tyndall measured the capacity of CO_2 and water vapor in the atmosphere to absorb infrared radiation and hypothesized that variations in the concentrations of these "greenhouse gases" could account for changes in climate, including glacial periods.[28] Toward the end of the century, Swedish scientist and Nobel laureate Svante Arrhenius calculated that an eventual doubling of atmospheric concentrations of CO_2 from the burning of fossil fuels would increase the temperature of the lower atmosphere by from 4°C to 6°C, which is in the same general range as the recent projections of modern computer models.[29]

The possibilities for human-induced global warming received little further attention until the 1950s when Roger Revelle and Hans Suess of the Scripps Institution of Oceanography warned that humanity, by adding substantially to atmospheric GHGs, was unwittingly conducting a large-scale geophysical experiment on the planet that could have very serious consequences.[30] The International Geophysical Year, an eighteen-month period of international scientific cooperation sponsored by the International Council of Scientific Unions (ICSU),[31] led to important new atmospheric monitoring programs.[32] Stations set up at the Mauna Loa Observatory in Hawaii and the South Pole have documented a steady rise in atmospheric concentrations of CO_2. During the 1960s, the World Meteorological Organization (WMO) substantially strengthened its network for the collection, compilation, and distribution of weather data in the form of the World Weather Watch (WWW). From 1967 to 1981, WMO collaborated with ICSU on the Global Atmospheric Research Program (GARP) to further international research on atmospheric dynamics.[33]

The erratic weather of the 1970s offered confusing signals about whether the global climate was changing and, if so, in what direction and to what extent humans were a factor. Drought persisted in the Sahel region of Africa, and in other parts of the world a combination of drought, failed monsoons, cold winters, abbreviated growing seasons, and El Niño contributed to what became known as the World Food Crisis from 1972 to 1975. Unusually cold weather in the Northern Hemisphere during the latter 1970s led to speculation that the planet may be in the early stages of a new period of glaciation.[34] To assess existing scientific knowledge about climate change, both naturally occurring and human-induced, the United Nations convened the First World Climate Conference (FWCC) in Geneva in 1979, which was attended by 350 scientists from 50 countries, representing a variety of climate-related disciplines.[35]

Scientists have employed two approaches to investigate the impacts that rising concentrations of atmospheric GHGs could have on global climates. One such strategy looks to the past by using paleoclimatological methods, such as analysis of tree rings, sediments, and fossils to study prehistoric climates and the factors that have influenced them. Air bubbles trapped in ice cores up to two miles in length, which have been extracted from the ice sheets of Antarctica and Greenland, provide a record of climate and the chemical composition of the atmosphere over the past 420,000 years. Analysis of these bubbles has revealed that warming and cooling trends in temperature over this extended period closely paralleled changes in the amount of atmospheric CO_2.[36] This finding gives rise to concerns that the global temperatures will rise significantly over the next century and beyond as GHGs accumulate in the atmosphere.

A very different, but complementary, approach has been adopted by several major research institutions that have developed large, highly complex Earth system and climate models. Over time the climate models have been enlarged and refined to the point that they can predict past temperature trends reasonably accurately when data on levels of human pollutants is incorporated into the model. These models can then be used with increasing confidence to project future temperatures as concentrations of GHG continue to rise, with most forecasting a rise in global mean temperatures of several degrees centigrade over the century, unless steps are taken to sharply reduce the buildup of GHGs in the atmosphere.[37]

The 1980s saw the creation of several major international scientific projects to investigate climate change and its relationship to other components of the Earth system. Responding to recommendations from the FWCC, the WMO, in partnership with the United Nations Environment Program (UNEP) and ICSU, launched the multifaceted World Climate Program to compile a full range of data that may have a bearing on climate change, to determine the extent to which human activities may be affecting climate, and to study the impacts of climate change.[38] In 1986, ICSU inaugurated the Stockholm-based International Geosphere-Biosphere Program (IGBP) to engage the world's scientific community in a coordinated effort to research phenomena related to "global change," namely the ways in which human activities are having ever increasing and irreversible impacts on the functioning of the Earth system.[39] The IGBP is organized into projects that seek to expand knowledge about the key components of the Earth system—the atmosphere, land, and oceans—and the interfaces between them. ICSU and the International Social Science Council cosponsor the International Human Dimensions Programme on Global Environmental Change, which is a social scientific counterpart to the IGBP.[40]

In 1988, WMO and UNEP created the IPCC. Its purpose is not to conduct scientific research, but to assess and summarize available scientific research that provides knowledge relevant to induced climate change. The IPCC is structured around working groups that focus on (1) the science of climate change; (2) impacts, adaptation, and vulnerability to climate change; and (3) policy options for mitigating climate changes. The IPCC has issued major assessment reports in 1990, 1995, and 2001, which are designed to educate policy makers about the state of climate change science and to facilitate negotiations on international agreements to address the problem. The first of these IPCC reports informed the Second World Climate Conference, which was held in Geneva in 1990 and the first official negotiations on a climate change treaty that commenced early the next year. The second report, issued in 1995, gave impetus to negotiations on the Kyoto Protocol of 1997, while the third report, released in 2001, lent urgency to efforts to finalize the Kyoto Protocol at Marrakesh later in the year.

The IPCC assessment reports are widely viewed as the most definitive compilation of what is known about climate change and the conclusions of "mainstream" science on the subject. They are the products of a rigorous process that engages prominent scientists with relevant expertise to serve as lead authors, who are given responsibility for drafting sections of the reports. In preparing the drafts, the lead authors are instructed to take into account the diversity of peer-reviewed, internationally available scientific research on their subjects. These drafts are then subjected to reviews, first by other experts from a broad variety of organizations including industry, as well as by scientists from developing countries, and second, by the relevant governmental departments and ministries of the states involved in the process. Finally, summary reports for policy makers are written and reviewed on a line-by-line basis.[41]

The first two IPCC reports noted a modest increase in global average mean temperatures in the range of 0.3°C to 0.6°C over the past hundred years. However, whereas the 1990 report drew no conclusions about whether this increase could be attributed to human factors, the 1995 report suggested that there was now evidence of a "discernible human influence on global climate."[42] The 2001 report, which takes into account the unusually warm temperatures of the 1990s, noted a 0.6°C increase during the twentieth century and concluded that "most of the observed warming over the past fifty years is likely to have been due to the increase in greenhouse gas concentrations."[43] The IPCC reports also forecast future changes in global mean temperatures based on the projections of approximately twenty large computer models of atmospheric dynamics, some of which take into account other dynamic features such as oceans, sea ice, vegetation, and soils. The 2001 report projected a rise in global average mean temperatures of between 1.4°C to 5.8°C by 2100, which is significantly higher than the rise of 1°C to 3.5°C that was foreseen in the 1995 report.[44]

Despite a growing body of scientific evidence suggesting that humans are having significant impacts on the global climate, as compiled in the reports of the IPCC, there are a number of outspoken scientists who argue that the science on the subject is still not definitive. They point to the natural variability of climate over time and the remaining uncertainties about atmospheric processes acknowledged by mainstream scientists. Questions have been raised about the reliability of weather data used to plot global trends and the adequacy and the validity of the climate models that are used to project future trends. Anomalies that appear to run counter to the warnings of global warming have been noted, such as a slight cooling of temperatures in the lower atmosphere and recent observations of cooler temperatures and thickening ice at certain locations in Antarctica. They have also raised questions about the seriousness of the potential impacts of climate change. Some challenge the objectivity of the IPCC assessment reports, in particular in the selection of the authors responsible for drafting its reports, and contend that mainstream scientists exaggerate the threat of global warming to attract research funding.[45]

Scientists convinced of the seriousness of the threat of climate change have countered many of the arguments raised by the skeptics, for example, by offering explanations of the anomalies. While acknowledging that there are still numerous scientific uncertainties and puzzles on as complex a subject as atmospheric dynamics, they argue that enough is known about the potential human impact on climate to strongly suggest that human-induced global warming has already taken place and poses a very serious threat for the future. They further warn that to wait until the science of global climate change is substantially more definitive would significantly reduce the options available for addressing the problem.

The debate between scientists convinced of the seriousness of the threat of climate change, as conveyed in the reports of the IPCC, and the skeptics emphasizing lingering uncertainties has been highly contentious, especially in the United States. Have these exchanges given the American public a balanced and accurate picture of the state of climate change research? Or have the widely disseminated views of the skeptics confused the American public and given them an excuse for persisting in an environmentally destructive life style? How have international policy makers responded to mounting, but not entirely conclusive, evidence of human interference with the global climate system and the potential for catastrophic consequences?

THE INTERNATIONAL RESPONSE TO GLOBAL CLIMATE CHANGE

The pronounced global warming trend of the 1980s lent greater urgency to addressing the threat of human-induced climate change. An international meeting of scientists convened by WMO, UNEP, and ICSU in Villach, Austria, in 1985 issued a report warning that an increase in global mean temperatures of from 1.5°C to 4.5°C was "highly probable" if concentrations of GHGs in the atmosphere were to double, as was projected to occur by 2030.[46] The Villach report encouraged policy makers to consider options for limiting, if not reducing, emissions of GHGs. In June 1998 Canada hosted the nongovernmental Conference on the Changing Atmosphere, which called upon the industrialized nations to reduce their emissions of CO_2 by 20 percent from 1988 levels by 2005. Over the next two years, a spate of ministerial-level conferences held around the world discussed strategies for addressing climate change, the most important of these being the Ministerial Conference on Atmospheric Pollution and Climate Change in Noordwijk, the Netherlands, in November 1989.[47] The prenegotiation stage of international discussions on climate change came to a climax at the Second World Climate Conference in November 1990, which called for negotiations to begin on a climate change treaty.[48]

The United Nations General Assembly promptly established the Intergovernmental Negotiating Committee (INC) through which representatives from 135 countries drafted the FCCC, which was ready for adoption at the Earth Summit held in Rio de Janeiro in June 1992.[49] The negotiations on the FCCC were often contentious. The members of the European Community and most other Western industrialized nations argued that the first major treaty on climate change should include a timetable for the developed countries to limit their GHG emissions.

The United States was strongly opposed to incorporating a timetable for mandatory emission limits into the first treaty on climate change. Instead, it proposed a more modest agreement that would promote additional scientific research and international information exchanges to further clarify the nature and seriousness of the threat posed by climate change. To the extent that limits were considered, the United States argued for a comprehensive approach that would encompass all of the principal GHGs, so that credit would be received for reductions in CFC emissions previously mandated by the 1987 Montreal Protocol, which controls substances that threaten the ozone layer.[50]

Throughout the climate change negotiations, the Group of 77 (G-77)/China coalition of 130 countries repeatedly reminded the developed countries that they were responsible for a disproportionately large share of the CO_2 that had been added to the atmosphere since the dawn of the industrial revolution. Thus, the coalition has steadfastly insisted that the developed countries take the first significant steps toward mitigating the problem of climate change by cutting back on their emissions of GHGs. These developing countries have been highly unified in rejecting calls, especially from the United States, to limit their own much lower levels of emissions. The small-island and low-lying states, which are especially vulnerable to rising sea levels, have been the most aggressive advocates of mandatory cutbacks in GHG emissions by the industrialized countries. To strengthen their voice in the climate change negotiations, forty-three of these countries (which account for only 5 percent of the world's population, but more than 20 percent of the membership of the United Nations) formed the Alliance of Small Island States (AOSIS).[51] By contrast, the OPEC members among the developing countries, led by Saudi Arabia, have been concerned that their economies would be severely damaged if an international treaty mandating limitations on CO_2 emissions caused developed countries to cut back substantially on their imports of petroleum. Thus, the oil-exporting countries were among the few countries that aligned with the United States in resisting binding reductions of GHGs. They also asked to be compensated for any loss of exports resulting from an international agreement to limit GHGs.

As the Earth Summit approached, compromises were struck on key provisions of the framework treaty, which made the version of the FCCC weaker than many of the developed and AOSIS countries and environmental NGOs would have preferred. The treaty did establish a goal of stabilizing GHG emissions at "a level that would prevent dangerous anthropogenic interference with the climate system," but didn't define what is meant by "dangerous." In deference to the United States, it specified no binding targets and timetables, but the developed countries listed in Annex I (principally the industrialized countries of Europe and North America), were to "aim" at returning their GHG emissions to 1990 levels by the end of the decade and to provide detailed information on what measures they had adopted to achieve this goal. Developing countries were to provide inventories of their

GHG emissions and removals of sinks, but they were not expected at this stage to limit their emissions.[52]

The FCCC entered into force in 1994 upon ratification by fifty countries. A year later the first of a series of annual Conference of the Parties (COP1) was held in Berlin to consider additional steps to address global climate change. It was still not possible to reach an agreement spelling out a timetable for mandatory limits on GHG emissions, due primarily to the continued opposition of the United States, now represented by the Clinton administration. The delegates did, however, adopt a nonbinding statement, known as the Berlin Mandate, which acknowledged that the commitments contained in the FCCC were not adequate to effectively address the problem of global warming. The mandate also called for negotiations to begin promptly on a supplemental protocol to be ready for COP3, scheduled for Kyoto in 1997, by which the Annex I countries would commit to a timetable for limiting their GHG emissions.

Meeting through the night at the close of COP3, delegates hammered out an agreement that become known as the Kyoto Protocol of 1997, which supplements the FCCC adopted five years earlier.[53] The protocol's most notable feature is the binding commitments of the Annex I countries to reduce their emissions of six GHGs (carbon dioxide, methane, nitrous oxide, hydrofluorocarbons, perfluorocarbons, sulphur hexaflouride) by an average of 5.2 percent from 1990 levels by the period 2008 to 2012. These commitments are "differentiated," meaning that the Annex I countries agreed to individualized emission targets that would take into account their unique circumstances. Thus, the members of the EU (formerly the European Community) and several other European countries agreed to reduce GHG emissions by 8 percent, the United States by 7 percent, and Canada and Japan by 6 percent, while the Russian Federation, Ukraine, and New Zealand committed to emissions no higher that their 1990 levels. Norway and Australia would restrain increases of emissions to 1 percent and 8 percent, respectively. Developing countries were still not expected to accept any limits on the their GHG emissions.

The willingness of the United States and several of the other Annex I countries to adopt a schedule for binding limits contained in the Kyoto Protocol was contingent on the inclusion of several so-called flexible mechanisms, also known as Kyoto mechanisms, which offer the possibility of less costly ways for achieving the emission targets. One such flexible mechanism was the alternative of offsetting part of a country's commitment to limit GHG emissions by enhancing carbon sinks, such as by net increases in forest cover that sequesters carbon. Another would allow emission trading, whereby states whose emission levels were lower than their target level could sell credits to countries that were not achieving their commitments domestically. States could also receive credits for investments in "joint implementation" projects that resulted in emission reductions in other Annex I countries. Likewise, they could contribute to a clean development mechanism (CDM) that would enable them to receive credits for emission-saving projects they financed in developing countries. Finally, under a bubble provision, countries could agree to a collective emission target, but decide among themselves on differentiated responsibilities for achieving that target. The EU has opted for the bubble option and negotiated differentiated emission-reduction targets among themselves, with Germany and the UK bearing a disproportionate responsibility for achieving the group's commitment to an 8 percent reduction.[54]

When the Kyoto Protocol was originally drafted at COP3 in 1997, numerous contested details regarding rules for implementing the flexible mechanisms were left to be resolved at future COPs. As negotiations resumed, a small bloc of developed countries known as the Umbrella Group, which included the United States, Australia, Japan, Canada, and Norway, argued against limits on the use of the flexible mechanisms. The EU and its member states held out for limits on the application of the flexible mechanisms to ensure that all of the Annex I countries, including the United States with its high per capita GHG emissions, would take steps to curb their domestic emissions, rather than achieve their targets primarily through the less expensive options in other countries. Furthermore, the United States insisted that developing countries also make commitments to limit their increasing GHG emissions, as the Clinton administration responded to strong opposition to the Kyoto Protocol in the Senate for exempting developing countries from binding emission limits. The G-77/China bloc persisted in its position that the Annex I countries should take the first substantial steps to address global climate change.

Little progress was made toward resolving the remaining contested issues at COP4 in Buenos Aires in 1998 and COP5 in Bonn in 1999. However, the delegates did agree to make a concerted effort to finalize the protocol at COP6, scheduled for The Hague in 2000, so that ratifications could proceed and the parties take steps to comply with the approaching dates for achieving their emission limits. At COP5, the United States, still represented by the Clinton administration and with some support from the other Umbrella Group countries, pushed a new proposal that Annex I states should be given credit for preserving existing carbon sinks, such as forests and range land, toward fulfillment of their emission limits. In its original form, the Kyoto Protocol would grant such credit only for documented increases in the size of carbon sinks. As the COP6 meetings proceeded, the United States scaled back its proposal by reducing the amount of emission reductions that could be fulfilled by existing sinks, but the EU remained staunchly opposed to allowing any such credits. Thus, the initial phase of the COP6 talks ended in stalemate without finalizing the Kyoto Protocol.

The COP6 meetings in November 2000 were not officially adjourned to leave open the possibility of reconvening them the next year. In the meantime, the new American administration of George W. Bush dealt a major blow to the Kyoto process by declaring that the United States would not implement the protocol because it was "fatally flawed" in that it did not include emission limits for the developing countries and the emission reductions would be too costly for the United States to implement. COP6 was resumed in July 2001 with the United States present, but in the background.[55] Even though its allies from the Umbrella Group—Japan, Canada, Russia, and Australia—were hesitant about going forward without the United States, significant compromises were struck on most of the remaining contested issues. Later in the year, the final details were worked out during intense negotiations at COP7 held in Marrakesh, Morocco (box 4.1).

In its final form, the Kyoto Protocol is considerably weaker than the

Box 4.1 Key Provisions of the Kyoto Protocol As Finalized at Marrakesh

- Annex I (developed) countries commit to reduce combined emissions of six GHGs (carbon dioxide, methane, nitrous oxide, hydrofluorocarbons, perfluorocarbons, and sulphur hexafluoride) by 5.2 percent from 1990 levels by 2008–2012.
- Emission targets vary by country: most European countries (–8%), United States (–7%), Canada and Japan (–6%), Russia and Ukraine (0%), Norway (+1%), Australia (+8%), Iceland (+10%).
- "Bubble": Groups of states may collectively commit to an emission target, while agreeing among themselves to differentiated responsibilities for achieving the target (adopted by the EU).
- Enhancement or conservation of carbon sinks, such as forests or crop or rangeland, may be used to offset GHG emissions.
- Flexible mechanisms permit countries to acquire emission reduction credits for reductions achieved in other countries by:
  - *Emission trading:* purchasing credits from other Annex I countries that have more than achieved their emission target.
  - *Joint implementation (JI):* financing projects that reduce emissions in other Annex I countries.
  - *Clean Development Mechanism (CDM):* investing in emission-avoiding projects in developing countries.
- A fund is to be established to assist developing countries to fulfill treaty obligations and adapt to impacts of climate change.

European and G-77/China countries and environmental NGOs would have preferred, as significant concessions were made to the remaining Umbrella Group countries to gain their approval. Among these were rather generous allowances for conserving existing carbon sinks, such as forests and crop and range lands, which for Russia was increased from 17 million metric tonnes (mmt) to 33 mmt of carbon per year at Marrakesh. Moreover, the Annex I countries are not limited in their use of flexible mechanisms to achieve emission targets, which would allow Annex I countries to purchase large numbers of emission credits from Russia and Ukraine, whose emissions are well below 1990 levels because of their sharp economic downturns. Such permits are known as "hot air," because the emissions would have occurred even in the absence of the Kyoto Protocol. The Europeans prevailed in their insistence on tough enforcement provisions that would require Annex I countries to reduce their future emissions by 1.3 tons for every ton of emissions that exceeded the limits for the period 2008 to 2012.[56]

The determination of most of the parties to the FCCC to go forward with finalizing the Kyoto Protocol at Bonn and Marrakesh, despite its rejection by the United States, is a remarkable development. Nevertheless, the future of the Kyoto Protocol is very uncertain. To enter into force, the treaty must be ratified by fifty-five countries, including those responsible for at least 55 percent of the CO_2 emissions of the Annex I countries in 1990. Given that the United States accounts for approximately 35 percent of these emissions, its continued rejection of the protocol would mean that ratification by all of the other significant emitters is needed. By early 2003 more than one hundred countries had ratified the treaty, but Russia's anticipated ratification was still needed to satisfy the emissions requirement. Will the other Annex I countries be willing to make the sacrifices necessary to reduce their emissions and possibly disadvantaging many of their producers in international trade, if the United States persists in playing the role of free rider in international efforts to combat global climate change? Or will the unity of so many countries in proceeding with the Kyoto Protocol pressure the United States to accept the agreement?[57]

CONCLUSION

The international efforts to address the threat of global climate change are an interesting case study in the interplay between science and policy. The climate change topic is notable not only for the sheer magnitude of the scientific resources that have been devoted to the subject by a broad variety of disciplines, but also for the extent to which these efforts have been facilitated internationally through programs sponsored by WMO, UNEP, and ICSU. Moreover, the climate change issue is unique for the major investment that has been made through the IPCC to engage a large number of leading scientists in a continuing assessment of the state of climate change research and to issue reports designed to inform international negotiations on the subject. Science has played a critical role in alerting the world community to the distinct possibility that human activities are releasing gases into the atmosphere that could significantly alter the global climate in ways that would trigger a myriad of other environmental changes, which in turn would have economic and social impacts. The negotiations leading to the FCCC and the Kyoto Protocol were responses to the warnings of the scientific community of the seriousness of the threats posed by climate change.

While science has called attention to climate change and triggered its placement on international agendas, it has played less of a role in dictating the provisions of the international treaties that address the problem. Even if fully implemented, the Kyoto Protocol will only slow the growth in concentrations of GHGs in the environment. Under the best of circumstances, including the full participation of the United States, the protocol would achieve no more than a 5.2 percent reduction in the GHG emissions of the developed Annex I countries from 1990 levels by 2008 to 2012. If the Annex I countries take full advantage of the flexible mechanisms and allowances for existing carbon sinks, the reductions in their emissions will be substantially less. In either event, global GHG emissions will continue to rise because of increases in developing countries, albeit at a slightly slower pace than without the protocol. Thus, while the Kyoto Protocol is important as a first step toward an international climate change regime, it can do little toward achieving the substantial cutbacks that would be necessary merely to stabilize concentrations of GHGs in the atmosphere, much less reduce them. Will the international community be able to negotiate the much stronger international agreements that are needed to more adequately address the threat of global change?

NOTES

1. Kevin Trenberth, "Stronger Evidence of Human Influences on Climate: The 2001 IPCC Assessment," *Environment* 43(4) (May 2001): 13.
2. Trenberth, "Stronger Evidence."
3. Intergovernmental Panel on Climate Change (IPCC), "Summary for Policy Makers," *Climate Change 2001: The Scientific Basis* (2001): 7, at www.grida.no/climate/ipcc_tar/syr/index.htm, accessed 15 May 2003.
4. IPCC, *Climate Change 2001*, 5.
5. World Meteorological Organization, "WMO Statement on the Status of the Global Climate in 2001," WMO Press Release #670, 18 December 2001, at www.wmo.ch/index_en.html, accessed 16 May 2003.
6. Trenberth, "Stronger Evidence," 13.

7. For a listing of reported harbingers of global warming, see IPCC, *Climate Change 2001*, 2–3.

8. UN Department of Economic and Social Affairs, Population Division, *World Population Prospects: The 2000 Revision*, ESA/P/WP.165 (28 February 2001), v, at www.un.org/esa/population/publications/wpp2000 (click on highlights.pdf), accessed 18 July 2003.

9. IPCC, *Climate Change 2001*, 5.

10. For a summary of potential impacts of global warming, see IPCC, "Summary for Policymakers," *Climate Change 2001*, at www.grida.no/climate/ipcc_tar/wg2/005.htm, accessed 17 July 2003.

11. IPCC, *Climate Change 2001*, 16.

12. IPCC, *Climate Change 2001*, chapter 11, paragraph 2.3, available at www.grida.no/climate/IPCC_tar/wg1/416.htm, accessed 19 July 2003. Collapse of the West Antarctic ice sheet is possible. A recent, sudden breakup of the sea-borne Larsen-B ice shelf off Antarctica (that would not affect sea-level) may portend the collapse of land-based ice sheets (that would increase sea-level); Eric Pianin, "Antarctic Ice Shelf Collapses into Sea; Scientists Split on Global Warming Role," *Washington Post*, 20 March 2002, A03.

13. See Chris Bright, "Anticipating Environmental 'Surprise'," in Lester Brown et al., *The State of the World 2000* (New York: W. W. Norton, 2000): 22–38.

14. National Research Council, *Abrupt Climate Change: Inevitable Surprises* (Washington, D.C.: National Academy Press, 2001), at www.nap.edu/books/0309074347/html, accessed 18 July 2003.

15. IPCC, *Climate Change 2001*, 16.

16. Elinor Ostrom, *Governing the Commons: The Evolution of Institutions for Collective Action* (New York: Cambridge University Press, 1990): 2.

17. Garrett Hardin, "The Tragedy of the Commons," *Science* 162 (1968): 1243–48.

18. See Marvin S. Soroos, *The Endangered Atmosphere: Preserving a Global Commons* (Columbia, S.C.: University of South Carolina Press, 1997).

19. IPCC, *Climate Change 2001*, 12.

20. Seth Dunn, "Can the North and South Get in Step?" *World Watch* 40(6) (November/December, 1998): 23.

21. Eugene B. Skolnikoff, "Same Science, Differing Policies: The Sage of Global Climate Change," MIT Joint Program on the Science and Policy of Global Change, Report #22 (August 1997): 11–13.

22. See Fred Pearce, "Greenhouse Wars," *New Scientist*, 19 July 1997, 38–43.

23. See Dunn, "North and South," 19–27.

24. *Congressional Record*, 25 July 1997, S8113-S8139.

25. Paul G. Harris, "Understanding America's Climate Change Policy: Realpolitik, Pluralism, and Ethical Norms," Oxford Centre for the Environment, Ethics, and Society, Research Paper #15 (June 1998): 60–5.

26. *New York Times*, 15 May 2000.

27. See the Pew Center's Web site at www.pewclimate.org/belc/index.cfm, accessed 18 July 2003.

28. William W. Kellogg, "Mankind's Impact on Climate: The Evolution of Awareness," *Climate Change* 10 (1987): 113–36.

29. Svante A. Arrhenius, "On the Influence of Carbonic Acid in the Air upon the Temperature," *Philosophical Magazine and Journal of Science* 41 (1896): 237–76.

30. Roger Revelle and Hans Suess, "Carbon Dioxide Exchange between the Atmosphere and the Ocean, and the Question of an Increase in Atmospheric CO_2 During the Past Decades," *Tellus* 9 (1957): 18–27.

31. ICSU has been renamed the International Council for Science, but retains its original acronym.

32. Wallace W. Atwood, "The International Geophysical Year in Retrospective," *Department of State Bulletin* 40 (1959): 682–89.

33. See Marvin S. Soroos and Elena Nikitina, "The World Meteorological Organization As a Purveyor of Global Public Goods," in Robert Bartlett, Priya Kurian, and Madhu Malik, eds., *International Organizations and Environmental Policy* (Westport, Conn.: Greenwood Press, 1995): 69–82; Arthur Davies, *Forty Years of Progress and Achievement: A Historical Review of WMO* (Geneva: World Meteorological Organization—No. 721, 1990).

34. See John Gribbin, *What's Wrong with the Weather? The Climatic Threat of the 21st Century* (New York: Charles Scribner's Sons, 1978).

35. Robert M. White, "Climate at the Millennium," *Environment* 21(3) (April 1979): 31–3.

36. J. M. Barnola, D. Raynaud, Y. S. Korotkevich, and C. Lorius, "Vostok Ice Core Provides 160,000-year Record of Atmospheric CO2," *Nature* 329 (1987): 408–14.

37. See Kevin E. Trenberth, "The Use and Abuse of Climate Models," *Nature* 386 (March 1997): 131–3.

38. See the program Web site at www.wmo.ch/indexflash.html, accessed 18 July 2003.

39. Thomas F. Malone, "Mission to Planet Earth: Integrating Studies of Global Change," *Environment* 28(8) (October 1986): 6–11, 39–42.

40. See the IHDP Web site at www.uni-bonn.de/ihdp, accessed 18 July 2003.

41. More information on these procedures is available at the IPCC's Web site at www.ipcc.ch/about/app-a.pdf, accessed 18 July 2003.

42. IPCC, "Summary for Policy Makers," *Climate Change 1995: The Science of Climate Change* Intergovernmental Panel on Climate Change, Second Assessment Report (1995), 2–3, at www.ipcc.ch/pub/sarsum1.htm, accessed 18 July 2003.

43. IPCC, *Climate Change 2001,* 13.

44. IPCC, *Climate Change 2001.*

45. See William K. Stevens, "Global Warming: The Contrarian View," *New York Times,* 29 February 2000, F1.

46. WMO, *An Assessment of the Role of Carbon Dioxide and of Other Greenhouse Gases in Climate Variations and Associated Impact* (Geneva, 1985).

47. Soroos, *The Endangered Atmosphere,* 189–92.

48. See Jill Jager and H. L. Ferguson, eds., *Climate Change: Science, Impacts, and Policy: Proceedings of the Second World Climate Conference* (Cambridge, UK: Cambridge University Press, 1991).

49. See the text of the FCCC online at http://unfccc.int/resource/conv/index.html, accessed 15 July 2003.

50. Harris, "Understanding America's Climate Change Policy," 41–8.

51. See the AOSIS Web site at www.sidsnet.org/aosis, accessed 15 July 2003.

52. For a detailed analysis of the FCCC, see Daniel Bodansky, "The United Nations Framework Convention on Climate Change: A Commentary," *Yale Journal of International Law* 18 (1993): 451–558.

53. Text of the Kyoto Protocol is online at http://unfccc.int/resource/docs/convkp/kpeng.pdf, accessed 17 July 2003.

54. For an analysis of the original Kyoto Protocol, see Herman E. Ott, "The Kyoto Protocol: Unfinished Business," *Environment* 40(6) (July–August 1998): 17–20, 42–5. See also David G. Victor, *The Collapse of the Kyoto Protocol and the Struggle to Slow Global Warming* (Princeton, N.J.: Princeton University Press, 2001).

55. *New York Times,* 28 March 2001.

56. For a report on the COP7 meeting, see http://unfccc.int/cop7, accessed 18 July 2003.

57. In February 2002, the Bush administration announced a plan for the United States that would reduce CO_2 emissions relative to gross domestic product by 18 percent by 2012, primarily through voluntary measures encouraged by $4.6 billion in tax credits. This proposal has been criticized by European leaders because it would allow American emissions to continue rising. *New York Times*, 17 February 2002.

5

Political Responses to Changing Uncertainty in Climate Science

Neil E. Harrison

Although there is disagreement about the nature of the relationship between science and politics, it is generally assumed that there is a close causative connection between the two streams of events and that theories should seek to model that interaction. If the two streams do interact, it should be possible to identify changes in one stream that occur in sympathy with or response to changes in the other. Once identified, the nature of those responses may be investigated in more depth to illuminate the interaction of science and politics.

Building on the previous chapter by Marvin Soroos, this chapter focuses on the political response to changes in scientific uncertainty. In climate change, the three assessments by the IPCC, each eagerly awaited by the political community and widely accepted as representing an authoritative scientific consensus, document shifts in scientific knowledge and uncertainty. The politics of international negotiations shows a response immediately after each assessment. On the one hand, this narrative shows that many political decisions have been made almost entirely independently of scientific knowledge. On the other hand, a close reading of this narrative also reveals several instances in which policy makers' attempts to reduce political uncertainty influenced scientific methods and the resultant reports upon which policy makers relied. Two such occasions are highlighted.

An examination of the development of each assessment and the changes between them suggests that, though constrained by the demands of their profession, scientists have made an effort to explain complex science and its many uncertainties in a way that is most accessible to nontechnical political decision makers. Unfortunately, this may have opened science to many criticisms and to its instrumental use by defenders of "business as usual." Early in the climate saga, many scientists were advocates and drew conclusions not warranted by their knowledge when making collective statements. Once climate science became institutionalized, its influence on politics was formalized and lessened.

The final section assesses several theories and more ad hoc explanations of the interaction between science and politics. I show that none explains more than a small part of the climate change story and consider the reason for the unexpected disjuncture between science and politics that this case narrative uncovers.

SCIENCE PUSHES THE ISSUE

Before 1988, climate change was not on the political agenda of most countries or of the international community. Atmospheric scientists, who were becoming more alarmed at the social and ecological implications of their research findings, pressed for an international policy to mitigate anthropogenic climate change. They overstated their case, but by doing so persuaded most countries to accept in principle their proposed policy.

The first important statement by the science community that climate change was a potentially serious problem was made in February 1979. The FWCC conference report declared that humans were inadvertently modifying the global climate and predicted that patterns of change in climate, especially at high latitudes, would be likely "to affect the distribution of temperature, rain fall, and other meteorological parameters, but the details of the changes [were] still poorly understood."[1] In response, WMO authorized a World Climate Program to collect further data "to develop and exchange methods for using information on climate," to assess impacts of global warming on economic sectors, and to improve understanding of climate related processes.[2]

Since at least 1957, the United States had led research on atmospheric issues, much of it on the link between energy consumption and CO_2 emissions.[3] Spurred by the book *The Limits to Growth* and the energy crises of the early 1970s, the Carter administration began to press for international action.[4] A report linked increases in atmospheric CO_2 to global warming, assessed the warming effect of a doubling of CO_2 levels from 1.5°C to 4.5°C, warned that this warming may change the global climate, and called for the United States to assign high priority to the "carbon dioxide problem" in all energy planning.[5] Along with other global environmental issues, climate change had "become a major U.S. foreign policy interest."[6]

International scientific conferences began to call for an international policy on GHG emissions. For example, the report of the World Climate Program meeting in Villach, Austria, in 1985 concluded that "increasing concentrations of greenhouse gases are expected to cause a significant warming of the global climate in the next century," a controversial forecast not supported by substantial scientific understanding.[7] In U.S. congressional hearings, a leading climate scientist estimated the probability of climate change to be as high as 99 percent and urged the United States to reduce its GHG emissions. Congress increased research funding and directed the EPA to report on the effects of climate change in the United States and on available policy options.[8]

Because the Reagan administration believed that environmental regulation reduced business efficiency, it had resisted international negotiations on stratospheric ozone depletion and ignored the findings of the climate research being funded by Congress over the objections of the White House. The administration's resistance was worn down beginning in 1983 when activists in the EPA leaked a scientific report on ozone depletion to nongovernmental organizations. The National Resources Defense Council then sued the government under the Clean Air Act. Responding to pressure from the international community, the EPA, and the U.S. public, in June 1987 President Reagan committed the United States to support an ozone protocol and authorized significant lobbying of the Soviet Union and Japan, which was instrumental in passing the Montreal Protocol.[9] President Reagan then ordered his science adviser to form the Committee on Earth Sciences (CES), comprising thirteen federal departments or agencies, to develop a national global change research strategy to include economic analysis of policy options.[10]

Some staffers in Environment Canada persuaded their minister to support a conference named "The Changing Atmosphere: Implications for Global Security" in June 1988. What began as a small scientific meeting grew rapidly, as first the Canadian prime minister and then Gro Harlem Brundtland, chair of the World Commission on Environment and Development (WCED), agreed to attend. The conference eventually attracted atmospheric scientists and technical experts, lawyers, ministers of government, economists, industrialists, policy analysts, officials of intergovernmental organizations, and more than three hundred journalists.[11]

The conference might have had little impact if it had not been for the hot dry weather in North America that had started in May after a dry winter and continued through the conference. News magazines published pictures of cracked earth, compared it to the dustbowls of the 1930s, wondered when the drought would end, and suggested a connection to global warming. Drought and journalists ensured publicity for the strongly worded conference report that compared climate change to nuclear war: "Humanity is conducting an unintended, uncontrolled, globally pervasive experiment whose ultimate consequences could be second only to a global nuclear war."[12] It summarized the scientific evidence behind this warning and called on governments to "work urgently towards . . . an international framework convention" and set an initial global goal of a reduction of CO_2 emissions by approximately 20 percent of 1988 levels by 2005.

The Toronto conference was hated by many industrialists and energy and industry ministries. A senior staffer from the UK Ministry of Energy complained:

> [I]t was grossly dishonest and devious since . . . it was presented as a scientific conference. It was very unscientific in its approach, in its organization. Conclusions bore no relation to the state of scientific knowledge. . . . It was a leap of assertion. I think it was based upon the premise that energy use, particularly fossil fuel use, is essentially sinful."[13]

Much of the data presented at the conference was later substantiated by an EPA report on the potential effects of climate change in the United States. Concluding that climate change was a real problem and that the impacts on the United States might be extensive, this report caused a storm of controversy in the Reagan administration.[14]

Politics After Toronto

Alarmed by the science summaries they saw at the Toronto conference and the strength of the conference statement, policy makers began to support policy action. One of the most surprising converts was Margaret Thatcher, prime minister of the UK since 1979, during which time the country had earned the title "dirty man of Europe." But on climate change, Thatcher was an outspoken advocate of a tough international climate policy and an early proponent of signing a climate convention at the Rio Summit planned for June 1992. Many countries, including most of the EU, also wanted to drop the two-step model of a convention that "defines norms, principles and decision-making procedures" and a later protocol that sets national emissions-reduction targets.

Thatcher had already spoken out in support of sustainable development saying, "No generation has a freehold on this earth. All we have is a life tenancy with a full repairing lease."[15] British diplomats were fond of explaining that she had been trained as a chemist and could understand the scientific method and some of the more arcane aspects of climate science. But Sir Crispin Tickell, the UK ambassador to the UN, who had published an early book on the climate issue, also had tickled her interest in the issue.[16] Because of recent electoral losses and plummeting popularity, she welcomed anything to improve her image on the environment, and talking "green" on climate change also would not affect current economic performance.

An informal task force established by WMO and UNEP to canvass countries and propose convention wording found more interest in setting emissions-reduction targets for a protocol than debating the principles needed for a convention. Along with many commitments, funding mechanisms, and administrative matters, only three principles, drawn from prior international environmental agreements, were proposed. The emphasis on early emissions-reduction commitments also reflected recent discussion on air pollution such as the Nordic Council's International Conference on Air Pollution, which required signatories to "pledge to bring about a reduction in emissions within an agreed period." An October 1989 background paper prepared for the Noordwijk Conference proposed that parties would commit to "measures to limit, reduce and, as far as possible, prevent climate change," support scientific research, exchange information, and develop and transfer technology.[17] The precautionary principle was suggested at a meeting of NGOs in February 1990 and was picked up later that year in the Second World Climate Conference report.[18]

The EU, Canada, and Australia were early and forceful proponents of commitments by developed countries to return their CO_2 emissions to 1989 or 1990 levels by 2000, still far less than the 50 to 60 percent reduction that science was predicting would be necessary to stabilize atmospheric concentrations. The 1989 Noordwijk Ministerial Declaration repeated the recommendation of the 1988 Toronto conference for a 20 percent reduction in emissions below 1989 levels by 2005. This relatively draconian proposal, likely to have caused short-term economic dislocation, illustrates that most developed states were, at that time, very concerned about climate change and believed that urgent action was needed. But the United States successfully blocked this early enthusiasm for binding commitments, insisting on further research.

THE FIRST IPCC ASSESSMENT

In December 1988, the UN General Assembly had authorized WMO and UNEP to form the IPCC to "provide internationally coordinated scientific assessments of the magnitude, timing and potential environmental and socioeconomic impact of climate change."[19] Its first assessment report (FAR), published in August 1990, estimated warming and its effects much as the FWCC had eleven years earlier. Working Group I (WGI), responsible for scientific assessment, reported substantial uncertainty about most important aspects of climate change and the carbon cycle. The policymaker's summary tried to convey this lack of knowledge by hedging many of the predictions that it was obliged to make. WGI was only "certain" of the natural greenhouse effect that allowed life to flourish on Earth. It "calculated with confidence" that some GHGs are more damaging than others and that atmospheric concentrations of GHGs adjust only slowly to emissions because of the effects of the oceans.

Though the report was widely accepted as authoritative, most of its other predictions were little more than partially informed guesses. Illustrating the many substantial gaps in knowledge of the carbon cycle and its effect is the impromptu comment from a leading climate scientist that between 1 and 3 billion tons of carbon was unaccounted for in the carbon cycle and that the "oceans guys" thought that this was possibly going into some unknown terrestrial sinks.[20] This was more than uncertainty: it was an acknowledgement of substantial ignorance. A survey of twenty-two leading active climate scientists found substantial agreement on most of the important issues raised in the IPCC reports, especially the significant uncertainties expressed in them. Their greatest concern was that uncertainties and the potential for climate "surprises" had been understated.[21] Although the IPCC was established on the ozone model to create an authoritative scientific consensus, industry and conservative critics interpreted the many uncertainties as meaning that no problem existed and that a "fraud" was being perpetrated on the world's citizens: without scientific evidence, there was no problem to remedy.[22] Some scientists, most of whom had not been involved in the process, disputed IPCC data, critiqued the general circulation models (GCMs) on which the IPCC relied as bad science (see box 5.1), and argued that while atmospheric concentrations of CO_2 had increased, there was no evidence that the effects would be harmful.[23] Industry representatives argued that the apparent correlation between atmospheric concentrations of CO_2 and global mean temperature did not constitute scientific evidence of warming and complained that any effective mitigation policies would destroy the global economy.[24]

A second line of attack was to impugn the IPCC process as tainted with politics.[25] They observed that there was some "negotiating" over the science and that some of it was "rammed through" WGI.[26] There were complaints that WGI summary reports did not accurately represent the many underlying reports, a view vehemently denied by the IPCC, and that "profound skeptics" had been intentionally omitted from the process and "shaken off or discarded."[27] A U.S. representative at the WGI meetings commented that "some people in the final writing of that report did a little bit of backing out."[28]

Some developing countries argued that their scientists had not been adequately represented in the IPCC because they had "invited the people that first of all they knew and secondly that they agreed with."[29] They felt that "deals were done [and] . . . a lot of things happened under the table that only

Box 5.1 Models and Policy Making

Early in the IPCC process, it had been decided to make GCMs (computer models of the relevant systems and their interactions) central to its work. Pure science usually requires theory based in explanatory knowledge, a series of empirically testable propositions that hypothesize the connections between background conditions, intervening events, and observed outcomes. A model shows how the world got the way it is and may predict its future, but does not explain cause-effect relations. The emphasis on prediction through GCMs was a response to policy makers' demands for a concrete framing of the problem in policy-relevant terms (for example, the nature, location, extent, and timing of harmful effects). Despite the defects in scientific knowledge and the limitations of the GCMs (that the FAR spelt out in some detail), the prediction scenarios they generated were critical to most policy makers' understanding of the science.

the big countries know what happened" and "so they stood back a little bit."[30] However, in the political negotiations, developing countries formally accepted WGI's report and argued that it showed that CO_2 emissions should be reduced.[31]

Political Response to the FAR

The many uncertainties in the FAR did little to dampen enthusiasm for early and effective mitigation. The first meeting of the Intergovernmental Negotiating Committee for a Framework Convention on Climate Change (INC 1) convened on a day of record heat in Chantilly, Virginia, in February 1991. Mostapha Tolba, executive director of UNEP, set an urgent tone by demanding a "more aggressive convention than Vienna."[32] He recalled the wording of General Assembly Resolution 45/212 that required any convention to be "effective" and to include "commitments." Bert Bolin, chair of the IPCC, told delegates that "climate change is most likely going to be a serious problem for the world within the next half a century."

Their public statements and private comments showed that most countries were alarmed and wanted to heed Tolba's call to action. The EU hoped for a new world order of sound resource management, peace, and a more equitable distribution of wealth. Austria called global warming an "atmospheric time-bomb" and demanded early global stabilization and, thence, reduction of emissions. For Australia, climate change was "the greatest global environmental threat of all," requiring an equitable approach and the "widest possible participation." China pointed to the scientific uncertainties and the need for the participation of all countries and argued for common, but differentiated, responsibilities. Germany proposed the precautionary principle, immediate global action, and a commitment to Noordwijk reductions by the developed countries. AOSIS, which "faced the frightening specter of accelerated and excessive climate change, global warming, and sea level rise," called for immediate and significant cuts in all GHGs, adherence to the precautionary principle, new and equitable funding mechanisms, the expeditious and equitable transfer of technology, the application of the polluter-pays principle, and commitments to energy conservation and efficiency requirements. Despite the small size and limited influence of the AOSIS states, the final convention reflected many of these points, if not as clearly or extensively as developing countries had hoped.

The early enthusiasm for an aggressive response to an accepted problem quickly faded as delegates concentrated on the minutiae of treaty writing. The United Nations FCCC, signed at the World Summit in Rio de Janeiro in 1992, was long on framework, but short on action. The primary cause of the rapid dissipation of interest was the intransigence of the United States.

At INC 1 the United States had agreed that climate change was "the most complex and critical environmental issue . . . an enormous unplanned experiment that is slowly changing the composition of Earth's atmosphere." It proposed an effective convention that called for emissions limits for all GHG emissions together (not just CO_2), further research, immediate no-regrets policies, a long-term perspective, and recognition of the special needs of developing countries.[33] Only four months later its position had changed: it now opposed any legally binding commitments.

Domestic politics drove the reverse in the U.S. position. After INC 1, the first Bush administration's enthusiasm for an effective convention and protocol fell as quickly as the president's popularity. Dogged by breaking his promise not to raise taxes, his popularity from the Gulf war fast disappearing, the country sliding into recession, and a presidential election fast approaching, Bush chose the economy over the environment, even though the effect on either might not be seen for decades.

President Bush had a good record on the environment: he had been instrumental to the passage of the 1990 Clean Air Act and had worked to mitigate Ontario's acid rain problem. But on climate change he relied primarily on economic advisors, including Michael Boskin, chairman of the Council of Economic Advisors; Richard Darman, director of Office of Management and Budget; and Chief of Staff John Sununu, a Ph.D. engineer.[34] They were "unconvinced that the scientific evidence for global warming justified the cost of mitigation," and because of "uncertainty over the consequences of some proposed actions, such as a carbon tax, the U.S. was flatly opposed to any targets and timetables." That is, these advisors valued economic growth and relative certainty of policy outcomes over the possibility of uncertain calamity. In all the discussions, the EPA was "on the outside looking in."[35] During the negotiations U.S. delegates had argued privately that its litigious domestic NGOs would sue if the United States agreed to a binding target and failed to meet it. Bush finally agreed to sign the convention after any effective legal mandates (beyond certain reporting obligations) had been removed.

The influence of government advisors also is clear elsewhere. Apart from Tickell's influence on Thatcher, mentioned above, since 1990 most British policy makers have relied on David Fisk, senior scientist at the Department of Environment, for interpretation of climate science reports. Within the UK government, he is often credited with maintaining his country's consistent support for an aggressive response to climate change. He has a rare skill, an ability to translate scientific data into language intelligible to policy makers and to render scientific uncertainties meaningful without inhibiting decision making.[36]

The election of President Clinton brought an immediate reversal of policy. Energized by Vice President Gore's advocacy of a global equivalent of the Marshall Plan to save planet Earth, in April 1993 Clinton accepted as binding the convention's suggested emissions limits and agreed to reduce U.S. GHG emissions to 1990 levels by 2000. To help achieve these reductions, the president proposed a carbon tax that was rejected by Congress and by many Democrats.[37] He then resorted to small technology transfer programs dedicated to increasing industrial energy efficiency.

THE SECOND ASSESSMENT

The IPCC issued several intermediate reports that suggested that ozone depletion and aerosol production might reduce warming, but the IPCC second assessment report (SAR) was anticipated as the next authoritative summary of the state of knowledge. Published in December 1995, it did not disappoint. The SAR expressed a greater certainty that global warming from human activities was real and dangerous and already underway. Except for a few hardened skeptics, climate change could no longer be dismissed as theory.

Between 1991 and 1995 understanding of the carbon cycle had increased: knowledge had advanced in four principal areas. First, the SAR was able to point to "identifiable physical mechanisms" that could be built into climate models, such as an enhanced hydrological cycle in a warming atmosphere and the greater warming of the land than of oceans.[38] Second, reflecting this increased understanding, GCMs could now model extensive ocean-atmosphere interactions (that affect rates of warming) and the potential for delayed warming after stabilization of GHG concentrations. Third, GCMs included the effects of anthropogenic sulphate aerosols that, though short-lived, tend to reduce radiative forcing and depletion of the stratospheric ozone (a GHG) layer. Finally, research had shifted from a focus on global mean temperature changes to modeling observed spatial and temporal climate patterns.

Four major areas of uncertainty still hampered researchers. First, data for temperature and climate patterns from recent centuries was limited and unavailable before the fourteenth century, making it difficult to measure a climate baseline to separate anthropogenic change from natural variation or "noise." The "magnitude and pattern of long-term variability" was still poorly understood. Second, the pattern over time of the effects of, and natural responses to, changes in GHG and aerosol concentrations and land use were not well understood. Third, estimation of future "emissions and biogeochemical cycling" of GHGs and aerosols was inadequate. Fourth, GCMs still poorly represented natural feedback cycles associated with "clouds, oceans, sea ice and vegetation." Other lesser uncertainties (for example, the poor "characterization" of the effects of rapidly increasing ground level ozone produced by vehicles) were identified throughout the science report. However, the report emphasized the long-term danger of continuing with business as usual and cautioned that there could be substantial climate "surprises."

> The SAR's most salient conclusion was that the ability to quantify the human influence on global climate is currently limited because the expected signal is still emerging from the noise of natural variability, and because there are uncertainties in key factors. These include the magnitude and patterns of long-term natural variability and the time-evolving pattern of forcing by, and response to, changes in concentrations of greenhouse gases and aerosols, and land surface changes. *Nevertheless, the balance of evidence suggests that there is a discernible human influence on global climate.* [Emphasis added]

The word "discernible" was not chosen lightly. The scientists recognized that policy makers needed the strongest possible statement of the nature and extent of the problem. But they were constrained by their training and their standing among their peers from expressing more certainty than the evidence clearly demonstrated. At the final negotiations, the scientists wanted to report that, for the first time, the majority of technical experts agreed that an anthropogenic warming signal had been read against the background noise of natural climate variation. But the available data was not conclusive, did not "speak for itself," as the signal was yet very faint. Gylvan Meira, the chair of WGI, asked for the English meaning *descernable*, the French word for something that is just barely perceptible.[39] For the first time, the IPCC recognized that the problem existed. But the careful negotiation of this important language attracted charges that politics had entered into the scientific process. Critics claimed that the IPCC had interpreted the data through values different from their own. That they attacked the IPCC for politics was, therefore, disingenuous, but their attack had a crude logic based on scientific principles: without conclusive evidence should the IPCC draw important conclusions that they can reasonably expect policy makers to rely on?

As with the FAR, critics of the report's conclusions complained about the IPCC process. In his remarks to INC 1, Bolin had referred to the FAR as containing "carefully balanced reports," suggesting that they were negotiated, which they were. Under conditions of pervasive uncertainty and incomplete knowledge, individual scientific reports may be contradictory. So, participating scientists negotiate the wording of the final reports, including the summary reports for policy makers, to give an appropriately "balanced" view of knowledge and uncertainty. This process, required by the rules of the IPCC, led to allegations that the SAR had been doctored to understate uncertainty.

Critics alleged that the scientific chapter of the final SAR had been altered for political reasons after its approval by the plenary. They noted that sections in the report approved in Madrid in November 1995 had been removed from the final report published in December and alleged that this led "policy makers and the public into believing that the scientific evidence shows human activities are causing global warming." The excised wording included the comment, "None of the studies cited above has shown clear evidence that we can attribute the observed [climate] change to the specific cause of increases in greenhouse gases."[40] IPCC officials retorted that the changes were made "in response to written review comments received from governments, individual scientists and nongovernmental organizations during plenary sessions of the Madrid meeting," as required by IPCC procedures "to produce the best possible and most clearly explained assessment of the science."[41] They insisted that the SAR's conclusions had not been changed, that the final version "draws precisely the same 'bottom-line' conclusion" as the original version, and that "[t]aken together, these results point towards a human influence on climate." They asserted that no uncertainties had been suppressed.

Political Response to the SAR

For four years the convention was worked over financing mechanisms and administrative details were elaborated, but no significant progress was made before the Second Conference of the Parties (COP 2) in June 1996, seven months after the SAR had been published.

Illustrating the division among the delegations, the plenary of COP 2 tried to make the IPCC responsible for defining "dangerous interference" with the global climate, prevention of which is the stated goal of the FCCC (Article 2). Bert Bolin refused, arguing that this was a political, not a scientific, issue. However important, it was not a political issue that the plenary felt able to resolve; all efforts to debate it had been opposed by the oil-producing countries, Australia, and Russia. Only two weeks later, primarily as a result of the unexpected intervention of the United States, there was a consensus around a broad definition of "dangerous interference," and the United States had accepted the principle of legally binding targets and timetables.

The Geneva Ministerial Declaration issued at COP 2 was the result of three days of ministerial discussions lead by the United States. It defined dangerous interference as atmospheric concentrations exceeding "twice pre-industrial levels" (about 550 ppm of CO_2 by volume). The resulting "acceptable" rate of warming was about 2°C, which would increase mean sea level by about 50 centimeters, and "significant, often adverse, impacts on many ecological systems and socio-economic sectors." The declaration also committed developed countries to "quantified legally-binding objectives for emissions limitations and significant overall reductions within specified timeframes . . . with respect to their anthropogenic emissions by sources and removals of greenhouse gases not controlled by the Montreal Protocol."

The declaration was a coup for U.S. diplomacy. Through extensive lobbying and a forceful speech in the first ministerial plenary session by Timothy Wirth, undersecretary for global affairs, Department of State, the United States had catalyzed this unexpected advance in the negotiations. Although it had always opposed legally binding emissions-reduction targets and timetables, the U.S. delegation denied that this was a policy u-turn.

The United States radically changed its position at COP 2 because a window of opportunity had opened in domestic politics. President Clinton persuaded the public that Republicans who had won control of Congress in 1994 had waged the "most aggressive anti-environmental campaign in our history."[42]

The environment was an important priority for nearly three out of four voters, including a majority of Republicans and 61 percent of GOP women. Clinton and Gore were reelected and a Democratic administration again tried to make environment central to foreign policy. In April 1996 Secretary of State Warren Christopher called climate change one of the most "significant long-term environmental and diplomatic challenges facing the world," announced a new "green" foreign policy, and made an international agreement on further cuts in GHGs a top priority.[43] This led directly to the diplomatic coup at COP 2.[44]

Other developed countries also had changed their positions between 1992 and 1996. For example, in 1991 Australia had voluntarily adopted an emissions-reduction target comparable to that of the European countries. But by 1996 it was working to block progress in the negotiations. Since 1992 its political and economic situation had changed. A conservative government had replaced a left-wing Labor government, and huge new energy resources in western Australia were being developed to satisfy rapidly expanding energy demand in Asia. The conservative government preferred economic growth to climate mitigation and found that reducing emissions was impossible.

A conservative government also had been elected in New Zealand. In an effort to exploit the country's large natural and managed forests, the government now pressed to expand the uses of sinks in the calculation of emissions reductions. Russia, which was expanding its oil production and still thought that it might benefit from warming, joined Australia, New Zealand, and the oil-producing states in opposing the EU and the United States. In the final plenary session it questioned the SAR and spoke out against the declaration.

Canada had been an active policy entrepreneur, advocating progressive solutions in both the ozone and early climate negotiations. But by 1996 it had receded into the background, embarrassed by its inability to implement a national policy. To the chagrin of Ottawa, the province of Alberta, rich in gas and oil, had been able to veto implementation of much of the federal government's ambitious Green Plan of 1990.[45]

The developing countries continued to view the climate issue as a problem caused by the developed countries. More concerned about development, they still hoped to use the issue to generate "new and additional" financial resources for development (and secondarily for environmental protection), despite being continually rebuffed. But the debate at COP 2 was primarily among the developed countries with the G-77 and China watching from the sidelines.

The Emergence of Sinks

After COP 2, the issue of carbon sequestration by land systems (sinks) became increasingly important. The 1997 Kyoto Protocol increased political pressure on developed-country governments to reduce the economic cost of meeting emissions targets. Article 3.3 of the protocol permitted parties to count towards their targets "direct human-induced land use change and forestry activities, limited to afforestation, reforestation and deforestation since 1990, measured as verifiable changes in carbon stocks in each commitment period." Developed countries led by Australia and New Zealand had inserted this language because they expected that they might achieve much of their Kyoto targets with only minimal changes to their forest management and land-use practices. The problem was finding a definition of forest practices and land-use changes that would constitute emission offsets under the protocol, as well as benefiting countries with widely different biomes. This political mechanism needed to be translated into science to permit "transparent and verifiable" reporting under Articles 7 and 8 of the protocol.

Because there is no single definition of a forest (also see chapter 7) and definitions of afforestation, reforestation, and deforestation are similarly subjectively responsive to local conditions (see table 5.1), the IPCC tried to develop a single definition of each concept. But it soon became apparent that "in applying the IPCC Revised Guidelines, different assumptions could cause a forest to be classified as either a source or a sink."[46] How soon after harvest should establishment (which includes planting and natural revegetation) occur to count as a human-induced land-use change? "If the . . . period is less than twenty years, countries would be able to convert natural forests to other land uses, begin a plantation scheme, and then declare these lands as 'reforested.' In this case, reforestation would lead to net emissions rather than sinks."[47] Under some definitions, planting after clear-cutting, as is common in temperate and boreal zones, might not constitute reforestation. Table 5.2 shows the range of carbon stock effects under different definitions.[48]

In 1998 Dr. Robert Watson, IPCC chair, had promised that the special report on land-use change and forestry to be published in 2000 would be "designed to provide scientific, technical, economic and social information that can [help] governments operationalize Article 3.3 of the Kyoto Protocol" and would be "policy relevant, but . . . not be policy prescriptive."[49] But several countries sought definitions that, *for them*, were tied to specific policies to increase carbon stocks. For example, the United States calculated that under its preferred definitions, "forest, cropland and grazing land management would result in average net removals of between 260 and 360 million metric tons of carbon equivalent (MMTCE) per year, with a central estimate of about 310 MMTCE per year."[50] As discussed later, sinks were almost the last issue agreed on for ratifying and implementing Kyoto.

THIRD ASSESSMENT REPORT

The third assessment report (TAR), published in 2001, included two important innovations in an attempt to more clearly convey the meaning of the detailed research and the reports from the three working groups (science, impacts, and policy responses). The IPCC included an executive summary titled "Synthesis Report Summary for Policymakers," structured as integrated responses to questions of interest to policy makers, and tried to categorize and quantify uncertainty.[51]

The TAR is confident that the observed warming is human induced: there was a "negligible" probability that the observed physical changes from this

Table 5.1 What Is a Forest?

Australia	A forest is an area dominated by trees having usually a single stem and a mature or potentially mature stand height exceeding two meters, and with existing or potential projected cover of overstorey strata about equal to or greater then 20 percent. Countries should be able to employ a definition of forest appropriate to their particular biophysical circumstances.
Austria for EU	Forest definition needs to be linked to data on the dynamics and equilibrium or time average values of carbon uptake and storage. Definitions in terms of crown cover would not be useful unless linked to carbon-stock data in this way.
Finland	Forestry land in Finland is grouped into three classes according to site productivity.
Iceland	A broad definition is needed; a narrow definition would introduce the danger of exclusion by parties of deforestation activities on the grounds that an area being cleared does not meet the definition of a forest. Restrictive definitions should be avoided. The capacity of a forest to sequester carbon is to be quantified by individual parties. The contribution of a forest should be determined by its capacity to remove carbon, not by arbitrary defining criteria such as tree height or cover.
Japan	Rigorous FAO definition has the following issues to be discussed: land that has the same carbon removal effect as forests, such as orchards and city parks, could be excluded from "forests."
New Zealand	Forest definition should take into consideration the context or purpose for which land is being managed. The issue for any definition will be how can the establishment (or removal) of a forest be distinguished from other land-use activities?
Philippines	Forests are ecosystems, which include all living organisms (flora and fauna), as well as nonliving components (litter, soils, water, etc.). This concept solves the issue of having a too-narrow definition of forests. More specifically, a minimum crown cover of 10 percent must exist in wild or natural conditions, and there must be absence of agricultural cultivation.
Samoa for AOSIS	To be defined by the IPCC special report.
Switzerland	Kyoto Protocol should not lead to incentives for deforestation and forest degradation. Sustainable forest-management practices that contribute to high carbon sequestration, preserve biodiversity and soil quality, and serve important socioeconomic ends (e.g., as shelterwood) should be recognized. The terms deforestation, afforestation, and reforestation need to be defined accordingly.
USA	Forest definition should be based on interpretations of Kyoto Protocol articles consistent with the level of commitment parties agreed to for the first commitment period. Interpretations of key LUCF terms should be based on sound science, should promote other environmental objectives related to land use, recognizing trade-offs and complementaries among environmental goals, and should create appropriate incentives.

Source: Based on UNFCCC Secretariat, "Note by the Secretariat: Matters Related to the Kyoto Protocol, Matters Related to Decision 1/CP.3, paragraph 5, Land-Use Change and Forestry, for the Fourth Conference of the Parties, 28 October 1998," FCCC/CP/1998/INF.4.

Table 5.2 Estimated Annual Developed-Country Carbon-Stock Changes, 2008–2012, from Afforestation and Reforestation under Different Definition Scenarios

FAO land-based 1	−759 to −243 MtC/yr.$^{-1}$
FAO land-based 2	−190 to +259 MtC/yr.$^{-1}$
FAO activity-based	+87 to +573 MtC/yr.$^{-1}$
IPCC definitions	+197 to +584 MtC/yr.$^{-1}$

Source: Based on IPCC Special Report, p. 12.
Note: MtC = Million tons of carbon equivalent.

warming could have occurred by chance alone. Impacts from warming, it concluded, would be negative and are already being felt. By 2100, in the absence of mitigation measures, atmospheric GHG concentrations of between 90 and 250 percent above preindustrial levels are probable. The TAR increased estimates of future warming to a range of 1.4°C to 5.8°C and increased its estimate of sea-level rise and many other impacts. Most impacts would be negative, and the negative impacts would accelerate as the rate and level of warming increased. More people would be negatively than positively affected, with developing countries and the poor everywhere hardest hit. Beyond 2100, plausible changes to natural systems would be irreversible. Reasons for concern now included "risks to unique and threatened natural systems," the possibility of abrupt and large changes in global or local climates, and extreme or catastrophic weather events.

To make the scientific data that it presented more useful for policy makers, the TAR quantified uncertainty and risk by attaching a quantitative probability assessment to each of its major pronouncements.[52] For example, far from barely discerning a changing climate, the TAR is "very confident" that the climate is already changing and will continue to change, with more hot days and fewer frost days in land areas, and that poorer states and people will suffer most. However, in a controversial move that appears to increase uncertainty, a special report asserted that the future is inherently unpredictable and assigned equal probabilities to all emissions scenarios.[53] This challenged the notion, supported by the environmental groups, that there is a "best" development path. The synthesis report commented that reliable prediction of future climates is not possible because the projected rate of warming is "very likely to be without precedent during at least the last ten thousand years."

The multipart questions in the synthesis report directly address policy makers' principal concerns. For example, the first question talks to the issue of "dangerous anthropogenic interference" and integrates data from all three working groups in the response. Again the IPCC explicitly noted that dangerous anthropogenic interference, avoidance of which is the principal purpose of the convention, must be a value judgment "determined through socio-political processes, taking into account considerations such as development, equity, and sustainability, as well as uncertainties and risk." Issues addressed in other questions include the causes and consequences of changes in global and regional climates, extreme weather events, the risk of nonlinear changes in climate, knowledge of atmospheric sensitivity to GHG concentrations, costs and benefits of mitigation interventions, and key uncertainties. Together with the probability measures, the responses to these questions give the most comprehensive and clearly stated assessment of the risks of climate change and the costs of mitigating it. Many significant uncertainties remain. But the synthesis report also argues that the risk from failing to implement mitigation measures early exceeds the risk of climate change turning out to be insignificant. It concludes that most of the necessary technologies are available, that they probably are cost effective, and that early intervention would probably lower overall mitigation costs.

Political Response to the TAR

Initial responses to the TAR were subdued; most states had been represented in the IPCC approval process (which had been much less divisive than that for the SAR) and were familiar with its major conclusions. Worry about the new U.S. administration's climate policy was initially alleviated in early March 2001 when the administrator of the Environmental Protection Agency told reporters at a G8 meeting (an annual summit of the leaders of the eight countries with the largest economies, excluding China) that "the climate review being undertaken by this administration does not represent a backing away from Kyoto." However, before the month was out, President George W. Bush had done just that.

On March 28 Bush announced that he was "unequivocal" in opposing the protocol because it was not in the United State's best interests and that he was seeking an alternate response. He rejected it as "fatally flawed" and unfair to the United States, the world's largest polluter, because poor countries are not required also to cut their emissions. Although the U.S. Department of Energy estimates that the United States could cut emissions by 75 percent—much greater than required by Kyoto—at no net cost, Bush insisted that the protocol would impose "draconian costs" on the U.S. economy.

The response from Europe was completely negative. The EU was furious, and the UK environment minister called the rejection of Kyoto "an issue of transatlantic foreign policy." Fearing that United States' withdrawal would kill the protocol, the EU asked for detailed U.S. demands and offered to negotiate changes to satisfy those demands. President Bush refused, saying that "first things first are the people who live in America. That's my priority." The EU considered, then dropped, retaliatory measures. Stating that he still believed that climate change was a danger, President Bush sought a "science-based" policy that would reduce U.S. emissions at minimal cost by "encouraging research breakthroughs that lead to technological innovation, and take advantage of the power of markets."[54] In July, Bush announced several small initiatives to improve scientific research, sequester carbon from the atmosphere, exchange debt for developing country forests, and improve scientific cooperation and policy coordination in the Western Hemisphere. The EU complained that this did not amount to a climate strategy, and President Bush was subjected to relentless attacks on a visit to European capitals. Not all was unity at home as he was attacked by opposition Democrats. A National Academy panel dominated by auto industry representatives proposed increased fleet fuel economy standards that Bush had publicly opposed.[55]

International Negotiations

Despite the U.S. rejection of the protocol, negotiations on its implementation continued. Since 1997, the international community had been searching for rules to turn political decisions in Kyoto into workable policies for countries to adopt. The Buenos Aires Plan of Action from COP 4 had been designed to produce those rules by COP 6 at the end of 2000. But that meeting (in The Hague in November 2000) failed to produce the hoped for result and was suspended when the EU refused to relax wording on the use of sinks (as advocated by the United States) in measuring national progress to Kyoto targets. By the time the IPCC issued its TAR, negotiators were dispirited. At the request of the George W. Bush administration, the next meeting (strictly a continuance of COP 6) was scheduled for June 2001. In the meantime, President Bush dropped his bombshell.

The U.S. rejection of Kyoto changed the dynamics of the protocol negotiations in two important ways. First, the EU took a clear leadership role, accepting the task of hammering an agreement out of the continual bickering. Second, countries like Canada, Australia, Japan, and particularly Russia gained more negotiating clout.[56] In Article 25 the protocol set the standard for its entry into force as ratification by fifty-five countries including countries representing 55 percent of CO_2 emissions from Annex I (roughly, developed) countries. With the United States gone, the EU, Russia, and Japan had to sign, and Canada and Australia also would probably be necessary.

To everyone's surprise, and the chagrin of the United States, the continued COP 6 in Bonn in July 2001 reached political agreement on all the important issues before it, in large part because the EU wanted an agreement more than specific rules. The rump of the Umbrella Group was able to extract favorably lax wording on sinks and emissions-trading principles, and the developing countries won increased funding. As demanded by Russia and Japan, the compliance mechanism—to punish transgressions—was relaxed. Preparation of a legal text was postponed until COP 7 in Marrakesh in November 2001.

At Marrakesh, negotiations followed the same pattern as in Bonn: the EU sought to broker a deal, and the remaining Umbrella Group countries, led by Russia and Japan, sought and won concessions to their particular interests. Because of its industrial implosion after the end of the Soviet Union, Russia's GHG emissions have fallen more than 30 percent.[57] At Marrakesh, Russia won a near doubling of the credits it could earn from sinks. But Russia remained unconvinced of the benefits of Kyoto: the United States's withdrawal might reduce Russia's potential income from trading emissions credits (needed by other countries to meet their Kyoto obligations) by up to 90 percent. The EU and Russia are considering a plan to limit Russian emissions credit in exchange for appropriate EU investment and to increase joint implementation projects.[58] Japan won concessions from the EU on the contribution of sinks to Kyoto targets and some slight loosening of the rules on Protocol compliance. Japan, which has historically taken its international commitments very seriously, was concerned it might not meet its agreed-on target. Despite a flat economy, Japan's emissions have risen since 1990, and it is struggling to meet its Kyoto goal of a 6 percent reduction by 2012. But its major concern was that it did not believe the Kyoto process should continue without the United States.

As of July 2003, the Kyoto Protocol teetered on the edge of entering into force. Australia has followed the United States and rejected its Kyoto commitments, and Japan, the EU, and Canada (despite threats from the energy-producing province of Alberta to sue) have ratified the protocol. Entry into force rests with Russia, which is expected to ratify.

Without a sea-change in domestic politics, it is not likely that the United States will reverse course. In February 2002 President Bush announced his long-awaited alternative to Kyoto.[59] It is a voluntary program that he expects to decrease U.S. energy intensity (energy used per unit of GDP) by 18 percent, as opposed to the decrease of 14 percent otherwise projected. With minimal government intervention, the U.S. economy achieved a 17 percent reduction during the 1990s.[60] Without an economic depression, this will result in a substantial increase in emissions by 2012, not the 7 percent decrease Clinton and Gore accepted at Kyoto. When the U.S. EPA acknowledged that human activities were likely to change the climate with potentially significant effects, President Bush dismissed the report as "put out by the bureaucracy."[61] He has also ignored a report by the National Academy of Sciences that warned of abrupt and significant changes in climate and argued that early mitigation would be advisable.[62]

COMPARING EXPLANATIONS

This final section assesses nine possible explanations for the interactions between science and politics in international climate policy. But the climate change saga is long and complex enough that the reader may envisage other explanations or develop elaborations of those discussed here.

Epistemic Communities

A community of concerned scientists may have influenced scientific statements and policy makers' preferences before 1991 and helped the international community to take up the issue. The frequent and strong advocacy of mitigation policies by "scientific" meetings shows that many scientists were interpreting incomplete data through their beliefs. This activism forced policy makers to take up the issue, and the Toronto comparison to nuclear winter alarmed many states, as seen in their INC 1 comments. But once science was institutionalized with the IPCC, scientific knowledge has been expressed in more measured terms, and its influence on politics has fallen dramatically.

In some cases, individual scientists have influenced countries' policies, as Sir Crispin Tickell is credited with influencing Prime Minister Thatcher, and David Fisk has "educated" several UK environment ministers about climate change. In countries as diverse as Thailand and Brazil, governments have relied on specialists to make sense of arcane technicalities.[63] But in the first Bush administration, many of these specialists were excluded from decision making, and the second Bush administration seems to have ignored them. In 1991 many heads of delegations were scientists (as Fisk for the UK), but since COP 1 in 1995, delegations have become increasingly dominated by government officials without scientific training. Since the early days of the convention negotiation, changes in scientific understanding have had little apparent effect on countries' climate policies.

Discursive Practices

The brief narrative given here prevents clear identification of discursive practices, but they may have influenced politics, if not science, between June 1988 and May 1991. At that time, the idea developed by scientists and promulgated through their meetings that global warming was a serious and immediate threat may have "disciplined" policy makers to believe that early mitigation was necessary. The early rush to combine a protocol of targets with the convention for signature at the 1992 Rio Summit gives circumstantial support to the power of discursive practices. After the promising early rhetoric, countries learned more about the uncertainties and the costs of mitigation and became more cautious, with several (e.g., Australia, Canada) reversing their support for early mitigation. The use of GCMs in response to political needs and the debate over sink definitions do not reflect the discipline of ideas so much as regulatory science (see "Mutual Construction," below). And since the FAR, there have been only minor changes in the structure or methods of the IPCC and no new scientific concept has reshaped political debate.[64]

Mutual Construction

As a boundary organization, the IPCC is designed to serve two masters: scientists and policy makers.[65] Although compromise is an essential component of its processes, it seeks to establish its reports as authoritative within both realms. The centrality of GCMs in the IPCC's assessments reflects the need to be both relevant and authoritative within politics, even at the price of attracting criticism within science.[66] Similarly, the political choice of definitions may distort scientific research on sinks, but is necessary to retain scientific influence on policy making. Although the IPCC has been responsive to political demands, it has protected the borders of science, as when Bolin insisted that defining "dangerous interference" was a political matter.

The concept of boundary organizations usefully describes the organizational reality of science as a hybrid activity. But it explains little of the actual processes that generate scientific influence. And, paradoxically, as the climate narrative has progressed, science's influence on politics has diminished, despite its policy responsiveness. Rather than science and politics mutually constructing each other, politics has been less influenced by science than it has used it.

Because political changes within states have become more important than IPCC reports, political processes have increasingly operated independently of scientific statements. President Bush has occasionally suggested that he is unconvinced that climate change is a problem and that more science is needed, but the primary reason he cited for rejecting Kyoto was the cost to the United States. Ignoring the National Academy of Sciences' (NAS) warning of possible abrupt and significant changes in U.S. climate, President Bush is confident that the United States can adapt to any changes as they occur, a view (shared by the Australian government, among others) that does not rely on science. He, like many government leaders, has ignored the precautionary principle.

In democracies, cautiously worded reports reflecting scientific uncertainty do not motivate energetic political responses like apocalyptic crises. The IPCC has protected its organizational existence by forcing consensus and excluding radical dissent. Consensus building (which environmental NGOs hoped would spur action) demands negotiation and dilutes the language to a carefully hedged least-common-denominator message that does little to stimulate action. And institutionalization of regulatory science in a boundary organization also has masked (or emasculated) political activism among climate scientists. The IPCC's scientific credentials continue to be questioned. The IPCC's reliance on GCMs has drawn criticism of its scientific credentials and distracted policy makers from the possibility of nonlinear climate change, and sink research has been circumscribed by political needs.[67] Finally, the TAR has been accused of making elementary statistical errors in forecasting global economic development.[68]

National Interests

Country choices in international negotiations are often assumed to reflect identifiable national interests.[69] But because of the long timescale, the many causes, and the largely nonquantified effects of global warming, national interests cannot be objectively defined in the same way that security of the state from attack might be. By 1992, most developed countries had assessed in qualitative terms the likely effects of climate change on their territory and trade. They generally believed that they had the technical ability and financial strength to adapt to a changing climate. A nonlinear change in the system could not be assigned any probability and was ignored.

Why, then, did the Europeans choose mitigation when the United States did not? The EU countries had some advantages from sharing the benefits of the elimination of dirty industries in East Germany and the conversion of much electric production in the UK from coal to natural gas. But by some measures (using bottom-up economic models that assume application of best available technology), the United States could reach its Kyoto goal at no net cost. Other benefits from reduced fossil fuel use, such as increased security from reduced petroleum imports, were discounted by the first Bush administration.[70] Except for Australia, which changed its policy partly in response to growth in resource extraction and energy production activities after 1992, there are few countries in which national conditions changed sufficiently to explain change in national policy.

Domestic Politics

Some theorists expect that domestic politics influences countries' negotiating positions and international negotiation processes. Some countries that changed their policies independently of science and national conditions were responding to domestic politics. For example, government changes in Australia and New Zealand, in both cases to more conservative regimes that valued markets and economic growth more highly, radically changed their climate change positions between 1992 and 1996.[71] The United States switched its position from opposition to support of mitigation policies with the 1992 election, to support of legally binding targets in 1996 and acceptance of Kyoto in 1997, to rejection of Kyoto after the 2000 election. Domestic politics also probably influenced Thatcher's early embrace of the climate change issue. However, changes from rightist to leftist governments in the UK and Germany or changes of political leadership in Brazil or India have not materially affected their policies. In some cases, public commitment to climate change mitigation is strong enough, as in Germany, for the issue to be almost permanently on the political agenda. In other countries, as in the United States, neither polls nor the voters' consumption choices indicate much interest in the issue. Differences in culture and political structures may partly explain these divergent responses.

Personal Beliefs

Leaders' personal beliefs may dominate the policy-making process if public apathy opens the necessary political space. Just as interpretation of incomplete and uncertain science depends on personal beliefs, the policy response depends on the value preferences of policy makers. Both Carter and Reagan could read the same data, but they interpreted it differently. Carter, the engineer, had responded to scientific evidence of climate change and the limits-to-growth argument with expanded research and proposals for international policy. Valuing economic growth and unilateral action over environmental conservation and multilateral collaboration, President Reagan brushed aside his predecessor's concerns, attempted to reduce the climate research effort, and temporarily withdrew from international negotiations on ozone depletion. As illustrated by the different choices of the first President Bush and Prime Minister Thatcher, the leader's beliefs may influence whom he or she listens to, which, in turn, determines what the leader hears. If the personal beliefs of leaders dominate policy choices, science will probably be used instrumentally to buttress prejudgments or ignored as irrelevant.

Crisis

Anything that mobilizes public concern can force political leaders to accept policies that they personally oppose. An unacknowledged motivation for the negotiators of the Montreal Protocol was the unexplained existence of the ozone hole. Pictures of a hole in the Antarctic stratospheric ozone layer were a tangible image of anthropogenic atmospheric damage. With all the issues crowding the international agenda and the many more domestic issues that vie for attention, an environmental issue attracts sufficient attention to advance serious policy making only when it vividly captures the public's imagination. A tangible and threatening image can motivate publics and policy makers in ways that carefully worded scientific reports cannot.

Climate change has no such image. The complexity of its causes and the pervasiveness of its effects prevent neat encapsulation in a tasty sound bite. Nearly every human activity—from farming to eating to work—increases the potential for climate change, and every human is affected. Pictures in U.S. newsmagazines of drought-cracked Midwestern earth electrified interest in climate change in 1988 when they were connected to the Toronto conference. Should climate change ever become nonlinear, only adaptive strategies would then be possible. But at least the danger would be a tangible problem for policy makers to debate, if it is not too late.

Fairness

A rational response by countries faced with significant uncertainty about the effects of their decisions is to distribute costs and benefits from international policy as fairly as possible.[72] There are three principal reasons why fairness has been largely ignored: it would allocate large costs to the developed countries before the problem was clearly understood; the developed countries dominate international environmental agreements; and the United States has held a veto (until 2001 it was commonly believed that the protocol would fail without the United States's participation). That fairness has not guided the climate negotiations shows that policy makers have not properly understood (or have conveniently ignored) the many scientific uncertainties and the possibility of significant abrupt changes. It is not clear if this is a failure of politics or of science.

An overall principle of fairness might be that the distribution of costs and benefits of international climate policy should be equal (in some way) "unless an unequal distribution . . . is to the advantage of the least favored" (the poorest and most likely to be harmed by a changing climate).[73] This may be implemented in at least five ways. First, emissions could be allocated equally by country. Although the UN operates under the principle of one country, one vote, allocating equal per-country emission rights cannot work when China has a population of more than 1.2 billion and St. Kitts and Nevis about 40,000. Second, mitigation costs could be allocated in proportion to gross domestic product, per capita income, or another "ability to pay" (comparable to "from each according to their abilities"), and benefits would be allocated (in terms of adaptation assistance) as technological and financial assistance from richest to poorest "according to their needs." However, convention and protocol provisions for financial and technical assistance, emissions trading, and joint implementation are insufficient to achieve these goals. Third, costs could be allocated to those who have gained the most from the pollution that caused the current warming: a stronger form of the polluter-pays principle in international law. The convention makes the developed countries primarily responsible and requires them to act first. But economic and trade agreements take priority, and Article 2 mandates that prevention of "dangerous interference" with the global climate "should be achieved within a time-frame sufficient . . . to enable economic development [for all countries] to proceed in a sustainable manner." Fourth, costs of mitigating future emissions could be allocated to countries in proportion to their expected baseline (business-as-usual) increases in emissions: those with the highest expected emissions growth rates would pay the most. This would disadvantage poor countries that are industrializing and only reduce the rate of growth in emissions, while the TAR advocates global emissions reduction.

Finally, each human could be allocated an equal portion of global carbon sinks as called for by India in 1991. Countries' emissions limits would be a function of the global average per capita target and their populations. To stabilize atmospheric concentration within a few decades, as was initially hoped for, would require an estimated 60 percent reduction in global emissions from current levels. But global population will increase about 40 percent by 2050. Thus, per capita emissions must fall even faster, imposing on the developed countries, and especially the United States, a huge cost, even if the best available technology could be widely implemented (see table 4.1). As a result, measures of "dangerous interference" are now being revised upward to more than double preindustrial atmospheric concentrations.

CONCLUSION

This case shows that there is no single, simple explanation for the interplay of science and politics in climate change. Some explanations have currency at different stages in the process. Rather than driving politics, did science—certainly after 1991—retard policy development by not being able to give policy makers a clear policy-relevant statement of a serious problem? Would a blind fear of the unknown have more effectively moved the international community to effective action—at the risk of incurring unnecessary costs—than has the IPCC? Unfair to the poor everywhere and ignorant of scientific warnings, inaction reduces current costs and disregards the sustainability principle of the convention. Is that why the international community—and particularly the developed countries—insists on using scientific research to guide international environmental policy? Is there an alternative?

NOTES

1. United Nations, World Meteorological Organization, *Proceedings of the World Climate Conference: A Conference of Experts on Climate and Mankind* (WMO—No. 537, 1979): 709.

2. United Nations, World Meteorological Organization, *WMO and Global Achievement,* (WMO—No. 741, 1990): 17.

3. National Research Council, *Energy and Climate* (Washington, D.C.: National Academy Press, 1977); and National Research Council, *Carbon Dioxide and Climate: A Scientific Assessment, Report of an Ad Hoc Study Group on Carbon Dioxide and Climate, Woods Hole, Mass., July 23–7, 1979* (Washington, D.C.: National Academy Press, 1979).

4. Donella H. Meadows, Dennis L. Meadows, Jorgen Randers, and William W. Behrens III, *The Limits to Growth* (New York: Universe Books, 1972).

5. United States, Executive Office of the President, Council on Environmental Quality, *Global Energy Futures and the Carbon Dioxide Problem* (Washington, D.C.: Government Printing Office, January 1981).

6. "U.S. Seeks International Response to Global 2000 Report to President," *International Environment Reporter,* 10 September 1980, 419.

7. Report of the International Conference on the Assessment of the Role of Carbon Dioxide and of Other Greenhouse Gases in Climate Variations and Associated Impacts, Villach, Austria, WMO—No. 661, 9–15 October 1985.

8. Interview with section head of the EPA, 27 August 91. Stephen Schneider, *Global Warming: Are We Entering the Greenhouse Century?* (New York: Vintage Books, 1989): 131; "Senators, Scientists Urge Policy Shifts to Avert Problems from Atmospheric Changes," *International Environment Reporter,* 9 July 1986, 241–2.

9. Richard Elliot Benedick, *Ozone Diplomacy: New Directions in Safeguarding the Planet,* enlarged ed. (Cambridge, Mass.: Harvard University Press, 1998): 65–7.

10. United States, Executive Office of the President, Council on Environmental Quality, *Environmental Quality: Twentieth Annual Report* (Washington, D.C.: Government Printing Office, 1990): 80–3.

11. Interviews with two senior Environment Canada officials, 13 December 1991.

12. "The Changing Atmosphere: Implications for Global Security," conference statement, Toronto, Ontario, Canada, 27–30 June 1988.

13. Interview with the responsible official at the UK Ministry of Energy, 30 July 1991.

14. United States, Environmental Protection Agency, Office of Research and Development, *The Potential Effects of Global Climate Change on the United States: Draft Report to Congress,* 2 vols., eds. J. B. Smith and D. Tirpak (Washington, D.C.: Government Printing Office, October 1988); interview with section head of U.S. EPA, 27 August 1991.

15. "Thatcher Urges U.N. Commitment to Climate Change Convention by 1992," *International Environment Reporter* (December 1989): 581; Mike Robinson, *The Greening of British Party Politics* (Manchester, UK: Manchester University Press, 1992): 178.

16. Matthew Paterson, *Global Warming and Global Politics* (London: Routledge, 1996): 34–5.

17. Canada and Malta, "Background Paper on Elements of an International Convention on Climate Change," October 1989 for the Ministerial Conference on Atmospheric Pollution and Climatic Change, with Particular Attention to Global Warming, mimeograph.

18. From my review of internal WMO correspondence, memos, and notes of meetings.

19. United Nations, General Assembly, 70th Session, 6 December 1988, Protection of Global Climate for Present and Future Generations of Mankind (A/RES/43/53).

20. IPCC First Assessment Report: Overview. Informal comments by Robert Watson, NASA at INC2, from my notes, 26 June 1991.

21. David H. Slade, "A Survey of Informed Opinion Regarding the Nature and Reality of a 'Global Greenhouse Warming,'" *Climatic Change* 16 (February 1990): 1–4. Without extensive knowledge of the relevant processes, which WGI did not have, no probability can be estimated for changes in climate.

22. For example, Alston Chase, "Global Warming Theory May Just Be Hot Air," *Denver Post,* 28 June 1992, D3; Rogelio Maduro et al., "The 'Greenhouse Effect' Hoax: World Federalist Plot," *EIR Special Report* (Washington, D.C.: Executive Intelligence Review, 1989); and Warren T. Brooks, "The Global Warming Panic," *Forbes,* 25 December 1989, 96–102.

23. Robert C. Balling Jr., *The Heated Debate: Greenhouse Predictions Versus Climate Reality* (San Francisco: Pacific Research Institute for Public Policy, 1992): 97–118. A senior Soviet scientist commented to me that warming would be good for his country.

24. Interview with the director of research, British Coal (15 July 1991). U.S. government representatives echoed these critiques (two officials from the Department of Energy and one from the Department of Agriculture on 27 August 1991, 29 August 1991, and 3 September 1991).

25. Discussions on 11 and 14 December 1991 with Australian officials, one of whom commented that the IPCC was "political science" and that the developing countries' beliefs and preferences were "overly represented."

26. The quotes are from interviews with officials from the U.S. Department of Agriculture (special assistant, 3 September 1991) and an official from the Ministry of Environment, Mexico (25 June 1991).

27. A senior official of Environment Canada commented to me (13 December 1991) that "some of the cautionary wording was omitted," and a member of the U.S. delegation suggested to me that the overview and the policymakers' summary for WGI were both intended to portray a greater consensus than actually existed. This view was further substantiated by interviews with the secretary of the IPCC (24 June 1991) and the deputy section head, UK Department of Energy (30 July 1991), and comments from two Australian delegates (11 and 14 December 1991).

28. Interview with section director, U.S. Department of Energy (29 August 1991).

29. Interview with official from the Ministry of Environment, Mexico (25 June 1991).

30. Interviews with official from the Ministry of Environment, Mexico (25 June 1991) and section director, U.S. Department of Energy (29 August 1991).

31. Interviews with delegates from Algeria (20 June 1991), Thailand (21 June 1991), India (22 June 1991), Mexico (25 June 1991), and Brazil (16 December 1991).

32. Quotes from statements to INC 1 plenary meetings are taken from printed copies of statements and my notes.

33. No-regrets policies are those that can be justified for other reasons, but also reduce emissions of GHGs. For example, the 1991 National Energy Strategy included policies that could be justified as reducing reliance on foreign oil, but also would reduce emissions (United States, Department of Energy, 1991).

34. Alan D. Hecht and Dennis Tirpak, "Framework Agreement on Climate Change: A Scientific and Policy History," *Climatic Change* 29 (1995): 371–402.

35. Interview with deputy assistant administrator, U.S. EPA (28 August 1991).

36. Comments of a British government scientist and a senior official at the UK Department of Environment, October 1997.

37. Hecht et al., "Framework Agreement on Climate Change," 371–402; Neil E. Harrison, "From the Inside out: Domestic Influences on U.S. Global Environment Policy," in *U.S. Climate Change Policy,* Paul G. Harris, ed. (New York: St. Martin's, 2000): 89–109.

38. The SAR can be found at the IPCC Web site at www.ipcc.ch, accessed 15 March 2003.

39. Private comments to me in November 1996 by Meira, independently corroborated by a scientist from the UK Department of Environment, who attended the meeting.

40. Frederick Seitz, "A Major Deception on 'Global Warming'," letter to the editor, *Wall Street Journal,* 12 June 1996, A16.

41. Benjamin J. Santer, "No Deception in Global Warming Report," letter to the editor, *Wall Street Journal,* 25 June 1996, A16. The letter was signed by forty scientists from eight countries who were lead authors of or contributors to the WGI. In response to the attacks, procedures were changed for the third assessment report published in 2001.

42. "Democrats Attack Republicans on Environmental Policy," ABC News Show: ABC World News Tonight, 6:30 P.M. ET, 26 February 1996, Transcript #6040-7 in Academic Universe, accessed 10 June 2000.

43. Barry Schweid, "Christopher Sets Out Ambitious Program to Protect the Environment," *Associated Press,* 9 April 1996, online in Academic Universe, accessed 10 June 2000.

44. According to Dr. Michael Harris, environmental affairs manager, ICI Halochemicals, this change in U.S. climate foreign policy was actually decided in the White House in December 1995, shortly after the SAR was published. Private conversation June 1996 (from my notes and observations at COP 2).

45. At the final preparatory conference for the Kyoto meeting held in Bonn in October 1996 a senior Canadian delegate privately expressed to me exasperation that Alberta was preventing Canada from continuing to take a lead on climate change mitigation.

46. From FCCC/SBSTA/1998/INF.1, based on J. Greenough, M. Apps, and W. Kurz, "Influence of Methodology and Assumptions on Reported National Carbon Flux Inventories: An Illustration from the Canadian Forest Sector," *Mitigation and Adaptation Strategies for Global Change* 2(2–3): 267–83.

47. FCCC/SBSTA/1998/INF.1, 7.

48. In 2000 developed country emissions were 3,421 MtC.

49. FCCC/CP/1998/INF.4.

50. UNFCCC Secretariat, "United States of America: Land-Use, Land-Use Change and Forestry United Nations Framework Convention on Climate Change, Subsidiary Body for Scientific and Technological Advice," Thirteenth Session, Lyon, France, 11–15 September 2000, Paper 9, FCCC/SBSTA/2000/Misc.6/Add.1. U.S. emissions for the 1990 baseline year totaled 1,340 MtC.

51. Intergovernmental Panel on Climate Change, "Climate Change 2001: Synthesis Report, Summary for Policymakers," available at www.ipcc.ch/pub/tar/syr, accessed 10 March 2003. The individual working group summary reports are also at the IPCC site.

52. The confidence scale used by WGI uses these probability ranges: VL = very low (<5 percent probability); L = low (5–33 percent); M = medium (33–67 percent); H = high (67–95 percent); and VH = very high (>95 percent). Unfortunately, the three working groups used slightly different measures, which increases confusion when the object was to simplify and clarify.

53. Nebojsa Nakicenovic et al., Special Report on Emissions Scenarios (IPCC, 2001), at www.grida.no/climate/ipcc/emission/index.htm, accessed 15 January 2003.

54. George W. Bush, Statement by the President, White House, Office of the Press Secretary, 13 July 2001, at www.whitehouse.gov/news/releases/2001/06/20010611-2.html, accessed 10 January 2003; Edward Alden, Nancy Dunne, and Robert Shrimsley, "Bush Repudiates Kyoto Pact," 28 March 2001, at FT.com, accessed 5 January 2003; FT Reporters, "Bush Says Kyoto Climate Treaty Would Harm the US," 29 March 2001, at FT.com, accessed 5 January 2003.

55. "Clueless on Global Warming," *New York Times,* 19 July 2001, A24, at www.nytimes.com, accessed 17 July 2003.

56. These countries were the most important remaining members of the Umbrella Group (the non-EU OECD countries, Russia, and the Ukraine).

57. Christiaan Vrolijk, "A New Interpretation of the Kyoto Protocol: Outcomes from The Hague, Bonn and Marrakesh," Briefing Paper No. 1, The Royal Institute of International Affairs, Sustainable Development Programme, April 2002.

58. Joint implementation under Kyoto allows one country to assist another through financial and technological investment to reduce its emissions.

59. George W. Bush, "President Announces Clear Skies and Global Climate Change Initiatives," speech on 20 February 2002 at the National Oceanic and Atmospheric Administration, at www.whitehouse.gov/news/releases/2002/02/20020214-5.html, accessed 5 January 2003.

60. United States Environmental Protection Agency, "U.S. Climate Action Report 2002," at www.epa.gov/oar/globalwarming.nsf/content/ResourceCenterPublications USClimateActionReport.html, accessed 5 January 2003.

61. "Blaming the Bureaucracy," *Washington Post,* 12 June 2002, A30.

62. Richard Alley et al., *Abrupt Climate Change: Inevitable Surprises* (Washington, D.C.: National Academy of Sciences, 2001).

63. Neil E. Harrison, "Heads in the Clouds, Feet in the Sand: Multilateral Policy Coordination in Global Environmental Issues," Ph.D. dissertation, University of Denver, Denver, Colorado, 1994.

64. Karen Litfin describes how the concept of chlorine loading simplified complex science, became the dominant way of framing the scientific information, and ener-

gized political activity in *Ozone Discourses* (New York, Columbia University Press, 1994): 131–4.

65. David H. Guston, "Boundary Organizations in Environmental Policy and Science: An Introduction," *Science, Technology and Human Values* 26(4): 399–408; Clark Miller, "Hybrid Management: Boundary Organizations, Science Policy, and Environmental Governance in the Climate Regime," *Science, Technology and Human Values* 26(4): 478–500.

66. In *Pandora's Hope: Essays on the Reality of Science Studies* (Cambridge, Mass.: Harvard University Press, 1999), Bruno Latour argued that science obtains its influence by satisfying the needs of nonscientists. This "regulatory science" is never pure.

67. Because they are based on historical relations, GCMs cannot effectively model acceleration of negative and positive feedbacks into a nonlinear system response.

68. See letters from Ian Castles and David Henderson to the IPCC chair at www.lavoisier.com.au/papers/articles/IPCCissues.html, accessed 15 March 2003.

69. For example, Detlef Sprinz and Tapani Vaahtoranta, "The Interest-Based Explanation of International Environmental Policy," *International Organization* 48(1) (Winter 1994): 77–106.

70. United States, Department of Energy. *National Energy Strategy: Powerful Ideas for America,* 1st 1991–1992 ed. (Washington, D.C.: Government Printing Office, 1991).

71. Discussions at the Bonn negotiations in October and November 1997, with the chief of delegation for Australia of the Australian government and a lobbyist representing GreenPeace, New Zealand.

72. Oran R. Young, "The Politics of International Regime Formation: Managing Natural Resources and the Environment," *International Organization* 43(3) (Summer 1989): 349–75.

73. John Rawls, *A Theory of Justice* (Cambridge, Mass.: The Belknap Press, 1971): 303.

III

SCIENCE AND PRECAUTION

In the two case studies in this part, science worked to raise the issue as a problem, but politics could see no problem that needed a solution. Chapter 6 shows how Ontario's government failed to appropriately interpret scientific evidence of acid deposition until other, mainly political factors, changed how they constructed their interests. Chapter 7 discusses how policy makers failed to create an international institution to regulate global forests because they focused on the absence of international effects of deforestation and ignored evidence of its international causes. In both cases, scientific evidence was overlooked or misinterpreted because the prevailing political paradigm obscured its meaning. In one case, political changes incited a reevaluation of scientific evidence that resulted in a change in policy. In the other, changes in politics only reduced the possibility of political action. In both cases, the idea of precautionary action was irrelevant because science was not interpreted as giving a warning of an environmental problem.

In chapter 6 Munton investigates the progression of scientific studies of acid deposition in Ontario, Canada. He finds that both science and politics were influenced by the prevailing paradigm: that any airborne pollution came from the Sudbury, Ontario, smelters and that regulation of the smelters would impede economic development. Although he was not specifically testing the effects of discursive practices, at this point in his elaboration of a complex history, the case study would seem to support that model. But this only draws attention to the causes of the paradigm shift that took place in 1977 and 1978. How important were the previous scientific studies in effecting that shift? Or had there been a shift in social paradigm? Did the public now value the environment more and economic development less? Perhaps the media had effected this change by popularizing the science (that the provincial government had long ignored) in the process of making news over the destruction of the bucolic lakes that were Toronto's playground. Then again, were changes in government an opportunity to reevaluate interpretations of scientific data?

This case also raises questions about the practice of science. Although science is often portrayed as a free and open discourse, here the discoveries of Swedish scientists only accidentally became available to Canadian researchers, despite publicity about acid rain at an international conference. Another curiosity in this case study is the continual blindness of scientific researchers to the possibility of long-range air pollution. Because they were not aware of the Swedish research, this may perhaps be forgiven, but some early Canadian studies had suggested that U.S. sources might be implicated. However, researchers in the 1970s invariably concluded that the Sudbury smelters were the source of observed acid deposition. How much influence did the widespread belief that Sudbury was the source of all significant air pollution in the province influence scientists' interpretation of their findings? Did this belief, to which government officials also subscribed, also influence the design of later studies that restricted their usefulness? Do policy makers always try to reduce uncertainty as epistemic community theories suggest, or only when they want to make policy for some other, usually political, reason?

Like Sherlock Holmes's mystery of why the dog didn't bark, chapter 7 is interesting for what was widely expected to happen, but did not. It shows that early interest in a global forest agreement by environmentalists, international organizations, and many countries fizzled as discussions stretched out and scientific reports continued to track rapid deforestation. Scientific data shows that deforestation is occurring across much of the globe and that its effects on ecosystems and social systems can be significant. Why did opposition to a global forest treaty prevail? Countries used diverse arguments, including sovereignty and the cost of implementation, in opposing an agreement. Developed countries like Canada and the United States wanted the freedom to manage their forests for production and balked at the increased development aid needed to assist developing countries manage their forests sustainably. Developing countries wanted the export earnings from clear-cut logging of tropical forests, and logging interests were more powerful than the indigenous forest populations. But that analysis is simplistic: some developing countries wanted a treaty to require sustainable management of tropical forests. If biodiversity deserves an international treaty, why do forests, in which much of the world's biological diversity blossoms, not? Given the failure to create an international forest policy, where did interest in international forest policy initially come from? Who would benefit from an international forest policy and who would not?

But that is not the whole story. The authority and accuracy of scientific assessments of deforestation were undermined by the absence of a global research program and by methodological differences among the available scientific studies. Thus, there was no unifying scientific statement of a global deforestation problem and its international causes and effects. Was the absence of a global forest research program a scientific or political choice?

For two reasons, it is unclear if the effects of deforestation deserved an international treaty. First, the most immediate effects of deforestation are usually local, in the region or the country. Second, global consequences had been excluded from the forest debate and subsumed under the carbon cycle problem in climate change, which raises the problem of dividing the global ecosystem into multiple issues. Is it politically practical, but scientifically disingenuous, or is global science an impossibility that must be ignored? Thus, should issues be carved out of the whole on the basis of their scientific practicality or their political convenience?

An important aspect of any debate on global deforestation was missing: the international causes of tropical deforestation. In keeping with other environmental treaties, causes were barely considered by either scientists or policy makers; debate focused on regulating the international effects of environmentally damaging activities rather than removing underlying causes endemic in social and economic structures. Why were the international causes of deforestation ignored? If the international consequences of deforestation are immaterial, are its causes irrelevant?

6

Using Science, Ignoring Science

Lake Acidification in Ontario

Don Munton

The Canadian province of Ontario made a name for itself during the 1980s as a firm and committed advocate of combating "acid rain."[1] Government ministers made speeches about the dangers of acidic deposition in the province's wilderness and recreation areas and about the need for Canada and the United States to deal with this problem in a cooperative manner. The same ministers made proselytizing trips to the United States to raise American consciousness, targeting both policy makers in Washington and the masses in the heartland. Ontario joined Ottawa in funding lobby groups to push for emission controls in Canada and the United States. The province even went to an American court along with some northeastern U.S. states to try to force the U.S. federal government to address the transboundary air-quality problem. Finally, the government of Ontario cracked down on the major provincial sources of acid rain, forcing significant emissions reductions. But that was the 1980s.

Ontario had not always been so keen to spread the word or take action to combat acidification. In earlier decades, the provincial government had studied the matter, but taken no effective action to deal with the problem or even indicated any desire to do so.

This chapter is the story of the transition. Most industrial countries adopted policies to reduce acidic deposition during the 1980s and, thus, most underwent some sort of policy change.[2] The conventional story in environmental-policy studies is that of how policy objectives are pursued and perhaps realized. In contrast, the focus here is on how Ontario's policy objectives were redefined and how scientific knowledge shaped this policy development.

Two sets of questions are central to this study. First, how did the Ontario government come to recognize in the late 1970s that acidification was a problem requiring a policy response? And what role did environmental science and other factors play in the transition? Second, was there scientific evidence of an acidification problem in Ontario prior to the late 1970s? If so, what was the governmental response to such earlier evidence and what does that response tell us about the role of science in environmental policy making? The process of answering these two sets of questions will resemble more that of peeling the layers of an onion than doing a traditional, historical-chronological analysis. The various factors that contributed to the conversion to a pro-acid rain control position will be examined in turn, starting with the scientific knowledge. Initially, the focus is on the late 1970s, but it later shifts to the 1960s and early 1970s.

Both policy watchers and scholars have tended to take as a given the relatively early response of Ontario and other governments in Canada to scientific evidence of ecosystem acidification. In contrast with the prolonged debates in the United States (and, for example, in the UK) during the 1980s about the nature and seriousness of acid rain, these questions were essentially settled in Canada by 1980. To be sure, the debate over the precise control policies to be adopted by Canadian governments took a while longer to resolve. That relatively early official acceptance in Canada of the scientific evidence pointing to acid rain as a major environmental problem has masked an illuminating story, however. The trail of scientific research on acidification, particularly in Ontario, actually extends back much further than is normally assumed. So, too, does the trail of government consideration of this evidence, although not necessarily the record of government actions taken. While not lacking scientific studies, earlier Ontario governments generally did not respond to the evidence provided.

The full history of acidification research in Canada is a long and complex one—too long to be related fully here. While acidic deposition has a range of negative impacts, on lakes and streams, fish, forests, crops, materials and human health, the focus of the present chapter is on one aspect of the overall acidification problem, albeit the one that received the most attention in Canada during the 1980s, that of lake acidification.[3]

ONTARIO AND "ACID RAIN," 1977–1985

When did the Ontario government first recognize acidification as a major environmental problem in need of action? Or, to put the question only slightly differently, when did it recognize acidification as a problem that seriously affected the province's interests?

This recognition was not merely one step, but a critical step for Canada. Under the Canadian constitution, and since 1867, provincial governments are responsible for natural resources. The acid rain control programs eventually adopted in Canada were thus provincial programs. As the largest and most impacted province, Ontario was the key to the Canadian response to acid rain.

Canadian federal environment minister Romeo LeBlanc first placed acid rain on both the domestic Canadian agenda and the Canada–United States agenda in June 1977. In a speech to the Air Pollution Control Association, a North America–wide organization, he colorfully characterized acid rain and other airborne pollutants as an "environmental time bomb."[4] Noting that its sources were to be found in both the United States and Canada, LeBlanc acknowledged that "we have both been negligent in this area. What we have allowed to happen, innocently enough perhaps, is a massive international exchange of air pollutants, and neither party to this exchange is free of guilt."[5] LeBlanc urged action in the form of an international agreement. Such an accord eventually emerged, but it took fourteen years to bring to fruition.

At the time acid rain was not a household word. It was not even a problem that many scientists and governmental officials understood. The 1977 meeting LeBlanc addressed was in all likelihood the first at which most of those attending had heard the phrase "acid rain."[6] Nor at the time was this a problem about which environmental NGOs were then exercised. The situation soon changed.

Ontario was the first Canadian provincial government to embrace the acid rain issue. A few months after the LeBlanc speech, the province's environment minister, Harry Parrott, proposed to his federal counterpart a joint Ontario-federal research program on acid rain.[7] Most other provincial officials across Canada were strongly inclined toward disinterest. Parrott made the first ever presentation on acid rain to the Canadian Council of Resource and Environment Ministers in 1978, outlining the results of his province's monitoring program. Recalls Parrott, "Most of them looked at me like I was out of my mind."[8] The Ontario government took up the challenge. In 1978 the province established the Acid Precipitation in Ontario Study (APIOS). The premier, William Davis, wrote to Prime Minister Pierre Trudeau in February 1979, calling for immediate action by Canada and the United States to attack acid rain.[9] The next month, at a major environmental conference in Toronto, Minister Parrott termed acid rain "a top priority for our Government."[10] A few months later, in April 1980, the minister indicated he would be announcing an acid rain control program imminently, a promise he made repeatedly over the next few weeks.[11] The program was a bit longer in coming, but it came.

A new control order limiting sulphur dioxide (SO_2) emissions at the International Nickel (Inco) smelter near Sudbury was issued in September 1980. These new requirements were actually less stringent than an earlier 1970 control order, the targets of which had still not been met. Inco did reduce emissions somewhat, but mainly through production cuts brought about by low nickel demand and low prices. More meaningful Ontario regulations requiring a drastic cut in smelter emissions took a few more years. After a change in governing party, Ontario inaugurated its own acid rain control program in 1985 and formally joined with the federal government and other provinces in reducing SO_2 emissions in eastern Canada by approximately 50 percent.

In sum, from 1978 the government of Ontario was rhetorically committed to the view that acidification was a problem, and from 1980 it was publicly committed to taking action to deal with it. Having perceived a need to respond to the threat of acidification in the late 1970s, Ontario did not waver from that perception or "construction" of its interests. Or, at least, it did not do so for the next decade.[12]

CONSTRUCTING ONTARIO'S INTERESTS ON ACIDIFICATION

The question here is why the critical shift in perceptions and eventually the policy changes happened at all. What prompted this "construction" of interests?[13] While LeBlanc's speech helped initially to draw attention to the issue, it was not the critical factor behind the Ontario shift. The necessary conditions lay elsewhere. As is usual in such matters, there was a constellation of factors at work. Scientific research was one key element. The set of factors included as well a triggering development; a receptive minister, or at least one willing to learn; an aroused press; a critical threatened resource; a concerned public; and a foreign bad guy.

SCIENTIFIC KNOWLEDGE

The APIOS study launched in 1978 and destined to run for the whole of the 1980s was not Ontario's first effort at researching acidic deposition. For most of the 1970s the environment ministry had led an interdepartmental effort termed the Sudbury Environmental Study (SES). The elusive goals of this lengthy project were to establish the impact of the emissions from the two Sudbury-area nickel smelters. Inco's was the larger of these; the smaller was owned and operated by Falconbridge Mines. Both tapped the enormous and rich nickel deposits of the Sudbury basin.

One basic component of the SES involved measuring the acidity levels of hundreds of lakes and assessing their fish stocks. Notably, it was this work that provided Environment Minister Parrott with a figure of 140 acidified lakes in Ontario that he used to support his policy. What the ministry kept fairly quiet at the time was the fact that these 140 lakes were not scattered throughout the province. All were in the Sudbury area. That is to say, based on scientific knowledge at the time, Ontario's acid rain problem was a Sudbury-area problem. Nor was Parrott then reporting a recent discovery. Ministry scientists had been documenting the damage to these very lakes for some time. The SES work, despite its preoccupation with Sudbury, actually found evidence of serious, albeit less advanced, acidification in lakes up to one hundred kilometers away.

An unstated objective of the SES had been to provide a sound scientific basis for requiring reductions in Inco's emissions. In this respect, the SES proved an utter failure. The government scientists involved all agreed in 1979 that they still lacked the scientific information to justify these reductions. They were not able to make authoritative descriptive statements of the sort that "acidic precipitation in northeastern Ontario is due to the emissions from Sudbury *x* percent of the time."[14] A major reason for this failure was the serious underfunding of the entire project. Another was that much of the initial SES work had actually been directed toward studying ecosystem remediation, rather than studying deposition patterns or developing models to assess control options. The remediation research sought to determine if the Sudbury area lakes could be stabilized by adding lime to them to offset their acidity—in effect, Tums for the lakes. While this work was never well publicized or explained, liming northern Ontario lakes was a way to address certain symptoms of SO_2 deposition and to put off tackling the basic cause, the smelter emissions.[15]

The SES did provide an important piece to the acid rain puzzle. Its scientists discovered, to their surprise, that precipitation acidity downwind from Sudbury did not decline significantly during a six to seven week shutdown of the Inco and Falconbridge plants during 1978. As stated in an Ontario Ministry of Environment (MOE) paper, "even in the absence of smelter operations there is a significant acid loading in the Sudbury area associated with precipitation."[16] The implications were fairly obvious, but nevertheless stunning. Much of the acidic deposition in this heavily polluted region, and possibly elsewhere in the province, must therefore be coming from sources other than the Sudbury smelters. As transboundary studies would soon show, it was coming, of course, from the United States via long-range transport.[17]

Discovery of acidification damage to areas of Ontario very distant from the smelters came not from the SES, but from an unrelated research project. In 1974, as part of a study called the Lakeshore Capacity Project (LCP),

Ontario's Ministry of Housing had established a research station in Dorset, Ontario, northeast of Toronto, and more than a hundred kilometers from Sudbury. The overall aim of this project was to gauge the impact of cottage development on the rural environment (that is, to determine how much cottage development could be encouraged and allowed before there was a serious negative environmental impact). Curiously, but serendipitously, the project's data collection regime included not only water-quality measurements, but also precipitation monitoring. These data showed both an increasing acidity in the lakes and significant acidic deposition from the atmosphere.[18]

At that time, Tom Brydges was a senior aquatic scientist in the environment ministry and one of those responsible for the liming-remediation projects under the SES. While he had talked with Scandinavian scientists about acid rain in Sweden and Norway, he did not originally believe it could be a widespread Ontario problem. Indeed, like many others, Brydges was at first highly skeptical of the idea that American air pollution was causing acidification problems in Ontario. He recalls thinking that this "was the biggest scam I had ever heard. No way!"[19] The Dorset-LCP precipitation-monitoring results, however, convinced him of the reality of acid rain in Ontario. After seeing only the first six months' worth of data, he was converted.[20] Brydges became a proponent of the acid rain cause within the environment ministry. The SES and LCP-Dorset findings, now buttressed by references to the work of scientists in Sweden and Norway, became the basis of the briefings Brydges and others gave Parrott when he became the minister.

The results of the two projects fitted well. The Dorset results pointed to a genuine problem beyond Sudbury, while the SES findings showed how serious the problem could become; it also pointed to long-range transport and, thus, suggested American sources of SO_2. Proof of the long-range transport hypothesis was soon to come from atmospheric research carried out by federal government scientists. What could not have escaped attention from the Sudbury-Dorset results were the serious implications for fish stocks of acidified lakes. The environment ministry scientists knew very well the connection here. What they had not discovered themselves already (a considerable amount), they knew from Scandinavian scientists and from the recent work of two University of Toronto scientists. Richard Beamish and Harold Harvey, Beamish's Ph.D. dissertation supervisor, had published an article in 1972 in the *Journal of the Fisheries Research Board of Canada*, the premier scientific journal of its type in Canada.[21] That article provided very strong evidence of the hypothesized link between SO_2 deposition, acidification, and the loss of fisheries.

In 1964, zoologist Harvey had begun working in the La Cloche lakes area of Northern Ontario, about forty miles south-southwest of Sudbury—a region that had been made famous earlier in the century by the paintings of Canada's Group of Seven artists. Intending to study the introduction of kokanee (a landlocked salmon) to the lakes, Harvey and Beamish stocked George Lake with thousands of fry. They returned the following year and, to their surprise, found no survivors. The Beamish-Harvey article was, in effect, the detective story of the disappearance of these and other fish from the area's lakes.

The La Cloche lakes had once been known as having a good sport fishery, especially trout and burbot. Surveys conducted by the Ontario Department of Lands and Forests in the early 1960s had shown abundant lake herring and good stocks of trout. By the mid-1960s, the burbot seem to have disappeared. By late 1960s, the trout were not showing up, even when fisheries officials used gillnets in an effort to find them. By the early 1970s, the herring were gone too. A pattern had thus been observed, but not explained, before detective Beamish arrived from Toronto.

The most common and handiest reason for a declining sport or commercial fishery is overfishing. That traditional explanation, however, was not consistent with a number of facts in the La Cloche case. First, the decline in fish populations had been general. Both those species sought as sport fish and those not so fished had disappeared. "While it is possible to argue that anglers were responsible for the decline in numbers of lake trout," Beamish and Harvey doubted that anglers were responsible for the disappearance of species that are not the targets of recreational fishers. The evidence pointed not to overfishing, but to habitat problems.

One by one, possible hypotheses and explanations were discarded. Beamish eventually concluded that the most unusual characteristic of the lakes was their high acidity. It was well established in the fisheries literature that low pH levels were toxic to fish. (The pH scale is used to measure acidity ranges from fourteen, or highly alkaline, to zero, or highly acidic.) Acidification affects most fish species even at a relatively mild level of acidity, such as pH 6.0. Water with a pH of 5.0 or less is toxic to nearly all fish. Beamish found the pH of Lumsden Lake to be 4.4, down from a recorded 6.8 in 1961. The main finding of Beamish's dissertation, which he completed and defended in 1970, was the extent of acidity and its role in the demise of the fish. He showed the loss of fish species was preceded by failures in reproduction and suggested these failures were due to lake acidity.

Having determined the acidification of the lakes, Beamish set about to identify the source of the acidity. Natural causes were quickly ruled out: the low pH levels of the La Cloche lakes were not due to natural oxidation of pyritic (sulphur bearing) rocks. Nor were they the result of organic acids; bogs were rare in the Lumsden lake chain. That left anthropogenic sources, of which there were no obvious ones in the immediate area. The remoteness of the lakes and the lack of human habitation or nearby mining or other development ruled out surface effluents. The only possible pathway was the atmosphere.

Beamish consulted the scientific literature on atmospheric chemistry. He found two well-established facts. One was that SO_2 emissions from industrial plants could be transported considerable distances. The second was that, in the process, SO_2 was oxidized to sulphuric acid. He did not know then that smelter emissions themselves also contain significant amounts of sulphuric acid. Following this trail, Beamish began to take rainfall samples around George Lake in 1969. The following year, he was taking snow samples around both George and Lumsden Lakes. The results staggered the young scientist. The pH of the rainfall averaged 4.4, well within the acidity range. That of the snow was 3.4, or 10 times more acidic, roughly the acidity of orange juice. The lowest recorded pH, for some surface snow, was 2.9, about the acidity of vinegar. This was an astonishing discovery. There was no doubt this wilderness area was receiving highly acidic precipitation, especially in winter.

More extensive lake water acidity measurements followed to determine the geographic scope of the acidification problem. Beamish sampled twenty-two lakes in or east of the La Cloche area in 1971. Those pH values were then compared to earlier tests done by the provincial Department of Lands and Forests. In all but two cases, the pH levels had dropped over the previous decade or part thereof. These lakes, too, had become from ten to one hundred times more acidic. More lakes were sampled in the area north of La Cloche and west of Sudbury. Some of these lakes were on carbonate-bearing rock formations and were better able to buffer the acidic deposition. But even there, the overall trend was toward increased lake water acidity. Beamish and Harvey observed:

> A profound increase in acidity has occurred in these lakes in recent years. It is unreasonable to conclude that these lakes have maintained pH levels suitable for fishes for between two thousand and nine thousand years and that natural phenomena have produced large and widespread increases in lake acidity only in the past decade.[22]

But where was the acidic deposition coming from? To Beamish and Harvey the only answer was the Sudbury smelter complex, which at that time was spewing out thousands of tons of SO_2 per day. The prevailing winds in the area, however, were from the southwest, across Lake Huron, not from the northeast where Sudbury lay. Beamish sought meteorological data on wind patterns in the area from Transport Canada. This data showed that on average the winds blew into the La Cloche area from Sudbury one hour or one day out of every four. During the winter, winds from Sudbury were more frequent. This winter pattern seemed to explain why snow was particularly acidic. The two scientists, then unaware of longer-range transport, concluded the smelters were the literal smoking gun. But they were reluctant to point the finger. "This wasn't the sort of charge one made lightly back then," Harvey has said. "We were very, very scared that we could be wrong."[23]

They submitted their paper to the *Journal of the Fisheries Research Board.* The response was rapid. Within a month it had been accepted. One of the reviewers broke with the convention of anonymity and asked to be identified—ecologist Eville Gorham of the University of Minnesota. Gorham had himself done research work in the early 1960s on the environmental impact of the Sudbury smelter emissions in the immediate vicinity of the plants, when he had been affiliated with the University of Toronto.[24] This pioneering work had been largely ignored at the time by the upper levels of the Ontario government and had met with little response from fellow scientists. Both Gorham and the second reviewer for the journal emphasized the importance of Beamish's work. The latter said the information ought to be drawn to the attention of conservationists and industry and suggested the unusual step, in scientific circles, of publishing a condensed version of the article in a more environmentally oriented journal.[25]

Beamish and Harvey made various attempts to bring their findings about lake acidity to the attention of governments. The reaction was at best subdued. At one 1969 meeting with officials, there were "no questions or anything." "We gave our presentation and that was it," Beamish recalls. It was "like they had been told not to say anything."[26] The Ontario government did respond, in small measure, to their work. A modest precipitation sampling network and water-quality monitoring program in the La Cloche region was established by the Water Resources Branch of the Ministry of Natural Resources. This program focused primarily on the lakes studied by the University of Toronto researchers. It was given little attention by high-level officials, even in the MOE. Officials did watch much more closely to see if public attention was aroused by the Beamish and Harvey work. It was not.

Richard Beamish had put together most of the pieces of the acidic deposition puzzle. One element missing from this work was systematic atmospheric research. As fisheries ecologists, such was not within Beamish and Harvey's expertise. Moreover, while provincial ministries of environment did ambient air-quality sampling and deposition studies, broad-scale atmospheric research was well outside their normal bureaucratic missions. Weather forecasting and the attendant monitoring stations were the preserve of the federal government. It is only a modest digression from our focus on Ontario to consider briefly this aspect of the 1970s scientific work.

Partly in response to the warnings made by Svante Odén and Swedish government reports at the 1972 Stockholm environment conference, and partly in response to European studies begun under the Organization for

Economic Cooperation and Development (OECD), the Canadian federal government began the first regular acid precipitation monitoring in Canada in 1973.[27] Operated by the Meteorological Service (later the Atmospheric Environment Service of Environment Canada), this trial project was eventually expanded into an on-going Canadawide monitoring program (CANSAP). In the early years this network comprised only eight monitoring stations, and the data collection was somewhat erratic. With increased funding, which became progressively easier to acquire in a period of expanding government programs, the network was expanded and improved.[28]

CANSAP data became the critical basis for research by Douglas Whelpdale and Peter Summers of the Atmospheric Environment Service (AES) on the extent of acidic deposition in Canada. Whelpdale teamed up with an American scientist to combine data from the precipitation monitoring networks in both countries with what was known of North American weather patterns and the location of large SO_2 polluters. They demonstrated scientifically, for the first time, but with remarkable accuracy, the transboundary nature and extent of long-range transport and acid rain in North America.[29] They estimated that approximately half of the acid rain falling in Canada came from American sources hundreds and even thousands of kilometers away, just as Swedish and Norwegian scientists had showed the extent to which acidic deposition in Scandinavia originated in the UK and continental Europe.

A joint Canada–United States scientific study released in October 1979 showed there was substantial scientific agreement on the validity of these estimates of the transboundary movement of sulphur compounds between the two countries.[30] It also showed that total American emissions of SO_2 were five times greater than Canadian ones, and American emissions of nitrous oxides ten times greater. While the United States overall had larger emissions and produced most of the transboundary air pollution, and the American contribution to Canada's acidification was significant, it was also clear that Canadian emission sources were part of the domestic problem.

TRIGGERING EVENT

The Ontario government's first SO_2 control orders had been placed on the two smelters in the Sudbury area in 1969 and 1970. While the companies, Falconbridge and Inco, were required to reduce emissions substantially, as well as to take other measures, the orders had nothing explicitly to do with acidification or acid rain. The objective was to improve ambient air quality in the Sudbury urban area, which had been a chronic problem for decades. The control orders were thus more human-health related, rather than ecosystem related. Falconbridge conscientiously set about to reduce its emissions through operational and process changes. These were eventually completed in the early 1980s, but not before the company sought and obtained a temporary reprieve from the Ontario government.[31] Inco built its famous 1,250-foot tall "superstack," but made no real effort to reach the mandated reduction in its SO_2 emissions from 3,600 to 750 tons per day.

It was soon clear to provincial officials that Inco was not going to reduce its SO_2 emissions by December 1978, as required. Inco quietly, but officially, informed the government in early 1976 of its decision not to try to do so.[32] The government waited until December 1977 to acknowledge publicly that Inco would not meet the reductions required by the 1970 control order. There was relatively little internal government debate about the wisdom of accepting Inco's inaction.[33] Environment ministry officials knew they could not substantiate the magnitude of the reduction mandated in 1970 should Inco fight the order, and it was assumed that Inco would do that.

The problem here lay in part with the SES research project. While it had made some important discoveries over the 1970s, as noted above, it had not attained its primary regulative goal, that of providing a post facto scientific justification for the emissions reductions required by the 1970 control order. The other technical consideration was that the new tall stack Inco completed in 1972 had improved the local Sudbury air quality, thus accomplishing the main purpose of the original order.

The province's environment minister formally gave Inco a reprieve on July 31, 1978.[34] In contrast to the muted response in 1973 when Falconbridge had similarly been granted an extension on an earlier deadline, the response this time was anything but muted.[35] A chorus of public criticism arose. A small band of environmental activists, calling itself the Inco-stinko faction, invaded the offices of the provincial MOE and let off a sulphur candle. The provincial legislature took up the issue, and the opposition party made as much of the controversy as it could. Ontario's largest source of industrial air pollution, Inco had long been a target. The tenor this time, however, was different.

The ensuing political uproar needed to be quelled. The Inco affair, moreover, was not the first controversial decision of the environment minister at the time. Within weeks, on August 18, 1978, he was moved to a different post—after just seven months in the job.[36] His replacement was the aforementioned Harry Parrott, an orthodontist and former education minister. Parrott was barely ensconced in office when he promptly stepped on a political landmine. He unwisely suggested that acid rain was a "glamour issue" and indicated he was more concerned about such down-to-earth problems as landfill sites and sewage treatment.[37] The ensuing press coverage and criticism made clear that the matter of SO_2 emissions from Ontario smelters was triggering a broad reaction. For the Ontario public, acid rain was no longer a glamour issue.

RECEPTIVE MINISTER

Parrott quickly asked his officials to brief him on acid rain. They were well prepared to do this, and the new environment minister proved receptive. Parrott's lack of environmental background made for a steep learning curve, but his familiarity with some scientific concepts was a positive factor. As scientist Tom Brydges noted, "you didn't have to explain pH to him." The minister listened, and "suddenly it clicked." Parrott realized acid rain "was not some kind of trivial little thing that was going to go away," but rather "a major problem on our hands."[38] Within weeks Parrott led the effort to create the APIOS research program, and was writing to his federal counterpart to propose research cooperation. If there was a moment when the actual process of reconstructing Ontario interests on this issue began, this was it. The policy initiatives followed and APIOS facilitated further changes in the ideational structure surrounding acidification.

AROUSING THE PRESS

The media played an important role in the reconstruction or redefinition of interests. To appreciate this role, it is necessary to go back a few years from the late 1970s.

Concerned about the implications of their La Cloche findings, Beamish and Harvey did something rare in the scientific community. At Harvey's suggestion, they spent part of 1970 preparing an article for the mainstream media. Their aim: to publicize the problem of acidification in the La Cloche lakes. In 1970, however, the mainstream media were not ready for acid rain. The submission was turned down by the *Toronto Telegram* newspaper and subsequently by *Maclean's,* a monthly magazine. In both cases editorial policy was cited as the reason for refusing publication.[39] One of the editors involved, however, assured the two scientists that they were almost certainly incorrect in their assertions. If their story were true, he said, he would surely have heard about it already.

The article was finally published in July 1971 by the *Globe and Mail*, Canada's "national" newspaper, as an op-ed piece. The scientists warned that "fish have disappeared suddenly from several lakes" in the La Cloche region and argued that the cause was a deterioration in the quality of the water due to acidification.[40] They went on to attribute the acidification of the La Cloche lakes to the deposition of acids from sources such as the Sudbury smelters.

The *Globe and Mail* editor seems to have been uneasy about the Beamish and Harvey piece and felt obliged to print some sort of counter to their provocative charges. Thus, the Beamish and Harvey article appeared on the

Globe op-ed page side-by-side with excerpts from a speech by an Inco company representative. The Inco man attacked the proposition that the erection of the then recently completed Inco superstack could lead to significant long-range pollution. The new stack, he assured the *Globe's* readers, "is the best solution that we, or anyone else, can find within the limits of today's technology."[41]

There was no follow-up in the *Globe* or elsewhere at the time. The Beamish and Harvey article generated a few letters from concerned citizens, mostly from fishing camp operators in the La Cloche region itself. But few Torontonians, and fewer provincial politicians, had ever been to the distant and somewhat inaccessible La Cloche lakes. The problem must have appeared too far away to worry about.

The next *Globe* piece on acidification, and first by a staff writer, Peter Whelan, did not appear until two years later.[42] Whelan's 1973 article essentially reflected the Beamish and Harvey findings. He noted that lakes near Sudbury had been rendered essentially lifeless, suggested this was due to acidification, and attributed the condition to dilute sulphuric acid in precipitation originating from the Sudbury metal smelters. There was also no follow-up to the Whelan article.

The next items on acid rain did not appear until two years later, and then two years after that. A second Whelan story appeared in 1975. In February 1977, freelance science writer Lydia Dotto wrote a long *Globe* article on the general phenomenon of acid rain. Dotto cited damage to aquatic and terrestrial ecosystems, basing her observations on the work of American scientist Gene Likens and the work at the University of Toronto.[43] Again, she made explicit the link to the Sudbury smelters, but also discussed the phenomenon of long-range transport of air pollution. LeBlanc's speech came a few months later and attracted more attention. Once again, however, the story did not take off. As an Environment Canada official recalls, there was still "no big bang in the media . . . no one was running to jump on the issue."[44]

The media big bang finally came in 1978, just prior to the Inco reprieve in July. One of the few people outside of the environment ministry closely watching the results of its scientific work on acidic deposition was a young *Toronto Star* reporter, Ross Howard. On June 7, 1978, Howard published an article entitled "Rain of Pollution Killing our Resort Lakes." The *Star* ran it on the front page. The article drew a very strong link between acid rain and damage to lakes in the Muskoka region just north of Toronto. It was based on reports out of the SES and the LCP, and much of the information had come, quietly, from Tom Brydges. The article hit home—as it was intended to. "When Ross Howard put acid rain on the front page of the *Star,*" Brydges recalls, "that blew it wide open."[45]

The Howard article more than caught the attention of the government and others. It was, unusually, the subject of a ministry memo that same day, apparently aimed at rebutting the article and providing "talking points" for the minister and departmental spokespeople, should reporters ask.[46] More articles followed in the *Star* and elsewhere.[47] A media feeding frenzy developed. This then was the context in which one month later Parrott's predecessor announced his decision to give Inco the reprieve on its control order. The media interest helps explain the negative reaction that reprieve received.

The impact of the Howard article and subsequent events on the Canadian media's acid rain coverage can be judged by a simple quantitative content analysis. The source of the data is the *Globe and Mail*. More than any other Canadian outlet in the 1970s and 1980s, and certainly more than the Toronto-oriented *Star*, the *Globe* set the national agenda. Its coverage influenced the rest of the Canadian media, both print and electronic, as well as the political system. Based in Toronto, it also significantly sets the Ontario political agenda. It thus plays a role in Canada equivalent to that played by the *New York Times* in the United States, and it is the most appropriate basis for gauging media interest.[48] The pattern of acid rain coverage in the *Globe and Mail* is shown in figure 6.1.[49]

Perhaps the most common notion about the media is that their practitioners benefit by criticizing governments. There is an ongoing mutual interdependence of reporter and source that is the basis of a specialized social network encompassing the media, the bureaucrats, and the politicians. The operation of this network in the late 1970s told the Ontario government that a new set of preferences was developing in its domestic civil society. These were preferences that the media were both reporting and reinforcing.

This media attention was not merely a domestic phenomenon. In fact, the Canadian attention in 1978 followed some international media attention to the Sudbury and acid rain problems. In September 1976, a BBC television

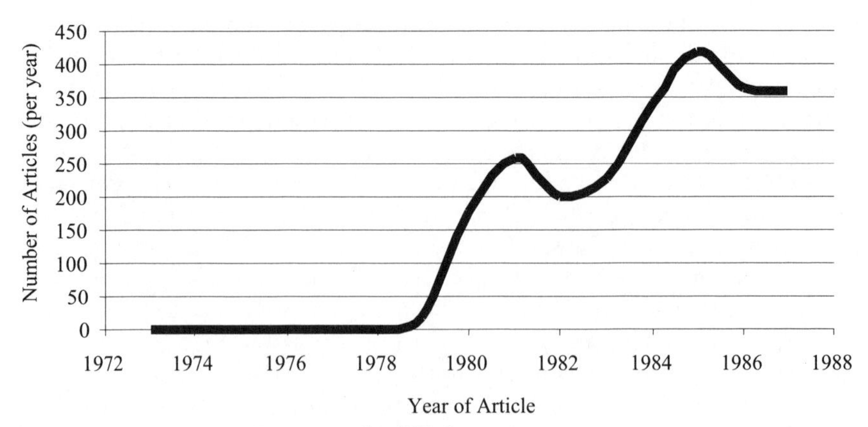

Figure 6.1 Newspaper Coverage of Acid Rain

crew had been in the Sudbury area reporting on the smelter emissions, acidified lakes, and the SES project. The following year, a crew from the NBC show "Weekend" was there doing a documentary on acid precipitation. The information that these foreign networks were coming had sent the environment ministry bureaucrats in Toronto scurrying. Apart from concerns about the public attention that might be generated, the ministry's fear seems to have been that someone in Sudbury might say the wrong thing.[50] In this case, however, the facts spoke for themselves.

A THREATENED RESOURCE

Compared with the Howard article in 1978, the 1971 *Globe* op-ed piece by Beamish and Harvey and the articles by Peter Whelan in 1973 and 1975 clearly had very little impact in terms of raising consciousness about acid rain. One likely reason is that environmental damage in the Sudbury area was old news. People in the Sudbury basin had been complaining for decades about local air pollution (what Inco and environmental officials still referred to as "fumigation incidents"), and even newly discovered damage of a new type was expected there. Moreover, such damage was, for some, including many in Sudbury, the price of well-paying jobs and a strong regional economy. In contrast, information about environmental damage to the seemingly pristine Muskoka area was new and thus news. Moreover, it was worrisome news about a highly valued resource.

The oldest and most favored cottage area for Toronto, the largest city in Canada and a city with hot, humid summers, the beautiful Muskoka lakes are close to sacred. Torontonians, at least those wealthy enough to have property in the Muskokas or lucky enough to have relatives or friends there, contend with bumper-to-bumper freeway traffic heading north on Friday afternoons and more bumper-to-bumper traffic coming back Sunday night, all in order to spend summer weekends at cottages in the area. Those cottages, many of them good-sized homes, sit on small, interconnected, picture-perfect lakes. The area is a boater's paradise. A Muskoka trademark is floating boathouses carefully constructed and painted to match their respective cottages. As one federal official has commented, "Muskoka is our special place."[51]

The Muskoka connection did not just bring acidic deposition close to home. For many people it was now a personal matter. It was, in Tom Brydges words, "a potent mix of viability of recreational industries, aesthetics, and threat to property value."[52] The cottagers were not the sort of environmental radicals who would have been turned away at cabinet ministers' doors; they generally held middle-of-the-road or conservative political values and were more like born-again conservationists than new-age environmentalists. They represented values that no Ontario politician could ignore.

Ross Howard's analysis of his own impact is interesting. His June 1978 article, he notes, had "colour . . . and all the elements": a recognizable "bad guy" (Inco), a "victim" (the Muskokas), "graphic elements" (dead fish), and "a minister on the ropes." Moreover, as Howard well understood, many of the upper management of the *Toronto Star* had summer cottages in endangered areas.[53] The Muskoka lakes had friends in high places, and not just at the *Star*. And, not surprisingly, the June 1978 article was far from Howard's last on the subject of acid rain and cottage country.[54] A social network involving influential members of Ontario society came into gear, working in tandem with the network involving the government and the media.

Amidst this growing controversy, one of Parrott's cabinet colleagues provided an insightful comment on the emerging acid rain issue. Frank Miller represented part of the Muskoka area in the provincial legislature (and would later become premier of the province). In 1978 he acknowledged there was a problem with acidic deposition, but wondered (out loud, to a local reporter) if society was ready to pay the price to fix it.[55]

It soon became clear to Miller and his colleagues that society was at the very least more willing than before to pay the price of reducing the emissions that led to acidification. Miller's likely off-hand comment itself offers us insight into the calculus at that time of the government and, for that matter, into the calculus of their political predecessors. They had consistently seen a distinct tradeoff in the interests at stake. "Fume" and "smoke" (often, acidification) damage to Ontario environmental resources was the cost Ontario society had to pay, and was willing to pay, for its strong economy and its mining and smelting jobs. As spontaneous, off-hand comments often do, Miller's exposes the heart of the pre-1980s thinking of Ontario and other jurisdictions. The prevalent value structure assumed the economic benefits of industrial production and industrial jobs in northern communities were to be preferred over the more esoteric values of environmental protection. That value structure, or ideational structure, was as strongly held by civic leaders and many workers in industrial towns like Sudbury as it was by the provincial politicians who supported the industries on which those towns depended economically.[56]

A CONCERNED PUBLIC

The Muskoka story and the subsequent announcement of the Inco reprieve did indeed hit home with an increasingly attentive public. Inquiries and criticism poured into the government.[57] The Muskoka Lakes Association wrote to the premier and other ministers about the Inco control order.[58] Cottagers asked, "What is happening and what will be done, now, about acid rain and dying lakes."[59]

That the public was aroused about the acid rain issue was evident both from the sustained pattern of letters to the editor and from the fact that the *Globe* was soon running them under a standing head on the newspaper's op-ed pages which always read, "Acid Rain."[60]

The message the Ontario government began to hear from the broad social network comprising the attentive parts of its own civil society was one of widespread anger and dismay that the Ontario government had known about this problem for seven years, but done nothing.

In response to letters from concerned citizens, the premier explained that the MOE "discovered 'acid rain' falling in the Muskoka-Haliburton area in 1976." Within months of confirming the problem, Davis assured one letter writer, more resources were applied and the findings were first made public in September 1977.[61] The premier made no mention, for example, of the decade-long SES project, the acidified lakes near Sudbury, and the experiments with liming those lakes.

Interestingly, the final version of the premier's letter differed in a significant way from the first draft prepared in the environment ministry. The latter had contained a fuller version of events. The MOE's draft had originally noted that "the Province has known about acid lakes in the extremely sensitive La Cloche mountain area for twenty years. The problem was thought to be very localized." It then went on to say that it was only during the past few years that the MOE realized the problem was more general.[62] The differences between the original draft and the version eventually sent suggest that the premier's office had belatedly recognized a political danger, one missed by the bureaucrats in the MOE. Trying to downplay the acid rain problem by arguing it had been around and known for a long time was problematic—even though the problem had been around and known for a long time. The officials and the government had been caught on the horns of a classic political dilemma. Either you admit you were not on top of a problem, or you admit you knew about it and did nothing. The government opted more for the former than the latter.

The earliest public opinion polls in Canada that asked about acid rain suggest a relatively high level of awareness and concern—at a time when the environment had declined as a priority issue from early 1970s levels. In September 1980 two in every three Canadians (66 percent) said they had recently heard or read about acid rain, and a year later the proportion was 76 percent. In the 1980 poll, moreover, most of the respondents (an impressive 60 percent) could offer a reasonably correct definition of what acid rain was or what its sources were. At the same time, the vast majority (78 percent) said acid rain was "very urgent" or "quite urgent."[63]

The acid rain story is not one of governments responding to environmentalists, environmental groups, or formal nongovernmental organizations, either international or domestic. The science was done, the media coverage began, and the public concern developed, all well before interest groups began to pick up on the issue. The first major sign of NGO activity came later. An "Action Seminar on Acid Precipitation" was held in Toronto in November 1979. Organized by a coalition of American and Canadian environmental groups, particularly the Federation of Ontario Naturalists, Friends of the Earth, and the U.S. National Clean Air Coalition, it was also supported by Canadian governments. The fact that the seminar was very well attended surprised the groups involved and may have been an additional sign to governments that acid rain had aroused public interest. The Canadian Coalition on Acid Rain, which assumed a high profile in the mid-1980s, was yet to be formed in 1979. It would not begin operations for another two years, and clearly had no role in Ontario's embrace of the issue.

A FOREIGN BAD GUY

Ross Howard, like Beamish and Harvey, Peter Whelan, and Lydia Dotto before him, had clearly pointed to the Sudbury smelters as the villain in the Muskoka story. The scientific research, however, was beginning to point elsewhere. The SES study during the smelter shutdown in 1978 and the atmospheric transport work of Summers and Whelpdale both suggested that Ontario was receiving considerable amounts of acidic deposition from the United States. As a result, the foreign bad guy image became both a factor encouraging the redefinition of Ontario's interests and a substantive element in the Ontario and Canadian perception of the acid rain problem.

In February 1979 Minister Parrott told a legislative committee that Ontario sources emitted less than three million tons of sulphur and nitrogen oxides compared to a total of thirty-nine million tons in the northeastern United States. Ministers' statements and government publications soon came to emphasize that much, if not most, of the acid rain in Ontario was coming from the United States. On this point the federal government agreed with Ontario.[64] The widely accepted image of acid rain as garbage dumped in Canada's backyard certainly encouraged the redefinition of Ontario's interests in the late 1970s. It is a common sociopsychological pattern that groups more quickly identify problems if they are caused by persons outside the group itself.

The transboundary nature of acid rain meant Ontario could not solve the acidification problem on its own. The political argument—"we cannot move unilaterally"—became a dominant theme in the rhetoric of Ontario politicians. This argument gave the government a solid reason not to demand reductions in SO_2 emissions from sources within the province. It also allowed Ontario to blame American sources for much of the problem and the federal government for not bringing the United States on board the acid rain express.[65]

While the line that Canada and Ontario had to work together with the United States rather than take a priori domestic unilateral action was both a politically useful and fairly constant refrain, it was also not an absolute stance. As early as October 1979, Parrott had announced Ontario was "prepared to act singly and in advance of other jurisdictions" to deal with acid rain. It eventually did so. With the 1984–1985 federal-provincial accords on acid rain control, Canada and Ontario finally moved first.

ASSESSING THE FACTORS

How did politics and science interact in the case of Ontario and the problem of acid rain and acidification? To answer this question, it is necessary both to establish that there was a redefinition of societal interests and to analyze the role science played in this process. This section considers the construction of Ontario's interests both before and after 1978 and the factors that may have played a part in changing the province's perceptions of its interests and its policies.

A host of factors encouraged a new construction of Ontario interests on acid rain during 1978 and 1979: scientific evidence, the triggering event of the regulatory reprieve to Inco, a change in environment ministers, an increasingly active media, an aroused public, and a foreign source for the problem. No attempt will be made here to weigh the relative potency of these factors. Such is at best an elusive, albeit common, epistemological quest. None of the factors here seem to be sufficient conditions, but some may arguably have been necessary conditions. The role of the science stands out in this respect.

A superficial consideration of the process through which Ontario reconstructed its interests on acid rain in the late 1970s might suggest that scientific knowledge played a critical role, perhaps the critical role. Such a conclusion would be in keeping with the thrust of the epistemic communities approach.[66] And a reasonable case could be made for this conclusion if one considers the various pathways that emerge from the above analysis. Scientific knowledge was provided at an important juncture to the relevant minister, Mr. Parrott. Scientific results also had a clear impact on media coverage and, through the media, on the public. Scientists like Tom Brydges were closely involved in the policy process and helped raise awareness among the public.

The rush to judgment here about the criticality of science, however, must be restrained. There are two reasons for caution. First, the very nature of the acidic deposition issue means necessarily that scientific information would be a part of any assessment of the interests at stake. Indeed, it is difficult imagining an environmental issue where science does not play a role. To find it being used in this or any specific case, then, is to be expected. It is not, in itself, proof that science played the critical role.

Second, and more seriously, the nature and timing of scientific discoveries need to be examined more closely. It is one thing for scientific knowledge to be produced and put to more or less immediate policy use. It is another matter if relevant scientific knowledge is produced, but does not have an impact on defining interests and deciding actions. What if key scientific facts central to the acidification issue in the late 1970s had already been discovered and presented to policy makers much earlier? In that case, scientific knowledge may have had at best a delayed impact, and perhaps little or no impact, at least prior to the eventual Ontario embrace of the acid rain issue in the late 1970s. In such a case it would seem that policy makers are perhaps not always receptive to scientific findings and that they do not always seek to reduce uncertainty about potential environmental problems.

Much was in fact known about acidification prior to the striking findings coming from Dorset in the late 1970s and, indeed, prior to the SES study being launched in the early 1970s. Knowledge of the environmental problems from smelter SO_2 emissions, including damage to aquatic and terrestrial ecosystems and to vegetation, goes back much further. A brief synopsis of some high points of this long and extraordinary history will perhaps suffice to make the point.

The provincial and federal government findings that grabbed people's attention in the latter 1970s were actually but the most recent part of a stream of evidence that had been emerging for years. The first solid scientific evidence on SO_2 damage to lakes and fish goes back a decade, to 1970–1971. Long before Minister Parrott began talking about those 140 acidified (Sudbury-area) lakes, Ontario government scientists observed lake acidification and fish loss in a wide area surrounding Sudbury. The resulting interdepartmental study was based on sampling from more than ninety lakes.[67] It pointed directly to SO_2 emissions from the smelters as the likely cause of the acidification. By coincidence, this research was being done at the same time Beamish was doing his similar work in the nearby La Cloche region and the same time that Svante Odén was publishing his first analyses of acid rain in Swedish scientific bulletins and a Stockholm newspaper. Odén's early reports, however, were published only in Swedish and received little attention in North America until the mid 1970s. In short, the 1970–1971 Sudbury Area Lakes Study (SALS) previewed the main lake acidification findings of the decade-long SES project and the many studies of acid rain–inflicted aquatic damage that followed in the 1980s.

The Sudbury lakes report was never made public, despite the fact that the responsible cabinet minister initially wanted it to be released. The conclusions of the original version of the report were nevertheless watered down within the Ontario bureaucracy to make the culpability of the smelters less clear. There was no immediate follow-up to determine the geographic extent of the damage from the smelter emissions. Thus, the generalized nature of acidic deposition in Ontario was left to be discovered a few years later.

Just as the APIOS study of the 1980s had been preceded by the SES and before that by the SALS, the latter was also not the first work of its kind. It had been preceded in turn by scientific work conducted by Eville Gorham, the assessor for Beamish and Harvey's article, who can be credited as the father of contemporary acidification research in North America. Gorham's early work on the acidification of lakes in the Sudbury area in the early 1960s followed on the heels of studies he conducted into acidic rainfall and acidic deposition in rural England and his native Nova Scotia.[68] He found circumstantial evidence in Nova Scotia of long-range transport from the United States.[69]

Gorham's fieldwork in Sudbury had been undertaken when he was a young professor at the University of Toronto. His coinvestigator and later coauthor was a scientist with the regional staff of the Department of Lands and Forests, A. G. Gordon. Gorham and Gordon found clear evidence in the Sudbury basin of connections between the smelter operations, SO_2, acidic deposition, and ecosystem impacts, in particular, negative effects on lake waters, vegetation, and soils.[70] Their work was at least a beginning toward a comprehensive conception of what twenty years later came to be popularly, if misleadingly, termed "the acid rain problem."

The Gorham-Gordon research was conducted in accordance with the normal openness of academic science, and the results were published in scientific journals. But it was also a quasiofficial study, given the involvement of a government department. For both reasons the work became well known to others within government.[71] But the Gorham-Gordon findings, like those of the later SALS study, did not lead the Ontario governments of the day to a serious reexamination of their perceived interests, let alone to consideration of regulatory action.

Although the Ontario government was well aware of this body of scientific evidence of acidification damage in the early 1970s, it deliberately set it aside and ignored suggestive evidence that acidification in the province was more than just a Sudbury problem. And it did so during a time of emerging public awareness about the environment. At the time, and over the previous half-century or more, the province had been locked into a definition of its interests that favored economic growth over environmental concerns. Indeed, the prevailing definition of its interests not merely gave strong preference to the mining and smelting industry, but consciously ignored its negative impacts. When the government finally began to consider acid rain controls in 1980, particularly controls on the province's nickel smelters, the policy shift thus reflected a considerable change in the province's perception of its economic and environmental interests.

To summarize: What was the role of science in this redefinition of provincial interests? Acid rain was identified in 1978–1979 both as a major environmental problem and as one in need of solution. This perception emerged from a bruising process in which new scientific findings, backed by changing norms in civil society and mobilized by media coverage, overcame government inertia and resistance to tackling the acid rain problem. The government recognized that the resources at stake were important ones and that the public was aroused. In short, acidification, or acid rain, was perceived to be a serious problem, and one that had to be addressed, both by domestic regulation and by international negotiation. The science of the latter 1970s contributed to a reconstruction of interests and to the adoption of acid rain controls in Ontario and Canada as a whole.

Much of the scientific evidence available in 1978–1979 was not, in fact, new knowledge. As shown here, the Ontario government was aware of considerable evidence on acidification a decade or more earlier, much of the evidence from its own scientists. It had ignored this evidence. It was able to do so in part because senior officials ensured that the available evidence did not get wide public circulation. The fate of the SALS work thus suggests that good science alone may not be sufficient. Governments can and will ignore scientific findings, especially when those findings seem to undermine firmly established perceptions and interests.

CONCLUSION

A basic question thus arises. Why did the scientific evidence of acidification during the late 1970s influence policy, whereas the science ten and twenty years earlier did not do so? Various explanations come to mind.

A sociological approach would emphasize changes in societal values and the changed nature of the social interactions between government and civil society. In contrast to earlier decades when governments avoided actions that would impact on major corporate interests, the Ontario acid rain controversy of 1978–1979 provides evidence of a popular belief in the late 1970s that environmental values had to be considered along with jobs and economic growth. If leaders, like other individuals, are by nature "group animals," then it follows that Ontario leaders of the day would feel a need to take a position on acid rain that placed them within, rather than outside of, their social group. They were, in addition, more predisposed to favor the preferences of their group over those of outsiders (such as the sources of pollution in the United States that led to long-range transport). The policy shift, therefore, may have occurred because the values of Ontario citizens changed and because political leaders both recognized and responded to these changes.

More traditional political explanations would focus on objective provincial or societal interests. By the late 1970s it was clear that Ontario ecosystems were threatened by acid rain. This problem was to at least a significant extent caused by emissions from outside the province's own boundaries. Thus, it was in the province's material interest to seek reductions in acid rain–causing emissions from all relevant sources, if only to share the cleanup burden.

Another explanatory perspective would point to calculations of political self-interest by those in power who wished to remain there. Acid rain had become a general provincial political issue for Ontario as a whole in a way that the regional environmental problems of Sudbury had not. Acid rain was widely perceived to be an urgent and important problem, one that had to be dealt with—or at least the government believed it had to be seen to be dealing with it. The influence of possible electoral punishment may well have been particularly strong given that the Ontario government in 1978–1979 lacked a majority of seats in the provincial legislature. These political factors were simply not present earlier in the decade.

The present chapter does not argue that one of these sets of influences (changing social values, objective societal interests, or perceived political interests) was paramount. Perhaps all were operating.[72] Conclusions are for readers to draw. Two questions merit reflection as we contemplate explanations.

First, what is the role of scientific research implicit in the above explanations and how is scientific knowledge disseminated? Citizens did not simply wake up one morning with a full realization of the dangers posed by acidic deposition. The societal explanations offered above for the policy shift assume dissemination to the public (likely through the media) of scientific knowledge about the causes and impacts of acidification. Whether or not this is a valid assumption may depend on one's perspective.

The second question concerns the acidification problem and perceptions thereof. Was it the material conditions of environmental damage that directly or indirectly led the government's policy shift? Or was it a change in perceptions and how interests were constructed? While the actual environmental damage done to Ontario lakes is impossible to assess precisely in retrospect (such as the extent to which lakes were acidified), the actual aquatic conditions were arguably more likely a near constant over the 1970s than they were a worsening factor or, in other words, a variable. Ontario was being damaged in the early and mid-1970s by pollution from both its own sources and from the United States, just as it was in the 1980s. If so, did the policy changes result from recognition of the actual environmental damage being done? Or from shifts in what the Ontario government perceived and wanted?[73] What it wanted after 1978 was arguably not the same as what it wanted before 1977, and for that matter, what it had wanted for decades earlier.

What seems clear is that the reconstruction of Ontario's interests on acidification in the late 1970s was very rapid. During the same period, the province's objective material interests did not change; certainly, they did not change as quickly as did the subjective definition of these interests. In other words, the reality mattered more after it became clear to the provincial government that some relatively new environmental values were challenging longstanding economic ones. The Inco-Muskoka issue, reinforced by the media and public inputs, perhaps made that value shift clear, even to the most determined proponents of the old ideational structure. In the end, though, any shift to new policies and practices comes about by an act of will. If the Ontario acidification case was in large part a struggle between new ideas and enduring economic interests, it would seem the new ideas won. How important those ideas were in reinforcing the science—or science in legitimating these new ideas—is another matter.

The long history of scientific studies in Ontario would seem to raise questions about at least one of the assumptions of the so-called epistemic communities approach to explaining environmental policy making. That approach assumes that policy makers seek to reduce uncertainty about environmental problems, that they thus welcome and support scientific studies that shed light on these problems, and that they then act on the basis of these findings. Contrary to these basic assumptions, scientific advice provided to the Ontario government prior to 1978 was not solicited and was ignored. Scientists were not welcomed into policy circles, but excluded from them and marginalized. Far from seeking to reduce uncertainty, as predicted by the epistemic model, the Ontario government wrapped itself in what uncertainty it could find and, when that failed, simply ignored and buried the evidence on the damage from acidification. Consistently over that long period, the government arguably turned a deaf ear to the scientific advice it was receiving in order not to place any significant burden of pollution control on its valued mining and smelting industries. Despite the mounting evidence, Ontario's construction of its resource and environmental interests remained largely unchanged through this entire period, until the very late 1970s. Governments faced with competing values and difficult political dilemmas may undertake scientific studies, but do not necessarily welcome the results of these studies and do not always act on them. Why they do so and when they do so is what the reader must now determine. Whether or not the findings of this case study can be generalized to other cases where science is brought to bear on environmental problems remains to be seen.

NOTES

The University of Northern British Columbia provided support for this research. Caroline Grey at the Archives of Ontario and Paulette Dozois and Catherine Bailey at the National Archives of Canada facilitated access to various records. Suzanne LeBlanc and Mary Ellen Kelm provided valuable questions and comments.

1. "Acid rain" refers to precipitation that is more acidic than normal. But rains are only part of the process, as snows and fogs can also be highly acidic, and much deposition of acidic compounds occurs in dry as well as wet form. "Acidic deposition" is thus a more accurate term. The general environmental problem is that of ecosystem acidification. See, for example, Canada-United States Air Quality Agreement Progress Report (Ottawa and Washington, March 1992): 28. Although the famous Trail smelter case of the 1930s is usually characterized as a problem of "sulphur fumes," it was arguably the first international dispute over transboundary acidic deposition. John Read, "The Trail Smelter Dispute" *Canadian Yearbook of International Law* 1 (1963): 213–29.

2. For a comparison of the North American and European experiences with acidic deposition, see chapter 8 by Kenneth E. Wilkening in this volume.

3. For a review of these effects, see: Environment Canada, *1997 Canadian Acid Rain Assessment,* 5 vols. (Environment Canada: Ottawa, 1997); "Acidic Deposition: State of Science and Technology," Summary Report of the U.S. National Acid Precipitation Assessment Program, September 1991; Canada, Federal/Provincial Research and Monitoring Coordination Committee, *The 1990 Canadian Long-Range Transport of Air Pollutants and Deposition Assessment Report,* 8 vols. (Environment Canada: Ottawa, 1990).

4. Victor Malarek, "Pollutants in the Air Are a Time Bomb, LeBlanc Says," *Globe and Mail,* 21 June 1977.

5. Reported in the *Toronto Star,* 21 June 1977. There were some follow-up stories to this speech, but no significant public reaction. See, for example, Frank Jones, "Acid Rain from U.S. Battering Canada," *Toronto Star,* 23 October 1977.

6. The term was introduced in North America in the early 1970s. See G. E. Likens, F. H. Bormann, and N. M. Johnson, "Acid Rain" *Environment* 14(1) (1972): 33–9; C. V. Cogbill and G. E. Likens, "Acid Precipitation in the Northeastern United States," *Water Resources Research* 10(6) (1974): 1133–37; G. E. Likens and F. H. Bormann, "Acid Rain: A Serious Regional Environmental Problem," *Science* 84 (1974): 1176–79.

7. Letter, H. Parrott to Len Marchand, 24 August 1978. Ministry of Environment, Acid Rain, 1 January–31 August 1978; RG 12–80; Archives of Ontario, Toronto (hereafter, AOT).

8. According to Parrott, Ontario was "the trend setter." "We were way out in front of the feds and the other provinces in 1978," (Dr. Harry Parrott, interview with author, April 1992).

9. Letter, William Davis to Pierre Trudeau, 21 February 1979; Ministry of Environment; Acid Rain, 1 January–31 March 1979; RG 12–80; AOT.

10. Barbara Baker, "Acid Rain Is Suddenly a Popular Issue," *Globe and Mail,* 5 November 1979, 35.

11. Rosemary Speirs, "Parrott Promises Plan for Reducing Acid Rain," *Globe and Mail,* 10 April 1980, 5; Michael Keating, "Parrott to Set Acid Rain Controls after Talks Next Week," *Globe and Mail*, 15 April 1980; Michael Keating, "Parrott to Set Acid Rain Limits Next Week," *Globe and Mail,* 22 April 1980.

12. In recent years, the Ontario government has drastically reduced its commitment to monitoring acidic deposition and has taken no new measures to reduce it further beyond the 1985 reductions. That, too, is another story that cannot be pursued here.

13. The notion of actors "constructing" their interests is developed by John Ruggie, *Constructing the World Polity* (London: Routledge, 1998); and Martha Finnemore, *National Interests in International Society* (Ithaca: Cornell University Press, 1996). Finnemore equates interests to "what states want" and the preferences they have. This is the sense in which that term is used here.

14. Minutes, SES Directors Meeting, 3 October 1977; Minutes, Technical Meeting, 28 September 1977; Ministry of Environment; RG 12–1-1 vol. 1 1977 Box 1; AOT.

15. The stated goal of these remediation research programs was "to develop a reclamation technique applicable to the acidified lakes in the Sudbury area," and they were initiated in 1973–1974 (Sudbury Environmental Study, 1978, Cont. 3, TB 27; RG 12–63; AOT).

16. Ontario MOE, "Notes on Acidic Precipitation," prepared for the Standing Committee on Resource Development, February 1979. This fact was understandably highlighted by Inco, for example, in a letter to the editor by Stuart Warner, an Inco vice president (*Globe and Mail*, 14 September 1978, 6).

17. The scientific basis of the long-range transport argument and of LeBlanc's claim that the United States was the source of some of Canada's acid rain was work by Canadian and American scientists discussed in more detail below. Senior Environment Canada officials saw prepublication versions of some of this work and appreciated its political importance.

18. P. J. Dillon, D. S. Jeffries, and W. A. Schneider, "Effects of Acidic Precipitation on Precambrian Freshwaters in Southern Ontario," April 1978; Ministry of Environment; Acid Rain, 1 January–31 August 1978; RG 12–80; AOT.

19. Tom Brydges, interview with author, 12 June 1992.

20. Brydges interview, 7 February 1992.

21. R. Beamish and H. Harvey, "Acidification of the La Cloche Mountain Lakes, Ontario, and Resulting Fish Mortalities," *Journal of the Fisheries Research Board of Canada* 29(8) (1972): 1131–43.

22. Beamish and Harvey, "Acidification of the La Cloche Mountain Lakes."

23. Harold Harvey, interview with author, October 1990.

24. E. Gorham and A. G. Gordon, "Some Effects of Smelter Pollution Northeast of Falconbridge, Ontario," *Canadian Journal of Botany* 38 (1960): 307–12; E. Gorham and A. G. Gordon, "The Influence of Smelter Fumes upon the Chemical Composition of Lake Waters near Sudbury, Ontario, and upon the Surrounding Vegetation," *Canadian Journal of Botany* 38 (1960): 477–87; E. Gorham and A. G. Gordon, "Some Effects of Smelter Pollution upon Aquatic Vegetation near Sudbury, Ontario," *Canadian Journal of Botany* 41 (1963): 371–8.

25. Letter, L. W. Billingsley, associate editor, *Journal of the Fisheries Research Board of Canada*, to R. J. Beamish, 21 March 1972; R. Beamish personal files.

26. R. Beamish, interviews, 7 February 1992 and 12 July 2001.

27. S. Odén, "Acidification of Air Precipitation and Its Consequences on the Natural Environment," (title translated from the Swedish) *Dagens Nyheter,* 24 October 1967, Bulletin No. 1, 1968, Ekologikommitten Statens Naturvetenskapliga Forskeningsrad, Stockholm; Organization for Economic Co-operation and Development, *Co-operative Technical Programme to Measure the Long-Range Transport of Air Pollutants* (Paris, OECD Environment Directorate, 1972).

28. It is worthy of note that Environment Canada and the U.S. EPA, despite their similar policy mandates, are somewhat dissimilar organizationally. To oversimplify, Environment Canada began and remains more of a scientific organization than a legally oriented regulatory organization in terms of staffing and leadership. The EPA, in general terms, is the reverse.

29. P. W. Summers and D. M. Whelpdale, "Acid Precipitation in Canada," *Water, Air, and Soil Pollution* 6: 447–455; J. N. Galloway and D. M, Whelpdale, "An Atmospheric Sulphur Budget for Eastern North America," *Atmospheric Environment* 14(4) (1980): 409–417.

30. Canada–United States Consultation Group on the Long-Range Transport of Air Pollutants, "The LRTAP Problem in North America: A Preliminary Overview," (Ottawa and Washington, 1979).

31. Because of the plant's proximity to the Sudbury airport, Falconbridge could not reduce ambient air pollution by increasing the height of the emissions stack. Having made expensive process changes, Falconbridge failed to meet its 422-tons-per-day emissions limit because its chosen technology did not work effectively. The MOE rewrote the Falconbridge control order in 1973, extending the deadline for compliance to May 1979. To meet the requirement, the company again modified its smelting process. By 1984, Falconbridge's process changes had achieved a reduction of 75 percent from 1970 levels.

32. Memo, W. J. Gibson to E. Piche, 8 March 1976; Ministry of Environment; SES-03 Branch Reps Meetings 1975 [no file number], RG 12–1-1 Box 3, AOT.

33. "Sulphur Dioxide Control Order to Be Changed to Suit Inco," *Eco/Log Week,* 2 December 1977. Minutes of Industrial Abatement–SES Technical Committee, 12 April 1977, found in file: Ministry of Environment; RG 12–1-1 Box 3 (no file number) SES-03 Branch Reps Meetings 1977, AOT.

34. Press release, 31 July 1978; Ministry of Environment, file: Sudbury Environmental Study, 1978, found in file RG 12–63 Cont 3, TB 27, AOT.

35. "Inco Pollution Deadline Extended 18 Months," *Globe and Mail,* 20 February 1973, 3.

36. *Nature Canada*, October–December 1978: 25.

37. Robert Sheppard, "Parrott Places Sewage Before 'Glamor,'" *Globe and Mail,* 5 September 1978, 4.

38. Interview with Tom Brydges, 12 June 1992.

39. Richard J. Beamish, personal correspondence, 1971 (R. Beamish personal files).

40. Richard J. Beamish and Harold H. Harvey, "Why Trout Are Disappearing in La Cloche Lakes," *Globe and Mail,* 26 July 1971.

41. R. R. Saddington, "How One Company Is Trying to Combat the Problems of Pollution," *Globe and Mail,* 26 July 1971.

42. Peter Whelan, "Lake Looks Healthy but Has Indigestion," *Globe and Mail,* 18 October 1973.

43. G. E. Likens, F. H. Bormann, and N. M. Johnson, "Acid Rain," *Environment* 14(1) (1972): 33–9.

44. Danielle Wetherup, interview with author, 15 February 1995.

45. Tom Brydges, interview with author, 12 June 1992. For example, the story was picked up the following day by, among others, a Sudbury television station, even though smelter emission stories had been notably absent on Sudbury TV news in the past (Ministry of Environment; "SES 05 Media Contacts," RG 12–1-1 Box 3 [no file number]), AOT.

46. MOE, "Acid Rain in Ontario," no date; *Acid Rain,* 1 January–31 August 1978, found in file: RG 12–80; AOT.

47. Ross Howard, "Cottagers Demand Acid Rain Answers," *Toronto Star,* 8 February 1979, A4; Ross Howard, "Acid-Laden Rain Killing Fish in Resort Areas," *Toronto Star*, no date.

48. Fred Fletcher, *The Newspapers and Public Affairs* (Ottawa, Ministry of Supply and Services for the Royal Commission on Newspapers, 1981).

49. The data collection is based on *InfoGlobe*, the electronic database of articles published in *The Globe and Mail. InfoGlobe* includes articles from November 1977 to the present. The data collection involved selecting any article (from 1977 to 1992) containing in its text the keywords "acid rain" or "acidic rain," "acid rainfall" or "acidic rainfall," and "acid precipitation" or "acidic precipitation." This search captured 3,333 articles, which included not only "hard" news stories, but also feature stories and editorials, as well as opinion ("op-ed") pieces. Some of these articles mentioned acid rain only in passing. *InfoGlobe* is based on the final Toronto edition of the *Globe*. The data collection was supervised by Adam Fenech and carried out by Heather Flett and Kim Hunter. The work was funded by a grant from IBM as part of the Harvard University–based project on The Press and Global Environmental Change.

50. Memos, September 1977; Ministry of Environment; RG 12–1-1 Box 3 (no file number) "SES 05 Media Contacts," AOT. This NBC coverage had followed on the heels of a Canadian Press Service story by Peter Michaelson, from Ottawa, on 27 August 1977 and requests from other media outlets for the SES report (Memo, E Piche to Barr, 6 December 1977; "SES 05 Media Contacts," found in Box 3 [no file number]), RG 12–1-1, AO T).

51. Ray Robinson, interview with author, 5 December 1991.

52. Tom Brydges, "Some Observations . . ." 1987, unpublished manuscript, 29.

53. Ross Howard, interview with author, 16 March 1995.

54. For example, see Ross Howard, "Acid-laden Rain Killing Fish." And he continued to question the ministry (Ministry of Environment: "SES 05 Media Contacts," Box 3 (no file number), RG 12–1-1, AOT.

55. Quoted in "Acid Lakes," *The Guide*, Huntsville, Ontario, 29 July 1978 found in file: Ministry of Environment; Acid Rain, 1 January–31 August 1978, RG 12–80, AOT.

56. By the early 1970s at least some of the citizens of Sudbury were as concerned about air pollution as job losses from regulation-induced production cutbacks. See

Letter, Michael Nash, Sudbury Environmental Law Association, to Hon. James Auld, 27 July 1973, found in file: Ministry of Environment, Air: Sudbury Air Pollution, 1973, file 511–2, Box 294 RG 12–45, AOT.

57. Ministry of Environment, RG 12–80, Acid Rain, January 1–August 31 1978, AOT.

58. Letter, 2 August 1978; Ministry of Environment; RG 12–80, Acid Rain, 1 September–31 December 1978, AOT.

59. Ross Howard, "Cottagers Demand Acid Rain Answers," *Toronto Star,* 8 February 1979, A4.

60. I am indebted to Michael Keating for this information.

61. Letter from Premier Davis, 1 May 1979, found in file: Acid Rain, 1 April–31 May 1979, RG 12–80, AOT.

62. Draft letter for Premier Davis, no date, found in file: Acid Rain, 1 April–31 May 1979, RG 12–80, AOT. Another letter (24 May 1979) in this same file refers to lakes being known to be acidic due to the smelters as far back as the 1950s. The source of that particular date remains unclear.

63. Canadian Institute of Public Opinion (CIPO), polls 442 (September 1980) and 455 (November 1981). This data was obtained from the University of British Columbia Data Archive and analyzed by the author.

64. Jim Foster, "Parrot [sic] Fears Flood of Acid Rain If U.S. Starts Using More Coal," *Toronto Star,* 14 July 1979; Victor Malarek, "The Acid Rain War: Showdown on Cuts in Stack Emissions," *Globe and Mail,* 6 August 1979, 14; Robert Sheppard, "U.S. Must Move First on Acid Rain Control," *Globe and Mail,* 26 September 1979.

65. "Ontario Active, Ottawa Lags on Acid Rain Problem," *Eco/Log Week,* 23 February 1979, 2.

66. See, for example, the special issue on epistemic communities of *International Organization,* 46(1) (Winter 1992).

67. There were two versions of the report, OWRC, "Preliminary Report on the Influence of Industrial Activity on the Lakes in the Sudbury Area," 1969–70; OWRC, "Preliminary Report on the Influence of Industrial Activity on the Lakes in the Sudbury Area," 1971, found in file: Ministry of Environment, RG 1–282–32, AOT.

68. E. Gorham, "The Influence and Importance of Daily Weather Conditions in the Supply of Chlorine, Sulphate and Other Ions to Fresh Waters from Atmospheric Precipitation," *Philosophical Transactions of the Royal Society of London* Series B, Biological Sciences, 24l(679) (1958): 147–78; and E. Gorham, "Factors Influencing Supply of Major Ions to Inland Waters, with Special Reference to the Atmosphere," *Geological Society of America Bulletin* 72 (1961): 795–840.

69. F. A. Herman and E. Gorham, "Total Mineral Material, Acidity, Sulphur, and Nitrogen in Rain and Snow at Kentville, Nova Scotia," *Tellus* 9(2) (1957): 180–3.

70. Gorham and Gordon, "Some Effects of Smelter Pollution Northeast of Falconbridge, Ontario," 307–12.

71. The SALS report of 1971 cites Gorham's papers (from 1958 and 1961) as well as the first article he coauthored with Gordon (in 1960).

72. One response to the latter argument might be that what Ontario was about to

do in reducing its emissions of SO_2 was not just a move toward pure self-interest. Ontario itself was a source of pollution to parts of the United States and, thus, any solution was going to be a movement toward a collective interest.

73. See Alexander Wendt, *Social Theory of International Politics* (Cambridge, UK: Cambridge University Press, 1999): 106.

7

Lost in the Woods

International Forest Policy

Radoslav Dimitrov

In the late 1980s, a forester described the international forest agenda as "a series of loudly trumpeted nonevents."[1] Today it could be more accurately portrayed as a series of loudly trumpeted events with no policy output. States have engaged in several global and regional initiatives to devise international policies for sustainable forest management, but negotiations have consistently failed to produce a forest treaty, despite a consensus among governments that the rate of deforestation is unsustainable. Multilateral scientific assessments of forest resources submit abundant information about the extent and causes of deforestation, but there is a remarkable paucity of data on its cross-border consequences, which are usually the main rationale behind environmental treaties. At the same time, the domestic costs of "green" policies are very high since forest utilization is a cross-sectoral issue that involves a number of socioeconomic interests. Few governments dispute that forests are important, and all agree that action should be taken, but many actors oppose the idea of an international treaty or that forest management should be coordinated at all. There are some existing legally binding agreements that bear tangential relevance to forest issues, but those have not been used to introduce international forest policy.

This chapter explores the intersection of science and politics in the context of global forest policy. It examines the role of extensive and increasing scientific evidence about the extent, causes, and consequences of forest loss in the recent attempts to form an international regime for forest management. Based on participant observation of several rounds of international forest negotiations, interviews with policy makers, and analysis of multilateral scientific assessments, the study explores the connection between shared knowledge and collective action in international initiatives on forest management. It generates questions about the origins of international concern with deforestation, examines the configuration of state preferences and the rationale behind the positions of states, and discusses the wisdom of the current international arrangement on forests. The discussion of the science of forest degradation focuses on scientific information contained in global multilateral assessments coordinated by international organizations. Such assessments are conducted on a global scale and, unlike national sources of information, are made available to all governments.

The case for managing forests at the international level has never been made either scientifically or politically. So far, research has not demonstrated that forest management policies in one country would significantly affect ecological or social systems in another. This raises a number of questions. Why do so many countries want a forest treaty? Where did interest in international forest policy come from? Who placed forests on the international agenda and why? How have political actors understood existing science, and how have they used it? Who would benefit from international forest policy? And why are countries that advocate a treaty unwilling to pay for its implementation? Perhaps treaty supporters are motivated solely by ethical concerns. Or perhaps their considerations have little to do with the consequences of deforestation or with forest management. The details of this complex issue weave a rich texture that provides clues to many possible answers to these and other questions.

THE SCIENCE OF FORESTS

A distinct feature of the forest problem is the absence of substantial scientific controversy. Forest degradation is not the subject of conflicting research results or competing scientific views, and policy makers do not debate existing empirical knowledge about it. This is not to say that our knowledge about deforestation is complete. There are significant gaps of information on important aspects of the problem that are well known in scientific and policy circles alike and are not a matter of debate in international deliberations.

Forest science is a well-established academic discipline, and forest research is institutionalized in policy making on both national and international levels. Specialized publications began to appear in Europe in the seventeenth century, and the first professional forestry school was established in 1768, in Ilsneburg, Germany.[2] Today, research on forest resources at local, national, regional, and global levels is carried out in diverse institutional settings by environmental NGOs, state governments, and intergovernmental institutions.[3] In addition to national research, international organizations coordinate multilateral scientific assessments of forest conditions on a regular basis. However, a group of some of the world's leading forest inventory specialists concluded that "the process of deforestation is such a complex process, involving physical, climatic, political, and socio-economic forces which are themselves very complex, that simple generalised models of forest change have so far not been developed."[4]

One major challenge in scientific assessments of forest resources is to standardize definitions and classifications of forest cover across countries and also across time. Forest density varies considerably, from thick cover to sparsely strewn individual trees, and experts do not agree on what constitutes a "forest." In a recent multilateral study, more than 650 definitions of forest types were assembled from 110 independent surveys in 132 countries.[5] The definitional disarray has, until recently, precluded a global estimate based on an aggregation of national data. Not until 1996 was a universal set of definitions developed at an international expert consultation in Kotka, Finland.[6] A 2000 study by the FAO was the first assessment to use a common global definition of forest: it considered a forest any area whose vegetation cover comprised at least 10 percent of the observable land area.[7] It was also the first to provide a consistent methodology for assessing forest change between two time periods. It relied on satellite imaging, as well as surveys of countries, and used statistical random sampling (a 10 percent sample) of the world's tropical forests through 117 sample units, and three time-point observations between 1980 and 2000. The study produced the first global forest maps that use satellite data (figure 7.1).

At the end of the twentieth century, forests covered 3,869 million hectares or about 30 percent of the world's land surface.[8] Their distribution today is highly uneven: three countries (the Russian Federation, Brazil, and Canada) account for more than 40 percent of all forest area. Of the total forested area, 95 percent are natural forests, and 5 percent are plantations created either through artificial afforestation or by conversion of natural forests. The FAO reported that global forests decreased by 11.1 million hectares between 1980 and 1990, and by 9.4 million hectares between 1990 and 2000, an annual depletion rate of 0.02 percent.[9] Thus, the rate of deforestation slowed down slightly in the 1990s.

Progress has been made in measuring the quantity of forests, but there are no precise estimates of their quality. The increase or reduction of forest area does not necessarily correspond to qualitative changes. During the 1990s, 1.5 million hectares of natural forests were converted to plantations each year, a trend that indicates reduction in forest quality, since plantations have lower biodiversity than natural forests.[10] While this may have ecological implications, the FAO reports that "it was not possible to make objective estimates of the extent or severity of [forest degradation] for most countries because

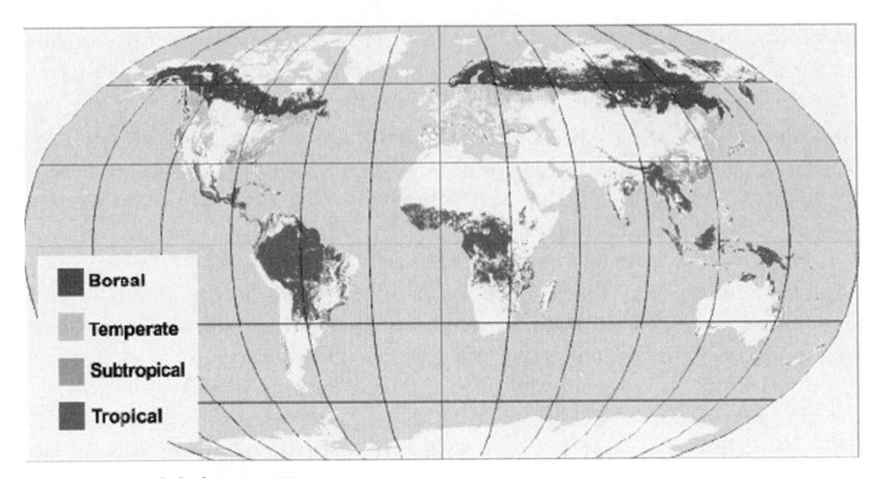

Figure 7.1 Global Forest Cover

Source: Forest Resources Assessment 2000, UN FAO.

of data limitations. Qualitative changes are reported in country reports and briefs, but it was not possible to derive globally valid statistics."[11]

Despite the gaps in empirically based knowledge—which are discussed below—no one disputes that the rates of deforestation and forest degradation are considerable. Similarly, their causes generate little scientific controversy. There is general agreement among scientists and policy makers alike that the main factors behind deforestation and forest degradation are human activities related to population growth and development: commercial logging, agriculture, clearing land for pastures, colonization programs, mining, hydroelectricity projects, and military activities. In addition, there are natural phenomena such as fires, diseases, and cyclones that can cause considerable damage to large tracts of forests.

By contrast, the consequences of deforestation are largely unknown by scientific standards. The benefits of forest cover are broadly accepted, but the ecological and socioeconomic impact of forest degradation is not verified by sufficient scientific evidence. On the one hand, there is general agreement that forest degradation implies loss of many of their ecological and social functions, including maintenance of water cycles, improvement of air quality, habitats for biological species, and stabilization of local microclimates. On slopes and in uplands, forests stabilize land, control erosion, and prevent landslides. Urban forests and trees mitigate air pollution, buffer noise, provide nontimber products, bestow psychological benefits, and, in arid and semiarid areas, they cool cities and protect them against sand and winds. At the same time, the empirical measurement of the value of forests is incomplete, and there is remarkable paucity of research-derived data on the functions of forests and, thus, on the derivative or indirect consequences of deforestation. The transboundary impacts of the problem are particularly uncertain: scientific reports explicitly acknowledge that global effects of deforestation on the climate or biodiversity cannot be measured with any degree of precision.

THE SOCIOECONOMIC CONTEXT OF FORESTS

Forests are sources of benefits in several socioeconomic sectors, such as commercial logging, international trade, agriculture, food security, energy consumption, and water management. Timber exports bring large revenues and logging is an important part of the economy, not only in developing countries, but also in countries such as Canada and the Russian Federation. In addition to timber production, forests are cleared for agriculture, to supply wood for household fuel, or to construct dams for flood control and hydroelectric energy production.

Commercial logging has received particular attention from policy makers, researchers and environmentalists. In national decision making, the principal value of forests is still measured in terms of their timber-producing capacity.[12] The FAO estimates that their economic value is $400 billion per year (that is, 2 percent of global GDP) and provides extensive tables with figures for each country.[13] In tropical countries, 11 million hectares were harvested annually during the 1990s. In industrialized countries, annual felling of timber was 1,632 cubic meters and actual removal was 1,260 cubic meters, implying significant waste and harvest losses.[14]

Healthy forests also generate employment, provide food items for subsistence, as well as tradable products, and support agriculture.[15] For example, in Jamaica, Thailand, Honduras, Sierra Leone, and Bangladesh in the 1980s, an average 12.3 percent of the population was employed in forest-based enterprises.[16] Maintaining forest cover on critical watersheds also safeguards water supplies for irrigation systems, and trees provide windbreaks for agricultural fields, protecting them from wind erosion and wind damage.

Large populations in developing countries depend on fuelwood for their energy needs. Burning wood meets 7 percent of the energy needs of the world as a whole, 70 percent of the needs in thirty-four countries, and more than 90 percent of energy consumption in thirteen, mostly African countries.[17] Forests are also sustainable sources of "nontimber forest products," including foods (fruit, honey, seeds, mushrooms), fibers (rattan, bamboo, and reeds), oils, rubber, gums and resins, waxes, ornaments, and pharmaceutical and cosmetic products.[18] For instance, in 1996 Tunisia exported 230 tons of essential oils worth $3.2 million, a considerable amount for this country. The average value of world trade in bamboo products is $36.2 million and of morel mushrooms between $50 and $60 million annually. Rattan is used throughout South East Asia in furniture, fish traps, mats, hammocks, and brushes. Yet, data comes piecemeal; no country estimates all its nontimber products, and there is no comprehensive global inventory.[19]

Activities that contribute to the depletion of forests are conditioned by a large complex of broader socioeconomic factors, such as government policies, population growth, poverty, and unemployment. Some studies find a high correlation between the change in forest area and the change in population density. Poverty and social inequality leave landless large populations of peasants, who clear forests to cultivate crops.[20] And because timber exports are important to economies, many governments encourage logging through investment incentives, overgenerous concessions of forests, business credits, and tax credits.[21] These incentive structures—in countries as diverse as the United States, Ivory Coast, Indonesia, China, and Brazil—often fail to value timber properly, while ignoring the value of nontimber forest products and their environmental functions. For example, between 1979 and 1982, the forest industries in the Philippines generated $1.5 billion in revenues, but the government earned only $170 million, that is, 11.4 percent of the profit.[22]

In thinking about forest management, it is important to bear in mind that the ownership of forests varies across regions. In Latin America, the Caribbean, and Asia, most forestlands belong to the state, and the control and administration of forests is in the hands of forestry and agriculture departments. In the late 1970s, the state owned practically all forests in Bolivia and Peru, over 80 percent in Brazil, Colombia, and Venezuela, and between 80 and 90 percent in Asia as a region. By contrast, most of the forests in English-speaking countries in tropical Africa and the Pacific are privately or tribally owned.[23] In Pacific countries, such as Papua New Guinea, clans and tribes own the forests, and the government must negotiate with them to use any forest resources. In the early 1980s, 82 percent of forests in Malawi belonged to traditional communities, and even today forests can be placed under protection or conservation only with the agreement of the locals.[24] To complicate matters further, in some countries, customary laws may conflict with formal laws and leave unclear who owns forests and controls their use.

At the national level, various policies for the protection of forests are in place, and their design and implementation vary considerably across the world. About 12 percent of the world's forests are in protected areas and another 2 percent (80 million hectares) are certified for sustainable management. More broadly, 89 percent of forests in industrialized countries and 6 percent of those in developing countries are under some type of management plan. Because of widely recognized problems with policy implementation and compliance, we do not know how much of these are actually managed sustainably.[25] "Paper parks" that exist only in theory are not uncommon.

DEVELOPING SCIENTIFIC KNOWLEDGE

While the literature on local and national forest resources has been extensive, comprehensive global assessments of all types of forests began only recently.[26] Although the FAO has carried out forest assessments since the 1940s, until the 1980s, none of the studies measured changes in forest cover over time or analyzed processes of deforestation. In 1982 the FAO and UNEP organized the first comprehensive study of tropical forests, within the framework of the Global Environmental Monitoring System.[27] This was a multilateral effort by several dozen experts from forty-five countries that used satellite imagery, as well as existing information from national governments and research institutions. Several multilateral scientific programs are now under way, assessing and monitoring forest resources to provide the international community with appropriate information. These include the Forest Resources Assessment (FRA) conducted every ten years by the FAO; the LANDSAT Pathfinder Tropical Deforestation Project, a collaborative research effort under the auspices of the NASA Goddard Space Flight Center, the University of New Hampshire, and the University of Maryland; and Project TREES by the Joint Research Committee of the Commission of the EU.[28] While each of these is well designed and rigorously pursued, only the FAO's projects provide truly global assessments of all types of forests in all ecological zones.[29]

Extent of Deforestation

Scientific assessments of the global rates of deforestation provide extensive data with a high degree of precision and are generally considered reliable. FRA 1990 consisted of two phases using different empirical methods: analysis of existing information from national inventories and statistical sampling of multidate data from high-resolution satellite imaging. The study concluded that between 1980 and 1990, global forests and other wooded lands had decreased by 11.1 million hectares, an annual rate of loss of 0.02 percent.[30] Apart from the global level, researchers found strong regional characters in the process of change. The rate of gross deforestation over the decade was 6.6 percent in Africa, 5.9 percent in Latin America, and a much higher 11.3 percent in tropical Asia. In Europe, there was a net gain in forest area of almost 2 million hectares, a continuation of a trend that had lasted for forty years.[31] Globally, there was a 230 percent increase in the harvested forest area between 1961 and 1990, and a 265 percent increase in the harvested volume, which suggests an increase in production efficiency or logging of denser forests.[32] The most recent global assessment, FRA 2000, was the first to use a consistent methodology and a common definition of forest cover.[33]

At the global level, gross deforestation for the period 1900 to 2000 was estimated at approximately 14.6 million hectares per year. At the same time, there was an expansion of natural forests and plantations of 5.2 million hectares per year. Natural forests (primarily in European countries) expanded by 3.6 million hectares per year, and deliberate afforestation totalled 1.6 million hectares annually. Plantations grew more than tenfold in two decades: from 17.8 million hectares in 1980 to 43.6 million hectares in 1990 to 187 million hectares in 2000. Today, China is the leader in plantation development, accounting for 24 percent of the global plantation area.[34] Net global deforestation for the period was 9.4 million hectares per year, lower than estimates of 11.3 and 13 million hectares for the two prior periods (1990–1995 and 1980–1990), respectively.[35]

While noting that the estimates are not fully comparable with earlier assessments because of dissimilar definitions used, the FRA 2000 assessment concluded that at 0.02 percent per year (and 3.6 percent for the entire decade), the rate of deforestation has slightly decreased since the 1980s.[36] Net deforestation rates were highest in Africa and South America, while the forest cover in industrialized countries remained essentially stable, and in Asia afforestation significantly offset the loss of forests (table 7.1).

Causes of Deforestation

Changes in forest cover can be natural or human-induced. While global data on the extent of forest changes from natural causes is incomplete (and global statistics on the extent and impact of forest fires nonexistent), those include fires, disease, plant parasites, and destruction by cyclones, particularly in the Caribbean, Central America, and South-East Asia.[37] In developed

Table 7.1 Forest Cover and Forest-Cover Change, 1990–2000 (in million hectares)

	Land Area	*Forest Area 2000*		*Change 1990–2000*		*Forests 1995*	*Change 1990–1995*
Region	*Million ha*	*Million ha*	*%*	*Million ha/yr*	*%/yr*	*Million ha*	*Million ha/yr*
Africa	3,008	650	17	−5.3	−0.8	520	−3.7
Asia	3,167	542	14	−0.4	−0.1	503	−2.9
Oceania	849	201	5	−0.1	n.s.	91	−0.1
Europe	2,276	1,040	27	0.9	0.1	933	0.5
North/Central America	2,099	539	14	−0.6	−0.1	537	−0.3
South America	1,784	874	23	−3.6	−0.4	871	−4.8
World Total	13,183	3,856	100	−9.0	−0.2	3,454	−11.3

Source: Food and Agriculture Organization, *Global Forest Resources Assessment 2000*.

countries, fires are a relatively minor cause of deforestation. The FAO reported that in the early 1990s the average forest area burned by fires in North America, Europe, and the former USSR was 4.26 million hectares, or 0.22 percent of the total forest area.[38] Natural fires often play a positive ecological role and enable forest regeneration. African woodlands, savannas, and the coniferous forests in Central America in particular are "pyroclimax," that is, they depend on recurrent fires for their survival.[39]

At an international expert meeting on forests, natural causes of forest degradation were not mentioned.[40] FRA 1990 mentions, but offers no figures on, forest change from diseases such as the *x*-disease of pines in southern California, the infestation of beeches by beech scale in western and central Europe, and the *Leucaena psyllid* in Asia and the Pacific islands.[41]

The main causes of forest degradation are considered to be human activities: commercial logging, agriculture, clearance for pasture, military activities, colonization programs, mining, and hydroelectricity projects such as the construction of dams.[42] One indication of the reliability of information about these causes is the fact that different studies arrive at similar estimates of the role of particular human-induced causes. For instance, according to a World Bank study, agricultural settlement is responsible for 60 percent of the loss of tropical moist forests,[43] and another study found that clearing land for agriculture accounts for nearly two-thirds of tropical deforestation worldwide.[44]

Although the FRAs by the FAO confirm that the main global cause of deforestation is agricultural activity, data reveals regional differences in the predominant causes of forest degradation. During the 1990s, the single most important cause of forest change in Africa was subsistence farming (the conversion of forests into small-scale agriculture); in Latin America, the main causes were large-scale cattle ranching and centrally planned operations such as resettlement schemes and hydroelectric projects; and in Asia the causes were shifting agriculture, permanent agriculture, and migration into new areas.[45]

There is more ambiguity about the role of background social factors that indirectly contribute to deforestation. Until recently, the FAO emphasized the impact of population growth and density on deforestation.[46] In its most recent assessment, however, it revised its position and concluded that existing studies show a weak and oversimplified connection between demographic factors and forest area.[47]

Consequences of Deforestation

The effects of forest degradation are closely related to the loss of functions that forests perform. The first ostensible attempt to assess the global nonwood benefits of forests was made during FRA 1990. In response to a questionnaire, national experts listed the following forest functions: wood production, water conservation, hunting, nature conservation, and recreation. Remarkably, they did not list the roles of forests in regulating climate or in maintaining biodiversity, nor did they include many nontimber economic benefits. The assessment relied heavily on descriptive accounts: for most of the listed nonwood benefits, countries did not provide, and the FAO did not report, concrete figures that would throw light on the negative impact of deforestation.[48] As the assessment's authors noted, "It was recognized that, inevitably, these estimates were subjectively arrived at by the correspondents responding to the questionnaire."[49]

No one questions the local benefits of forests. Healthy forests provide food items, tradable products, and employment, and serve ecological functions that support agriculture.[50] Maintaining forest cover on critical watersheds is essential for safeguarding water supplies for irrigation systems; and trees provide windbreaks for agricultural fields, protecting them from wind erosion and wind damage. Forests are also sustainable sources of a number of "minor forest products" (also known as nontimber forest products) such as fruit, seeds, mushrooms, fibers, rubber, gums, resins, waxes, pharmaceutical and cosmetic products, and ornaments.[51] However, there is little concrete data on nontimber values of forests. One review of existing research unequivocally concludes that information about their nonwood products and services is at a primitive level and that the problems with their estimate and quantification are overwhelming.[52] In 1995, the FAO submitted that there is no inventory of nonwood forest products.[53] Its first attempt to provide one was made during FRA 2000, when it collected all available information from all countries, only to conclude:

> Information on the resource base and on subsistence use of [nonwood forest products] is non-existent. . . . Despite their real and potential importance, national institutions do not carry out regular monitoring of forest resources or evaluation of the socioeconomic contribution of [nontimber products] as they do for timber and agricultural products.[54]

Individual countries provide figures on exports of particular products, yet both developing and developed countries lack comprehensive inventories. We know for instance that Portugal produced 135,000 tons of cork in 1995 worth $145.3 million.[55] Yet, few reliable estimates on nontimber products exist for most European countries, and even Canada could not provide comparable data for maple syrup, which is one of its main nontimber forest products.[56] Notably, many Asian countries are able to supply more data and have maintained national statistics for decades.

Multilateral assessments also do not quantify the connection between the loss of soil and deforestation. According to one estimate, 10 million hectares of arable land are lost because of erosion. Erosion hurts agriculture and also contributes to numerous other problems: siltation of harbors, disruption of sewage because of siltation, eutrophication of waterways, flooding, and increased water-treatment costs.[57] Forest degradation is known to bear relevance to these processes, but there are few estimates of how much it contributes to soil erosion. One of the FAO's reports includes a scant two-paragraph comment on the subject and offers no data or discussion of the socioeconomic consequences of such changes.[58]

Scientific knowledge that is considered least reliable by scientists and policy makers alike pertains to the impacts of forest degradation on global and regional biodiversity and on global climate change. Forests are rich in biodiversity. Tropical forests in particular are believed to provide habitat for half of all known species, even though they cover only 7 percent of the land surface of the planet. The general recognition of the importance of forests contrasts with the absence of concrete and reliable data on the number of existing species, the impact of deforestation on biodiversity, and most importantly, on the impact of biodiversity loss on human communities. FRA 1990 provided "indicative estimates" of the loss of tree species, but not of other species that reside in forests, and recognized that "the magnitude of such losses or the extent of degradation of biodiversity is unknown."[59] No country has complete data on the total number of its tree species. FRA 2000 reported that there is no global list of trees and no data on animal species living in forests. The FAO concludes that "there is no accepted methodology for linking these changes [in forest area] to their impacts on forest biological diversity. . . . This is especially evident when information is aggregated at the global level."[60] Scientists and policy makers can only assume negative socioeconomic impacts of forest-related biodiversity loss, but cannot estimate them. "On a scale of 0 to 10, we are somewhere between 0 and 1.5. That's how poorly we understand biodiversity."[61] Thus, the ecological role of forests as habitats of species has not been clarified scientifically. As to the impact of changes in biodiversity on societies, "They [scientists] can only speculate on that."[62]

With regard to climate, forests can either mitigate or accelerate climate change. On the one hand, when trees grow, they withdraw CO_2 from the atmosphere, and this offsets global warming. On the other hand, the burning of forests releases carbon and other GHGs such as methane, nitrous oxides, and carbon monoxide.[63] The FRA 1990 assessment did not generate data on its own, but cited an independent study that found that between 1980 and 1990, deforestation and degradation caused a release of 900 million tons of carbon.[64] In its second comprehensive report, the IPCC concluded that deforestation and forest degradation account for 23 percent of human induced CO_2 emissions.[65]

Despite such numbers from research, the science of forests' role in climate change is still considered incomplete. The senior science and technology officer for the FCCC summarized, "If you chop a tree, you know what happens to carbon—but what this is causing is largely unknown."[66] Ironically, global warming is expected to have a positive impact on the productivity and volume of forests: middle scenarios predict a 20 to 25 percent increase in net primary production, with a benefit to the forest sector of $10 to $15 billion.[67] The scientific uncertainty about the effects of the problem is so overwhelming that the newest assessments do not even try to address this aspect of the issue. FRA 2000 does not include any information whatsoever about the consequences of deforestation. A study of the forests' ecological functions was simply not part of the assessment.

Interpreting Forest Science

While the ostensible purpose of multilateral scientific assessments is to inform policy makers, forest assessments invariably conclude that the available data does not allow drawing reliable conclusions.[68] The FAO recognizes that "the essential needs of researchers and policy makers cannot be met satisfactorily" because of gaps in information and the considerable variation of data between countries.[69] The report from FRA 2000 pointed squarely at a contradiction between the policy emphasis on deforestation in the global forest agenda and the shortage of reliable information.

> Analysis of the information base shows that reliable information on forest area, forest change, volume and biomass was not available for the majority of the world's forests. . . . Comparable time series were absent in most countries, including many industrialized ones making precise estimates of forest change difficult at both national and global levels. The lack of fully compatible multidate information at the national level for most countries continues to be one of the biggest constraints in the global assessments.[70]

Even publications by pro-environment groups reveal that we cannot specify how important forests are. Comprehensive studies on the state of knowledge conducted by the International Union of Forest Research Organizations reveal a serious lack of data. One overview of existing research concludes that in view of efforts to set international forest policy,

> it is rather astonishing that so much data and knowledge are still missing. International organizations such as the FAO and the World Bank and aid organizations such as USAID, CIDA, SIDA, and Finnida [sic] have pumped billions of dollars into forestry activities over the decades, and yet we cannot answer some of the most basic questions concerning global forest resources and their functions.[71]

In summary, there is scientific uncertainty of various degrees over all aspects of the problem of deforestation. Despite recognized shortcomings, international assessments provide reasonably good information about global forest cover and the rate of deforestation, as well as undisputed principal understanding of the causes of deforestation. However, there is a marked paucity of information about the nontimber benefits of forests and about the consequences of deforestation. Even experienced and concerned foresters note "a lack of clarity about which tropical forests should be saved, how much, and for what reason."[72] The least reliable knowledge is on the shared, transboundary effects of deforestation. Most of the available data on socioeconomic and ecological benefits of forests are local and national in character.

INTERNATIONAL POLICY DEBATES

The international political agenda that addresses deforestation is decentralized and greatly fragmented. It consists of a large number of bilateral, regional, and global initiatives that involve a variety of actors: governments, international organizations such as the FAO, UNEP, and the UN Development Program, as well as a number of developmental banks including the World Bank. In addition, innumerable NGOs such as the Forest Stewardship Council and the International Union of Forest Research Organizations fund forestry-policy projects and actively engage in forest-related research, education, and training. However, there is no international agreement to provide coherent forest policy. At a major session of the UN Intergovernmental Forum on Forests (IFF), governments adopted consensus that there is no global legal instrument that deals with all types of forests in a comprehensive and holistic way.[73]

The international policy instrument that is most directly relevant to forests is the International Tropical Trade Agreement (ITTA) established in 1983 and renegotiated in 1994.[74] For three reasons, its design is inadequate for forest protection. First, the agreement is limited in geographic scope: it covers only tropical forests, while 80 percent of industrial wood comes from temperate and boreal forests.[75] Second, the ITTA is narrow in thematic scope: it is a commodity agreement that regulates trade in tropical timber and largely disregards the ecological functions of forests and their nontimber products. The 1994 amendments did add provisions for sustainable development and established the Bali Partnership Fund to assist countries with sustainability projects. It is difficult, however, to reconcile this objective with the primary objectives of the original treaty: to increase the processing of tropical timber, to promote the utilization of wood, and to boost wood exports (Article 1). Indeed, member states have rejected a proposal for certification of tropical timber from sustainably managed forests.[76] Finally, the agreement does not provide for mechanisms to monitor compliance by countries and to report on progress in improving forest conditions.[77]

Apart from the ITTA, there are several legal treaties that bear tangential relevance to forest issues, including the FCCC, the Convention of Biological Diversity (CBD), and the Convention to Combat Desertification (CCD). In addition, regional environmental agreements such as the 1978 Treaty for Amazonian Cooperation or the 1985 ASEAN Agreement on the Conservation of Nature and Natural Resources could provide a framework to coordinate national forest policies in a region.

Article 4(1)(d) of the FCCC requires the promotion of conservation of sinks and reservoirs of GHGs, including forests, and the rules agreed on at Marrakesh in 2001 for implementing the Kyoto Protocol reaffirm the role of forests as sinks for GHGs (primarily CO_2). Many joint implementation projects between developed and less-developed countries under the Kyoto Protocol have sought to prevent or reverse tropical forest degradation, but the Global Environmental Facility, which provides funds for implementing the FCCC, does not channel funding to forest projects.[78] International reports on forests recognize that the FCCC has had a minor impact on forestry since there is a wide range of forest-related issues that lie beyond its scope.[79] The CCD relies on regional annexes for implementation, and Article 8(3)(b)(i) of the African annex requires the development of national action plans to sustainably manage natural resources, including forests. Yet, forest policy does not figure in the main text of the agreement, even though forests are of central importance in combating the advance of deserts. In addition, the convention is considered impotent because no mechanism for funding implementation was established. The inadequacy of existing forest-related international agreements was highlighted by policy reviews that concluded that a comprehensive treaty on forests remains desirable.[80]

Negotiating a Forest Convention

The political move toward a global forest convention emerged in the late 1980s. In the context of proliferating environmental agreements and strengthening environmental values, the absence of a global policy on forest preservation began generating public pressures on governments. At a 1990 meeting in Houston, Texas, the G-7 discussed the problem, and at a press conference, U.S. president George Bush proposed starting negotiations on a global forest convention.[81]

What could Western countries gain from a treaty? Perhaps they had purely ethical concerns about environmental protection. Perhaps they were yielding to domestic public pressures fueled by environmental groups. It is also plausible that some of them were driven by trade-related concerns. Domestic groups had pushed the Canadian government and industries toward sustainable forest management, such as selective logging that increased production costs. This made Canadian timber exports more expensive and placed them at a disadvantage in the market competing with timber products from developing countries that did not face domestic pressures and could clear-cut their forests. Uniform international regulation could create a level playing field and make Canadian timber exports more competitive on the world markets. Besides, what could the West lose from a treaty? The focus of public concern was tropical deforestation, and the task appeared to be to change policies in developing countries. A treaty would not require significant changes in countries that already had strong policies, such as the Scandinavian countries who are renowned for their environmental traditions, or Costa Rica, one-fourth of whose territory is already in protected areas.

International deliberations on forest management have taken place within three institutional settings: at the 1992 UN Conference on the Environment and Development (UNCED) in Rio de Janeiro; in four sessions of the Intergovernmental Panel on Forests (IPF) between 1995 and 1997; and during four rounds of the IFF between 1997 and 2000.

Forest Discords at Rio

Sharp disagreements among states on the need for such a treaty prompted abandonment at the preparatory stage of a plan to include negotiations on a forest convention in the agenda for UNCED. While the United States, Canada, and European countries emphasized the principle of global responsibility in preserving forests, developing countries stressed their sovereign right to use their natural resources as they saw fit. They feared that a treaty would put limitations on their timber exports or oblige them to engage in selective logging that makes harvesting more expensive and their exports less competitive. Countries with large forests, such as Malaysia and India in particular, opposed the proposal from the very beginning, viewing it as a method to raise trade barriers. China, Brazil, and India argued that because the problem is essentially local in nature, it should be subject to national policy and not international regulation.[82] The Brazilian minister of the environment, José Goldemberg, stated that Brazil saw no need for an international forest convention unless the uncertainty about the connection between forests and climate change was dispelled.[83]

The Rio conference generated neither a legal instrument on forests, nor new funds for forest policies. The only output was a set of nonbinding "Forest Principles." At the last minute, participating states adopted an authoritative statement of principles on the management, conservation, and sustainable development of all types of forests.[84] The document, which is not legally binding, omitted references to the global or regional benefits of forests, and previous references to the interests of the world community were replaced with "values to local communities" [Preamble (f)]. Its insignificance as a legal instrument is reflected in the fact that a very small group of states cared to engage in negotiating the text.[85]

Interim Initiatives

After Rio, parties to the ITTA considered expanding the scope of the treaty to include boreal and temperate forests. The United States and the EU firmly objected, stressing that they considered changes in the character of the treaty unacceptable.[86] Such a position was considered by developing countries as a double standard: they saw the North as pressing them to take costly action to protect tropical forests, while being unwilling to reciprocate with temperate and boreal forests. A number of other initiatives were launched in the 1990s, including the Helsinki Process, aimed at developing nonbinding guidelines for forest management, and the Montreal Process on criteria and indicators for sustainable forest management. In addition to multilateral initiatives, at least thirteen unilateral and bilateral state initiatives were launched in the mid-1990s on various aspects of forestry.[87]

In the meantime, the activities of nongovernmental groups were beginning to alter forest practices worldwide and to affect governments' preferences regarding forest policy. Environmental groups such as the Forest Stewardship Council, established in 1994, encouraged sustainable forest management by issuing ecolabels to certify forest products from sustainably managed forests. They embarked on a double-track campaign to lobby companies such as furniture manufacturers to buy only timber harvested from certified forests; and to encourage individual consumers to buy ecolabeled products from such companies. People responded and the companies saw the opportunity to attract environmentally minded consumers. The new market for certified forest products began to affect the dynamics of the international timber trade. When by 1998, Malaysian logging companies had lost half of their European market, Malaysia changed its position on a forest treaty and began to view it as an opportunity to counter the NGO offensive. Like Canada, they believed that international standards negotiated by governments would be less stringent than standards promoted by NGOs.[88]

The IPF and IFF Processes

During debates at the UN Commission on Sustainable Development in April 1995, countries advocated an international dialogue dedicated exclusively to forests. To this end, they established a two-year ad hoc forum for discussion called the IPF and asked several international institutions to form an Interagency Task Force on Forests (ITFF) to provide information and technical advice. The ITFF was chaired by the head of the forestry department of the FAO and included the World Bank, UNDP, ITTO, UNEP, and the Secretariat of the Convention on Biological Diversity.[89]

The IPF consisted of four global conferences that took place between 1995 and 1997 and produced proposals for cooperation in financial assistance and technology transfer, forest research, developing criteria and indicators for sustainable forest management, and trade. States could not agree, however, on major issues such as the amount of financial assistance for forest policies in developing countries or the need for a global forest convention.[90] The UN General Assembly decided to continue the policy dialogue and established an ad hoc IFF, widely perceived as a mere continuation of IPF.[91]

The most controversial issue by far was the nature of a future international policy on forest management. The major options under consideration were coordinating existing treaties, strengthening existing nonbinding agreements, establishing a permanent governmental forum on forests, and creating a legally binding global forest convention. At the IFF's second session in 1998 (IFF-2), Canada and Costa Rica announced a separate initiative to assess the need for a global forest convention. The Costa Rica–Canada Initiative (CRCI) consisted of two global and eight regional expert meetings that involved six hundred experts from more than forty countries and organizations.[92] These sessions facilitated a free exchange of views, but divisions remained, and the initiative failed to achieve consensus on the desirability of a treaty.[93]

The bargaining process at the IPF and IFF was characterized by stagnation.[94] Throughout the eight global rounds of talks, the arguments offered on each side remained unchanged. A large group of countries advocated a global forest treaty: Canada, Scandinavian countries, France, Switzerland, the Russian Federation, Malaysia, Turkey, South Africa, Senegal, and the Czech Republic, among others. In their official statements, they all stressed the importance of environmental protection and the need for stringent action to ensure sustainable forest management. The EU pointed to the need for and absence of a holistic international approach to forest management. Canada, Costa Rica, Argentina, and Gabon, among others, emphasized that existing instruments do not adequately address the problems confronting the world's forests and supported initiating negotiations on a legally binding instrument.[95] Switzerland underscored the regional differences in forest resources and policy needs, and suggested creating a global framework convention with detailed regional protocols to be negotiated separately.[96]

The motivation of treaty supporters was most likely mixed. There were speculations among delegates that the Russian Federation saw a treaty as a way to remove forest policies from the jurisdiction of local and regional governments and gather them into federal jurisdiction. Some developing countries, as well as Eastern European countries, hoped that a treaty would channel development assistance into their forestry departments. Canada and

Malaysia were most likely betting on the prospect of a weak treaty that would fend off environmental criticism, while giving a green light to their logging industries by legitimizing environmental standards lower than those demanded by NGOs.

On the other side of the fence, the United States and Brazil were leaders of an anti-treaty coalition that included Australia, New Zealand, Japan, the United Kingdom, Mexico, India, Indonesia, China, and most other developing countries (which negotiated jointly as a group: the G-77).[97] This camp advocated a nonbinding arrangement. Regardless of its particular content, a nonbinding agreement would not entail any concrete policy obligations. Brazil repeatedly upheld the nonbinding Forest Principles of Rio as a comprehensive and sufficient instrument.

Reasons given for opposing a treaty also varied. At IFF-2, New Zealand expressed concern with the costs of negotiating a convention.[98] Others defended their sovereignty and rejected the idea of forests as global public goods. Everton Vargas, principal negotiator for Brazil, curtly stated, "Forests are not global commons, they are national resources." A U.S. delegate remarked, "Forests are inherently local, they are not global commons. The net effects [of deforestation] are too disaggregated."[99] Notably, no delegate objected to such statements. Japan, a major importer of tropical wood, did not want to see limitations on timber trade. Apparently, Indonesia and Brazil viewed a regime as potentially threatening, rather than facilitating tropical timber trade. Why did Indonesia and Malaysia hold opposing views provided that all of them have vast forests and rely on timber exports? Did they interpret the situation differently? The reasons for these differences are an interesting subject for further study.

Other countries had no firm preferences. At IFF-2 and IFF-3, Australia stated that all ideas should be on the table and supported rigorous analysis of all options for an international policy arrangement.[100] In private conversations, Australian delegates shared that "there are no serious reasons to have a forest treaty" and that "if someone comes up with a really good reason to have a treaty, we would consider it."[101] Yet, few concrete arguments for a treaty were offered. Even the most ardent proponents of a treaty acknowledged that forest issues could be handled effectively through unilateral action.[102] A Canadian negotiator, for instance, reasoned, "For an issue to trigger an international (policy) response, there have to be global dimensions, there has to be a problem that is shared . . . common impacts . . . Forests don't affect everyone in the same sense."[103] By IFF-4, Australia had made up its mind and openly opposed treaty negotiations. So did Gabon, a country with large tracts of tropical forests that originally had been a treaty supporter.

The United States had made a u-turn since the early 1990s, reflecting congressional skepticism of multilateralism in international relations. When negotiating text documents, the U.S. delegation opposed references to the need for a holistic and comprehensive international agenda for action on forests. Together with New Zealand and Brazil, it rejected references to an international policy "agenda" and preferred to speak of the less committal "dialogue." At IFF-2, the United States even opposed a proposal for an international framework for monitoring forest conditions. In general, it backed away from international obligations that would limit its freedom and draw criticism or legal action for noncompliance. Besides, any specific policy agreement would entail costs for implementation that are usually covered primarily by industrialized countries according to the scale of contributions to the regular UN budget. That would require the United States, Japan, Australia, and New Zealand to cover much of the costs for implementing a forest treaty.

Many countries, including South Korea, Peru, Colombia, Nigeria, and Gabon argued that it was premature to discuss a treaty given the lack of consensus on financial resources for policy implementation. At IFF-2 and IFF-3, China supported the establishment of an international arrangement, but stressed the need to include financial mechanisms for its implementation. At IFF-4, African countries such as Zambia and Nigeria indicated that they would consider joining an international forest treaty only if a global forest fund were created. They did not want to enter obligations without the funds to fulfill them. On the last day of IFF-4, the G-77 openly stated "we do not have a principal objection to a treaty if the money is provided to implement it."[104]

The industrialized countries, with the exception of Canada, were adamant that additional financial resources were not justified. At IFF-3, the EU opposed references to financing forest conservation in the decision text of the meeting. Even strong supporters of a treaty, such as Scandinavian countries, curtly stated that there was money already provided on a bilateral and multilateral basis and argued that this money was not being used properly and efficiently by recipient countries. Canada attempted to use the issue of an international fund as a trump card, suggesting throughout IFF sessions that discussions of a fund are contingent on the decision to create a treaty. It was obvious, however, that the vast majority of Western countries refused to pledge new financial resources for international forest policy.

One of the most remarkable aspects of the story is the position of environmental NGOs. Most NGOs present at the negotiations bitterly opposed an international convention. At UNCED in 1992 they were enthusiastic supporters of a convention, but their position had begun to evolve by the mid-1990s. They had gained confidence in their own ability to affect the behavior of governments and consumers of forest products. Witnessing the deep disagreements among governments and their refusal to commit new funds for forest policy, the NGOs believed that formal negotiations would take many years and that any resulting treaty would be a weak one. In formal statements at IFF, coalitions such as the Global Forestry Action Project, emphatically called for reliance on existing binding and nonbinding initiatives and stressed that negotiations of a treaty would set back forest policies by suspending current progress through existing arrangements.[105] During coffee breaks, they actively lobbied delegates with whom they were on a first-name basis after years of working together at similar meetings.

The denouement of the IPF/IFF process came at IFF's last session (IFF-4) in early February 2000. After long hours spent trying to reconcile opposing positions, consensus could not be reached, and the final decision amounted to rejecting the idea of a treaty on forests. The IFF decided on an international arrangement that would include two principal elements: a permanent UN Forum on Forests (UNFF) and a collaborative partnership among eight international organizations and secretariats of existing treaties to enhance coordination of forest-related activities. In order to appease the treaty supporters, the plenary agreed that after five years, UNFF would evaluate its own effectiveness and would "consider with a view to recommending the parameters of a mandate for developing a legal framework on all types of forests."[106] NGOs dubbed this part of the decision text the "Monty Python paragraph," and one delegate remarked, "In five years' time, a vast array of lawyers will spend large amounts of public money trying to interpret what the negotiators meant." The final text also indicated that the Forum discussed, but did not reach consensus on, whether to establish a global fund to facilitate the implementation of forest policies.

The United Nations Forum on Forests

The first session of this new nonbinding institution was held in New York on June 11–22, 2001.[107] Open debates and informal consultations were deadlocked over basic matters such as the content and structure of the UNFF's program of work. The differences that marred the meeting reflected the general stalemate in international forest policy making. The North and the South could not come any closer on the issue of financial resources for new forest policies. Industrialized countries maintained that additional funds are not justifiable, given funding already provided on bilateral and multilateral basis. Views also differed on the fundamental character of UNFF. Some saw it as a body for making policy at the international level, while many preferred a bottom-up approach that focused on national forest policies. On the last day, delegates still had no agreement on the overall purpose of UNFF's Plan of Action.

Seeking to bridge such differences, countries came down to the lowest common denominator, progressively stripping the international arrangement of substantive content. Arguments about the unique policy needs of each country, the importance of national priorities, and the difficulty of setting uniform targets were used to ensure that states are not bound by any decision of the UNFF. Developed countries succeeded in removing trade issues from the program of work. Thus, the North reserved its right to import timber products from developing countries while blaming the South for cutting trees. The United States walked a very thin line: they tried to portray the UNFF as a success in order to undermine arguments for a legal treaty; at the same time, they did their best to weaken its capacity to produce policy action. Throughout discussions, the U.S. delegation made numerous proposals to delete key paragraphs from draft texts, including references to financial provisions, targets and timetables, and concrete responsibility for monitoring and reporting.

The decision texts that emerged from the meeting are masterpieces of Machiavellian diplomacy. With meticulously chosen words, they contain all the right ideas but commit no one to do anything about them. On the national level, the adopted document allows countries to set their own priorities and not to report on progress if they do not want to. On the international level, the UNFF only "invites," but does not commit, organizations to mobilize resources and to address the needs of developing countries. In sum, the UNFF does not specify what policy targets countries should pursue, when to achieve them, how to finance them, or to whom and when to report the results.

When the decision to establish the UNFF was taken at IFF, no one seemed particularly enthusiastic about it. It was simply the only alternative to doing nothing at all. Few countries have an interest in an international institution with teeth, yet no government can afford the domestic political costs of not acting on a prominent and popular environmental issue such as deforestation. Some delegates at IFF-4 said in consternation that the sole reason for producing a final text with recommendations was to justify to their publics and governments the expensive three-year process of international deliberations. Thus, the UNFF promises to be the institutional excuse given by governments to their publics for not having an international forest policy.

KNOWLEDGE AND INTERPRETATION: WHAT DECISION MAKERS BELIEVE AND WANT

After years of research, intersessional meetings, and formal debate, there remains strong resistance to the extension of international forest policy to include a legal treaty. Governments agree to disagree on key questions of finance, technology transfer, trade, and the need for a legal convention. Since forest utilization is a complex cross-sectoral issue that affects a number of socioeconomic realms, it should not surprise anyone that the proposal for an international treaty on forest management encounters strong opposition. Yet, the configuration of state positions is somewhat baffling. The division between treaty supporters and treaty opponents is not based on either North-South lines or on types of forests: each camp includes rich and poor countries, some with tropical and others with boreal forests. Nor has the dividing line been between forested and nonforested countries. Who wants what and why?

Why are some states reluctant to take on international obligations? Are the political and socioeconomic costs of policy implementation too high? It seems that because the causes of deforestation are not the same everywhere, it is difficult to devise an equitable system of international obligations that satisfies the majority of states. Some states may hesitate to make commitments because they do not have the authority to control the use of their forests. What are the implications of scientific uncertainty about the transboundary consequences of deforestation? Who would benefit from an international system of state-to-state obligations? Did policy makers know what scientists knew? How did they use existing information?

Three caveats suggest caution in analyzing the role of scientific information in policy deliberations. First, the existence of scientific data published in books and journals does not necessarily guarantee that policy makers are aware of it. Some chief negotiators are foresters by profession, for example, a key member of the Norwegian delegations to several international meetings.[108] More often, delegations consist of government officials without scientific backgrounds. Forestry experts in national governments are critical to the process of transferring scientific knowledge to political actors since higher-level decision makers and government negotiators rarely read scientific reports themselves, but instead rely on their interpretation by experts in their own government agencies. Finland's decision makers and delegations, for instance, receive briefs by a large team of experts who follow scientific developments and review published assessments on national, regional, and global levels. These experts make summaries of reports, draw conclusions, and give input to the political decision makers in the Ministry of Foreign Affairs.[109]

Second, scientific input does not automatically shape policy decisions. Beyond the raw data and the conclusions of the scientists is the interpretation of those data by political actors. Given the same set of information, governments may interpret and use it differently. This interpretation may be an honest assessment of existing data and its implications. Or, it may be instrumental, using science to support prior prejudices. As science can be easily manipulated to serve the material interests of politically powerful actors, it is difficult to determine whether scientific knowledge is the genuine rationale for state preferences or whether it is used as justification for preferences that are motivated by other national concerns or personal agendas.

Third, in a more fundamental sense, particular knowledge does not directly lead to particular decisions because goals depend on values, and the interpretation of knowledge is mediated through these values. "Neither science nor logic can tell us where we ought to head."[110] Interests are formulated on the basis of the values that actors wish to maximize. Hence, even when there is consensus on scientific knowledge, there are differences in state preferences and negotiating positions. Scientific uncertainty creates additional opportunities for interpretation. It can be used to delay policy by arguing that policy should wait for a better information base. Alternately, it can be used to speed policy by arguing that we should take precautions even though we do not have complete knowledge of the problem.

All these caveats warrant the application of analytical rigor and methodological discipline in studying the interface between science and politics. Scientific knowledge does not exercise its role in a deterministic way. It does not dictate the particular choice of policy, but defines the range of policy options that are considered by political actors. Accepted scientific data sets the parameters within which political discussions take place.

Interviews with policy makers and national delegates in the forest negotiations reveal that their own perceptions closely reflect the state of existing scientific knowledge: they are generally aware of areas of scientific consensus and uncertainty. But government officials and negotiators tend to frame the forest problem in terms of their local and national benefits, and are dubious about or uninterested in transboundary consequences of forest degradation. All of the interviewees point out that the role of forests in global climate change and biodiversity are uncertain. They acknowledge a connection between forest change and biodiversity, but are not aware of any specific reports on biodiversity loss. Many of them question the link between forests and climate change. The UK's senior forest official for international policy commented, "Deforestation is seen as an international problem but in fact the links are not so easy to establish. . . . The connection between deforestation and carbon [emissions] still has to be confirmed and clarified."[111] A delegate from India stated, "Climate change is a transboundary issue but forests are not. What happens to one's forests will not affect other countries."[112] Even leaders of environmental NGOs openly state, "The impacts [of deforestation] are not global; there is no threat to health and human well-being. The consequences are local and national. Both the causes and the consequences are in the locale."[113]

Most remarkably, even political actors who advocate strong international policy openly recognize existing uncertainties. Canada ardently advocates an international forest treaty, but when asked about the scientific knowledge of the shared consequences of deforestation, one Canadian policy maker said, "Oh, it is all speculative. To my knowledge, there is no government that tries to ascertain [the shared consequences of deforestation]."[114] Another noted,

"The link between forests and climate change is somewhat weak, the knowledge is very much under construction."[115]

High-level officers in international forest institutions also acknowledge gaps of knowledge. The head of the secretariat of IFF commented, "There is a lot of mythology in [discussions of] biodiversity, a lot of fiction and few facts."[116] And the coordinator of NGO activity at forest negotiations remarked, "With forests, science is all conjecture. Both the causes and consequences of deforestation are debatable."[117]

Most participants in the international policy negotiations understand that scientific knowledge is incomplete. In its official final report, the IPF notes:

> Much attention is still given to timber and forest cover, whereas other goods and services provided by forests, such as fuelwood, the sustainable use and conservation and the fair and equitable sharing of benefits of biological diversity, soil and water protection functions, and carbon sequestration and sinks, as well as other social, cultural and economic aspects, are rarely covered and need to be considered.[118]

During discussions on forest research at sessions of the IFF, a number of delegates recognized the difficulty of evaluating nontimber goods and services such as biodiversity and recreation, and agreed on the need to strengthen research to inform policy. The group of developing countries, or G-77, and China called for research to develop valuation methodologies that take into account environmental, socioeconomic, ethical, and cultural considerations.[119] Suriname said that research priority should be on the functions and benefits of forests, while the EU preferred to focus on the causes of deforestation. New Zealand stressed the need to develop an approach to identify both the costs and benefits of sustainable forest management. All delegations concurred that such an approach is missing.[120]

Many appear unmoved by demands for an international policy. Opponents to a global forest treaty consistently argued that none was required because there were no global environmental effects of forest degradation. This reflects the prevailing foundation of international environmental law (built on concepts of sovereignty), which prohibits interference by a state with the environment of another state. Discussion of biodiversity and climate change to which forest degradation contribute were excluded as they were the subject of separate negotiations under the FCCC and the CBD. But why were the international causes of deforestation from international trade to industrialization and poverty not considered worthy of debate? Was it because they seemed to echo earlier developing-country demands for a new international economic order?

The story of the international debate on forest management raises many other important questions that may or may not have clear answers. Without serious scientific evidence of the transboundary consequences of deforestation, why was so much effort poured into international political initiatives? Could not these resources have been better used searching for sustainable forest policies that individual developing countries could adapt to their needs? Why was the UNFF created if it leaves everything for countries to do, lets them choose what they want to do, does not provide them with financial assistance to do it, and has no right to hold them accountable for the results of their (in)actions?

Most scholars analyze international environmental institutional formation and operation. But much can be learned from the failure to establish effective international environmental institutions. This chapter does not offer a single explanation for the failure of the international community to establish an effective international institution to preserve global forests. Its several insights into that failure have much to say about the uses of scientific information by political agents and its effect on the consequent international policy (and doing nothing is a policy choice).

CHRONOLOGY OF POLITICAL EVENTS IN INTERNATIONAL FOREST MANAGEMENT

1983 International Tropical Timber Agreement negotiated
1990 Industrialized countries propose negotiations on a global forest treaty
1992 Forest treaty negotiations lifted from the agenda of the Earth Summit
1992 Nonbinding "Forest Principles" adopted at the Earth Summit
1994 International Tropical Timber Agreement amended
1995 IPF created
1995 United States reverses its position and opposes a forest treaty
1997 IFF launched to replace IPF
1998 Malaysia reverts its position and supports a forest treaty
2000 IFF concludes with the decision not to pursue a global forest treaty
2001 A permanent UN Forum on Forests is created without a policy-making or implementation mandate
2001 A Collaborative Partnership on Forests is established among eight international organizations to coordinate relevant activities

NOTES

1. Jack Westoby, *Introduction to World Forestry: People and Their Trees* (Oxford, UK: Basil Blackwell, 1989): 165.
2. Westoby, *Introduction to World Forestry,* 77–81.

3. One sample of nongovernmental studies is Norman Myers, *Deforestation in Tropical Forests and Their Climatic Implications* (London: Friends of the Earth, 1989). See also reports of the FAO, as well as research conducted by the International Center for Forest Research (CIFOR) in Jakarta.

4. R. Päivinen and A. J. R. Gillespie, *Estimating Global Forest Change 1980–1990–2000* (Rome: FAO, 2000). This is a background document prepared for an international panel of experts convened to review methods for the Global Forest Resource Assessment 2000 by the FAO.

5. FAO, *Forest Resources Assessment 2000: Main Report* (Rome: FAO, 2001).

6. A. Nyyssönen and A. Ahti, eds., *Proceedings of FAO Expert Consultation on Global Forest Resources Assessment 2000*, research paper 620 (Helsinki: Finnish Forest Research Institute in cooperation with ECE and UNEP, 1997).

7. In other words, when observed from above, at least 10 percent of the land is covered by forest canopy. FAO, *Forest Resources Assessment 2000*, 363.

8. FAO, *Forest Resources Assessment 2000,* 1.

9. FAO, *Forest Resources Assessment 1990: Global Synthesis* (Rome: FAO, 1995): 7–8; FAO, *Forest Resources Assessment 2000*, 1. These figures pertain to net deforestation. Gross deforestation is higher, but is offset by natural or artificial reforestation.

10. FAO, *Forest Resources Assessment 2000,* 343.

11. FAO, *Forest Resources Assessment 2000,* 7.

12. Robert Repetto and Malcolm Gillis, eds., *Public Policies and the Misuse of Forest Resources* (Cambridge, UK: Cambridge University Press, 1988).

13. FAO, *State of the World's Forests* (Rome: FAO, 1995): 21–5.

14. FAO, *Forest Resources Assessment 2000,* 78. Note that existing data on commercial logging is not comparable across regions: data for one region is on harvest area and that for the other on harvest volume.

15. David Pimentel, Michael McNair, Louise Buck, Marcia Pimentel, and Jeremy Kamil, "The Value of Forests to World Food Security," *Human Ecology* 25(1) (March 1997): 91–120.

16. FAO, *Small-Scale Forest-Based Processing Enterprises* (Rome: FAO, 1987), cited in Pimentel et al., "The Value of Forests," 104.

17. FAO, *Forest Resources Assessment 2000,* 34.

18. FAO, *Tropical Forest Resources* (Rome: FAO, 1982): 64.

19. FAO, *Forest Resources Assessment 2000,* 81–7.

20. A. S. Mather, *Global Forest Resources* (London: Belhaven Press, 1990): 250–51.

21. Robert Repetto, "Overview," in Repetto and Gillis, eds., *Public Policies,* 15–37.

22. Eufresina L. Boado, "Incentive Policies and Forest Use in the Philippines," in Repetto and Gillis, eds., *Public Policies*, 165–203.

23. Local ownership is more common in British-speaking African countries. In French-speaking countries, forest law is based on Roman law, according to which any unoccupied land without ownership documents belongs to the state.

24. FAO, *Tropical Forest Resources,* 51–3.

25. FAO, *Forest Resources Assessment 2000,* 346.

26. FAO, *Forest Resources Assessment 1990: Global Synthesis.* FAO Forestry Paper 124 (Rome: FAO, 1995).

27. FAO, *Tropical Forest Resources.* An earlier study of deforestation in the tropics was Norman Myers, *Conversion of Tropical Moist Forests* (Washington, D.C.: National Academy of Science, 1980), but it did not offer a quantitative assessment.

28. The LANDSAT Pathfinder Project carries out complete mapping of the areas it studies, but it monitors only moist tropical forests. To learn more about it and access its database, you can visit http://geo.arc.nasa.gov/sge/landsat/landsat.html, accessed 4 February 2002. Information about the TREES project is available at www.gvm.sai.jrc.it/Forest/defaultForest.htm, accessed 4 February 2002.

29. Abundant information about the FAO and electronic versions of the reports of its assessments can be found at www.fao.org/forestry/Forestry.asp, accessed 4 February 2002.

30. FAO, *Forest Resources Assessment 1990,* 7–8.

31. FAO, *State of the World's Forests,* 30.

32. The distinction between volume and area can be rather significant. While the area harvested in Asia was only 89 percent of that harvested in Latin America between 1986 and 1990, the volume of logs produced annually in Asia was 3.7 times that produced in Latin America. FAO, *State of the World's Forests,* 30, 52–4.

33. FAO, *Forest Resources Assessment 2000.*

34. FAO, *Forest Resources Assessment 2000,* 27.

35. FAO, *State of the World's Forests 1997* (Rome: FAO, 1997); FAO, *Forest Resources Assessment 1990.*

36. FAO, *Forest Resources Assessment 2000,* 11.

37. FAO, *State of the World's Forests 1995,* 22; also confirmed by FAO, *Forest Resources Assessment 2000.*

38. FAO, *State of the World's Forests 1995,* 21.

39. FAO, *Tropical Forest Resources,* 89.

40. Personal observation of an expert meeting of the Costa Rica–Canada Initiative, Ottawa, Canada, December 6–10, 1999.

41. FAO, *State of the World's Forests* 1995, 21–2.

42. One example is the Vietnam War, when 1.25 million hectares were spread with defoliants and more than 4 million hectares were damaged by shells. FAO, *Tropical Forest Resources,* 90.

43. World Bank, *The Forest Sector* (Washington, D.C.: World Bank, 1991): 31.

44. R. Rowe, N. P. Sharma, and J. Browder, "Deforestation: Problems, Causes, and Concerns," in *Managing the World's Forests: Looking for Balance Between Conservation and Development,* N. P. Sharma, ed. (Dubuque, Iowa: Kendall Hunt Publishing Co., 1992).

45. FAO, *Forest Resources Assessment 2000,* 6. The data on causes is consistent over time; similar estimates were given in FAO, *Survey of Tropical Forest Cover* (Rome: FAO, 1996): 58–60; and FAO, *State of the World's Forests 1997,* 20.

46. FAO, *State of the World's Forests 1997,* 21–22.

47. FAO, *Forest Resources Assessment 2000.*

48. See for instance, FAO's general discussions on nonwood benefits in FAO, *State of the World's Forests 1995,* 18–21.

49. FAO, *Forest Resources Assessment 1990,* 25.

50. Pimentel et al., "The Value of Forests," 106.

51. FAO, *Tropical Forest Resources,* 64.

52. Sten Nilsson, "Do We Have Enough Forests?" IUFRO Occasional Paper No. 5, International Union of Forest Research Organizations, February 1996.

53. FAO, *Non-Wood Forest Products,* Report of the International Expert Consultation on Non-Wood Forest Products (Rome: FAO, 1995).

54. FAO, *Forest Resources Assessment 2000,* 95, 81.

55. FAO, *Forest Resources Assessment 2000,* 94.

56. UNECE/FAO, *Forest Resources of Europe, CIS, North America, Australia, Japan and New Zealand—Main Report* (Geneva: United Nations Economic Commissions for Europe, 2000), ECE/TIM/SP/17. Canada submitted that production of maple syrup was valued at US$44.9 million in 1993 and weighed 15.3 million liters in 1995, but could not provide both the quantity and the value for either year. Table 10–8 in FAO, *Forest Resources Assessment 2000,* 95.

57. Pimentel et al., "The Value of Forests," 106–7.

58. FAO, *Forest Resources Assessment: Tropical Countries* (Rome: FAO, 1993): 54.

59. FAO, *State of the World's Forests, 1997,* 41.

60. FAO, *Forest Resources Assessment 2000,* 47–8.

61. Evelyn Trines, Science and Technology Program, Secretariat of the Framework Convention on Climate Change. Interviewed 10 December 1999.

62. Evelyn Trines, interview.

63. FAO, *Forest Resources Assessment 1990.*

64. R. K. Dixon, S. Brown, R. A. Houghton, A. M. Solomon, M. C. Trexler, and J. Wisniewski, "Carbon Pools and Flux of Global Forest Ecosystems," *Science* 263 (January 1994): 185–90.

65. Intergovernmental Panel on Climate Change, *Climate Change 1995* (Cambridge, UK: Cambridge University Press, 1996).

66. Evelyn Trines, interview.

67. Nilsson, "Do We Have Enough Forests?" 34.

68. N. E. Koch and M. Linddal, "Commentary on the Results of the Assessment," in *Benefit and Functions of the Forest,* vol. 2 of *The Forest Resources of the Temperate Zones: The UN-ECE/FAO 1990 Forest Resources Assessment* (New York: United Nations, 1993): 226–48.

69. FAO, *Forest Resources Assessment 1990,* 41.

70. FAO Committee on Forestry, *The Global Forest Resources Assessment 2000: Summary Report.* COFO-2001/INF.5: 10–11, 21.

71. Nilsson, "Do We Have Enough Forests?" 55.

72. Westoby, *Introduction to World Forestry,* 167.

73. Report on the first session of IFF.E/CN.17/IFF/1998/L.1.

74. ITTO, *ITTO: Ten Years of Progress* (Yokohama, Japan: ITTO, 1996); and Fred P. Gale, *The Tropical Timber Trade Regime* (New York: St. Martin's Press, 1998). The text of the agreement can be found in a Tufts University database, at http://fletcher.tufts.edu/multi/texts/BH837.txt, accessed 10 January 2000.

75. David VanderZwaag and Douglas MacKinlay, "Towards a Global Forests

Convention: Getting out of the Woods and Barking up the Right Tree," in *Global Forests and International Environmental Law*, Canadian Council of International Law (London: Kluwer, 1996): 1–39, at 12.

76. VanderZwaag and MacKinlay, "Towards a Global Forests Convention," 16.

77. Rob Rawson, member of the International Tropical Timber Council and assistant secretary of agriculture, fisheries and forestry, Australian Forest Industries Branch. Interviewed 5 December 1999.

78. Richard G. Tarasofsky, *The International Forests Regime. Legal and Policy Issues* (Gland, Switzerland: IUCN and WFF, 1995): 14.

79. FAO, *State of the World's Forests 1997,* 104.

80. O. Ullsten, S. M. Nor, and M. Yudelman, *Tropical Forestry Action Plan: Report of the Independent Review,* Kuala Lumpur, May 1990, 46–7.

81. Ans Kolk, *Forests in International Environmental Politics: International Organizations, NGOs and the Brazilian Amazon* (Utrecht: International Books, 1996): 145.

82. A. J. Grayson and W. B. Maynard, *The World's Forests—Rio + 5: International Initiatives Towards Sustainable Management* (Oxford, UK: Commonwealth Forestry Association, 1997): 27.

83. Kolk, *Forests in International Environmental Politics,* 156.

84. The full text of the forest principles can be found in Appendix A of the Canadian Council of International Law, *Global Forests and International Environmental Law* (London: Kluwer Law International, 1996): 353–9.

85. Kolk, *Forests in International Environmental Politics,* 158–9.

86. Kolk, *Forests in International Environmental Politics,* 161.

87. Grayson and Maynard, *The World's Forests*, 29–46, 56–68.

88. Gareth Porter, Janet Welsh Brown, and Pamela Chasek, *Global Environmental Politics,* 3rd ed. (Boulder, Colo.: Westview Press): 206–7.

89. Grayson and Maynard, *The World's Forests*, 28.

90. The final decisions text is available at www.un.org/documents/ecosoc/cn17/ipf/1997/ecn17ipf1997–12.htm, accessed 30 June 2001.

91. Resolution 1997/65 of ECOSOC.

92. Final report of CRCI, on file with the author.

93. Information on the final CRCI meeting and on IFF-4 is based on direct observation; the author was a rapporteur for the Canadian government and for the *Earth Negotiation Bulletin.*

94. A complete account of forest-related international negotiations, including statements by delegations, is provided by the *Earth Negotiation Bulletin* on their Web site at www.iisd.ca/linkages/forestry, accessed 5 February 2002. One would be hard pressed to find a more useful resource on current and recent environmental negotiations.

95. Statements and interventions by these delegations at all IFF sessions were seemingly endless reiterations of this line of argument.

96. Statements of Switzerland at IFF-4. Daily coverage of the meeting can be found at www.iisd.ca/linkages/forestry/iff4, accessed 4 February 2002.

97. Approximately 120 countries are members of the G-77. At UN meetings, they usually make joint statements and negotiate as a coalition.

98. Daily coverage of IFF-2 is available at www.iisd.ca/linkages/forestry/iff2.html, accessed 4 February 2002.

99. Jan MacAlpine, senior forest officer, U.S. State Department, Office of Ecology and Terrestrial Conservation. Interviewed 11 December 1999.

100. Daily coverage of IFF-3 can be found at www.iisd.ca/linkages/forestry/iff3, accessed 25 January 2002.

101. Rob Rawson, interview.

102. Mike Fullerton, Canadian Forest Service, Division for International Affairs. Interviewed 9 December 1999.

103. Michael Fullerton, interview; also confirmed by Richard Ballhorn, chief negotiator for Canada at IFF-4, interviewed 7 December 1999.

104. Statement of Nigeria, official spokesman of the G-77 during IFF-4, 11 February 2000.

105. IFF and similar UN meetings are open to NGOs, which are allowed to make formal statements and to attend negotiations in working groups, but not to participate in them.

106. Final report of IFF to the UN Commission on Sustainable Development (E/CN.17/IFF/2000), available at www.un.org/esa/sustdev/ecn17iff2000-sprep.htm, accessed 2 June 2001.

107. The account on UNFF-1 is based on the author's observation of the session.

108. Knut Øistad, deputy director general, Ministry of Agriculture, Norway.

109. Birgitta Stenius-Mladenov, director of the Political Department, Ministry for Foreign Affairs, Finland. Interviewed 7 December 1999.

110. Ernst Haas, "Is There a Hole in the Whole? Knowledge, Technology, Interdependence, and the Construction of International Regimes," *International Organization* 29 (1975): 827–76.

111. Mike Dudley, head of International Division, United Kingdom Forestry Commission. Interviewed 10 December 1999.

112. C. P. Oberai, inspector general of forests, Ministry of Environment and Forests, India. Interviewed 10 December 1999.

113. William Mankin, director of Global Forestry Action Project, a collaborative initiative of the Sierra Club, Friends of the Earth, and the Natural Wildlife Federation (interviewed December 9 1999); and Stuart Wilson, Forests Monitor, interviewed 2 February 2000.

114. Mike Fullerton, Canadian Forest Service, Division for International Affairs. Interviewed 9 December 1999.

115. Ralph Goodale, Minister of Natural Resources, Canada, interviewed 10 December 1999. Decision makers from Finland, France, the United States, and the UK had similar views.

116. Jag Maini, head of the secretariat of the IFF. Interviewed 9 December 1999.

117. Bill Mankin, interview.

118. Report of the IPF to the UN Commission on Sustainable Development, E/CN.17/1997/12, paragraph 80.

119. Joint statement at the second session of the IFF, in Geneva, Switzerland, from 24 August–4 September 1998.

120. Statement by New Zealand at IFF-3, in Geneva, Switzerland, from 3–14 May 1999.

IV

SCIENCE, IDEAS, AND CULTURE

It is commonplace to assume that ideas and culture (a bundle of ideas) affect politics, but do they also influence science? The four case studies in this part indicate that they do, and in more ways and more extensively than epistemic communities theory suggests.

Chapter 8 compares the science-politics interplay and its effect on policy in three regions: Europe, North America, and East Asia. Wilkening shows that scientific knowledge about acid deposition must be localized to account for local differences in emissions, dispersion, and the characteristics of affected ecosystems. However, the comparison of the three regions also investigates the relative influence of localized science and political or cultural factors.

This case uses three hypotheses to structure and guide the regional comparisons, but it also generates many questions. What factors determine the relative importance of local science and the political-cultural context? Does the importance of context reflect individual choices of leading policy makers or a dominant norm shared among policy makers? Were the European countries that have a history of cooperating (especially those that were members of the EC/EU) more accepting of an international solution than the countries of North America or Asia? If culture is an important factor, could other ideational factors also be influential? For example, did the personal values or political calculations of leaders play a role? For example, why did the first President Bush reverse the policy of his predecessor (in whose administration he was vice president). Was it because of some personal belief or the result of a calculation of political opportunity?

If cultural or other ideational factors are important, does that mean that science is irrelevant beyond outlining the nature of the problem? Or does science still have a role in the political debate, informing political positions and guiding policy?

Is the concept of critical loads a distortion of science or a necessary simplification to educate policy makers? What role did critical loads and the RAINS model play in European cooperation? Did it relieve policy makers of policy-making responsibility by substantially automating policy? And if it did, is this an appropriate role for science? Is it a useful shortcut of potentially endless political debate or a dereliction of duty by policy makers?

Chapter 9 follows the attempts of Australia and New Zealand to prevent Japan from implementing experimental fisheries research. Firestone and Polacheck exhaustively detail the scientific and legal conflicts over the management of the bluefin tuna fisheries of the Pacific. As it has in whaling, Japan has tried to use the experimentation provisions of regional and international agreements to increase its take of tuna, a fish highly valued by the Japanese. Australia and New Zealand initially opposed Japan over the design of the experiment and then tried legal means under the United Nations Convention on the Law of the Sea (UNCLOS).

Incomplete scientific understanding of the reproductive patterns and life cycle of the tuna could be used to justify additional "experimental" catches, but it also prevented agreement on the design of the experiment and obscured logical or scientific arguments against it. The scientific debate over Japan's experiment raises many questions. How can one assure the integrity of scientific experiments that potentially can improve the management of a common-pool natural resource, but that also may involve increased environmental consumption and substantial economic gain? Is any dispute ever only about science, as Japan claimed? Does scientific uncertainty mean that science should be discounted? Or should scientists interpret incomplete and contradictory data as best they can, and does such interpretation diminish science's authority or increase its usefulness? What standards should govern whether a proposed scientific experiment should be undertaken: where should the burden of proof lie? Who should judge the appropriateness of such experiments and what role should international institutions play in resolving scientific disputes of shared international resources?

The legal issues surrounding fisheries experimentation and governance of the common-pool resource were no less complicated. One international tribunal under UNCLOS accepted Australia and New Zealand's contention that the issue was not wholly scientific, that the tribunal had jurisdiction, and that Japan's experiment potentially endangered the fishery, in contravention of international agreements. But an arbitral panel considering a final settlement of the dispute found that neither it nor the other tribunal had jurisdiction because the regional Southern Bluefin Tuna Agreement limited the application of UNCLOS. With the proliferation of international environmental agreements, how does the interaction of global and regional international institutions affect the governance of common-pool resources? Is it possible to write rules that can anticipate every eventuality or should only general principles be used to distinguish right from wrong behaviors? Where the rules are vague or imprecise should science be considered a sufficient justification for states to interpret them to their benefit, and can it be used in this way?

There is little observable politics in this case. Politics played a role in some of the scientific arguments and probably guided Japan's push for an experimental catch program. Politics is likely to have played a role in some of the scientific and fisheries-management committees. But where and how does law fit on the science-politics continuum? At one level legal agreements are the embodiment of political choices. At another level the law, like politics, is a forum where actors go to get what they want. If one country can use provisions of legal instruments to their advantage in a way not foreseen by the drafters of that agreement, is that politics or just good legal maneuvering?

Disputed science is common in environmental issues; how do policy makers and the public choose whom to believe? That is the issue considered in Chapter 10. Carolan and Bell bring the international environment home to small town America by asking the question: whom do we believe when respected scientists contradict each other over complex science? This case shows that social relations play a role, taking over where rational judgment of empirical evidence becomes impossible. What are decision makers to do when, as is increasingly the case, "science" contradicts itself? What are they to do when group X has an "expert," who, using scientific evidence, states global warming is human induced, and group Y has an "expert," who uses similar scientific evidence to refute that claim?

When a distinguished scientist, considered one of the authorities on dioxin, proclaimed a small-town trash incinerator to be a major source of airborne dioxin pollution, dispassionate scientific analysis was not key in showing the errors in Commoner's work. Powerful social groups used other knowledge to persuade decision makers that Commoner was wrong. Does this mean that social action always dominates science or only when there is more unity in society than among scientists? Or does it mean that science is an inherently social activity in which blocks and groups may influence the interpretation of data, much as political choices are selected? Or is the reality somewhere in between? Social relations among funding-agency directors and researchers and authors and editors (and so on) marginally influence what knowledge is researched, published, and accepted, but the rules of blind peer review and replication still dominate.

The dioxin issue revolved around uncertainty in the data and the environmental and health risk of the pollution. What is the connection between risk and uncertainty? Are they positively correlated so that risk rises with uncertainty? Is risk higher because of an absence of knowledge or does fear only make it seem that way? As knowledge increases and unreasoning fear subsides, risk reduces as it is measured. As uncertainty is always present, certainty of knowledge (which is not equivalent to belief) must be absent. Does that mean that risk also is always present? If risks are always present in some form, how much new risk is acceptable? And is the determination of a "safe" level of risk a scientific or political question? Or can risk be high even with full knowledge of cause-effect relations? Knowledge of the causes of cancer does not reduce the risk that any individual may develop the disease. With new knowledge, technologies to reduce risk are possible, but the increased knowledge does not itself reduce risk.

This case is interesting for another reason: it was about the causes of pollution, but none of the contending social groups wanted the victims to be included. How much dioxin pollution is too much and who makes this decision—only the residents of Ames, Iowa? What about those in surrounding communities, or surrounding states, or neighboring countries? Here an apparently tacit exclusion of the potential victims was accepted by all sides. This simplified the issue and kept it local, but did it bias the interpretation of the technical issues and distort the political problem? Was this a small example of first-world arrogance?

Chapter 11 compares the international management regimes for two large bodies of water and their related terrestrial ecosystems: the North American Great Lakes and the Baltic Sea.

In principal, a holistic approach to ecosystem planning focuses on the scale of resource interactions and ignores politically contrived administrative boundaries. The goal is broad stakeholder participation in an integrated management approach without regard to political boundaries. The benefit of participation and integrated management is somewhat offset in practice by the potential for enlargement of the range of stakeholder interests and, therefore, of the number of disagreements. In the comparison between the two great water systems, the different outcomes may be explained by different conceptions of holism, by different structures and rules for implementing comparable concepts, or by differences in the implementation of comparable policies.

In both regions the progressive movement to a holistic view was a response to scientific evidence of ecosystem damage and interaction among ecosystems. But is it possible for science to be truly holistic and can environmental science meet political demands for holism? Or is it doomed to reductionism and issue-specific interdisciplinary research? Why did different ecosystem-management principles and practices develop in two scientifically comparable cases? Were the visions of ecosystem-management principles and practices different, or were they just implemented differently? Did the difference in ecosystem-management practices in the two regions lead to different outcomes for ecosystems and social systems?

Despite the spatial and physical differences between the two regional water systems, they enjoyed comparable governance structures, but with very different effects. In both, efforts were concentrated on hot spots of the greatest damage, but with more success in the Baltic region.

Has holism been more effective in Europe because of a cultural or historical acceptance of cooperation that is missing in North America? Or was the U.S. failure to perform appropriately the result of political structures? In the early years of the twentieth century, the U.S. government was an effective proponent of cooperation on Great Lakes water quality. As cooperation with Canada drew in ever more sources of pollution, did the multiple processes in federal, state, and local governments impede effective decision making, or was it only a problem of communication among jurisdictions? Is true holism in politics possible across borders and boundaries, and can environmental policy appropriately address challenges created by politically fragmented spaces?

8

Localizing Universal Science

Acid Rain Science and Policy in Europe, North America, and East Asia

Kenneth E. Wilkening

Europe, North America, and East Asia are the world's three regional-scale acid rain (or acid deposition, as it will be referred to here) hot spots (see figure 8.1); all three can be considered geographically and politically independent problems.[1] Coupled with the fact that acid deposition has a long history as an international issue—over thirty-five years in Europe, over thirty years in North America, and over fifteen years in East Asia—this offers an ideal opportunity to use comparative analysis to search for underlying patterns in the science-policy relationship and to examine hypotheses regarding that relationship.

One underlying pattern that stands out is the importance of localizing science to a region; in other words, rendering the general elements of acid deposition science applicable to a given region. Thus, we can consider two dimensions of scientific knowledge related to environmental problems, each with distinctly different political implications. There is a universal or generic (space/time-independent) dimension and a local or region-specific (space/time-dependent) dimension. Universal scientific knowledge, once constructed, diffuses from its place of origin and is applied by scientists to other regions, thereby creating region-specific knowledge. Thus, the universal becomes localized. The universal dimension is the vehicle for transmitting general environmental understandings; the local dimension is the place- and time-specific manifestation of the universal.

In the relationship between science and politics, it is overwhelmingly the

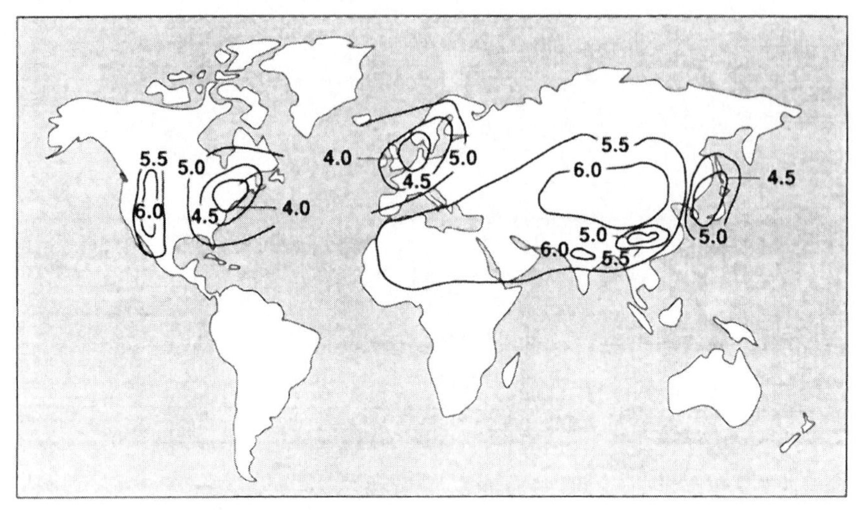

Figure 8.1 Acid Rain Hot Spots

Source: Neeloo Bhatti, David G. Streets, and Wesley K. Foell, *Environmental Management* 16(4), 1992. Used by permission.

localized dimension, not the universal, that interacts with the political circumstances of a region in defining an environmental issue and suggesting solutions. Recognizing the importance of the localization process allows us to frame and examine hypotheses related to the influence of science on the regional international policy process. I introduce three hypotheses, each addressing a different regional and science-related factor relevant to the international environmental-policy process to guide analysis of the acid deposition issue in the three regions. These are not the only hypotheses that can be formulated, but they highlight key elements of the science-politics interface.

- *Hypothesis 1: Variations in regional scientific knowledge explain differences in regional environmental policy.* If science influences politics, then it is reasonable to expect that regional knowledge (for example, differences in the sensitivity of regional ecosystems to acidic inputs) may explain divergence between regional environmental policies (see box 8.1). But if differences in scientific knowledge do not account for differences in regional policy, we must seek other factors, such as the differing ways in which political actors use local science. This generates a second hypothesis.
- *Hypothesis 2: Variations in how political actors use regional scientific knowledge explain differences in regional environmental policy.* Once

Box 8.1 The Effect of Differing Landscapes

East Asia's landscape possesses unique characteristics that differ from those of Europe and North America. In the northern parts of both Europe and North America, glaciers during the last ice age scraped the land, leaving behind thousands of lakes and a thin soil cover on top of the underlying granite bedrock. The lakes and soils of these areas are highly susceptible to acidification. "Lake cultures," too, factor into acid deposition politics. Places such as southern Sweden and Norway, southern Canada, and the upper Midwest in the United States can be called areas with a lake culture because of the sheer number of glacial lakes and the role these lakes have played in the histories of the peoples who lived there. Damage to these lakes from acidification draws rapid public attention. By contrast, glaciation played a relatively minor role in shaping the East Asian landscape. East Asia is not a land of lakes. However, it is a land of massive dust storms that loft huge quantities of alkaline particles into the atmosphere that neutralize acidic substances; it is a land of volcanoes whose acidic emissions rival anthropogenic sources; it is land with a temperate monsoon climate with winds that dramatically shift directions between summer and winter with expanses of ocean between countries. East Asia's vastly differing landscape may or may not make portions of it more prone to acidification. Resolving this unknown reinforces the necessity of creating local scientific knowledge.

constructed, scientific knowledge can be used in many ways by different political actors, such as politicians, bureaucrats, business groups, NGOs, mass media, and scientists themselves. What causes variations in the use of science? Two possible explanations are (1) there may be structural differences in political relations (within or between states) that influence the use of scientific knowledge in policy making, and (2) there may be differences in how scientific knowledge is understood and interpreted. This may occur as a result of the structural relations or may be the result of differences in the bundle of ideas that we call culture. This suggests a third hypothesis.

- *Hypothesis 3: Variations in the political and cultural contexts in which scientific knowledge is situated explain differences in regional environmental policy.* Larger circumstances or conditions, such as culture or the history of regional political relations or extraregional events, may influence policy and must be accounted for in explaining policy differences.

These three hypotheses are used to analyze the development of regional-scale, acid deposition–related international policies in Europe, North America, and East Asia. They and related questions are devices to distinguish the specific confluence of factors that influenced the development of policy in each region. Do variations in regional scientific knowledge alone account for differences in regional environmental policy? Do variations in how political actors use regional scientific knowledge account for differences in regional environmental policy? Do variations in the political and cultural contexts in which scientific knowledge is situated account for differences in regional environmental policy?

THE ACID DEPOSITION PROBLEM

Acid deposition is one of the oldest and most thoroughly studied international air-pollution problems. Decades of research have resulted in the creation of a relatively complete, yet evolving, universal or standard scientific model of the causes and effects of acid deposition phenomena. The three basic components of this model are (1) emission of acidic pollutants into the atmosphere, of which the major contributors are SO_2 and nitrogen oxides (NO_x), and their oxidized forms, sulfate (SO_4) and nitrate (NO_3), related primarily to energy production, in particular coal-fired power plants and petroleum-powered vehicles; (2) atmospheric transport, transformation, and deposition of emitted pollutants and their by-products; and (3) adverse impacts on humans and ecosystems once deposited (the most prominent of which are damage to freshwater ecosystems, forests, cultural artifacts, and human health).

The standard model contains universally applicable biogeochemical knowledge related to the acid deposition problem (i.e., related to emission, transport, deposition, and the impacts of acidic pollutants). However, to identify a specific problem in a given region, universal knowledge must be localized to the region. Thus, regional scientific pictures must be created. The chemistry governing the formation of acidic substances is the same everywhere, but the mix of emissions sources is different everywhere. The physics of atmospheric wind transport and turbulence is the same everywhere, but climatic patterns and air-mass trajectories are different everywhere. The general factors influencing soil formation are the same everywhere, but the specific factors that form a soil in a given location are different everywhere. The physiology of fish respiration may be the same everywhere, but fish species are different everywhere. In sum, the universal knowledge of the standard acid deposition model provides the template for localizing knowledge. Locally adapted knowledge details the unique character of an acid deposition problem in a region and stands as potential input into the region's decision-making processes.

Generally speaking, localizing acid deposition knowledge to a region involves answering four fundamental questions: (1) Does a regional-scale problem exist? (2) What are the sources of pollution emission (locations, characteristics, temporal trends, etc.)? (3) What are the source-receptor relationships (what sources deliver what quantities of pollutants to a given receiving location)? (4) What are the adverse impacts (type, extent, seriousness, etc.)? In order to answer these questions, scientists from multiple disciplines must ferret out and stitch together a comprehensive set of cause-and-effect relationships for the region. To do so, they generally engage in four basic research activities: (1) establishment and maintenance of monitoring networks, (2) development of emission inventories, (3) construction of atmospheric transport computer models, and (4) documentation and quantification of adverse impacts. Engaging in these activities in a region is an indispensable part of localizing acid deposition knowledge.

When these research activities and their influence on regional policy making are considered historically, stages of science and politics, separated by prominent events, are identifiable in each region (see table 8.1). The birth stage corresponds to the period leading up to the scientific acceptance (discovery) of a regional-scale acid deposition problem. The regime-formation stage corresponds to the period from discovery of the problem to formation of a transboundary air-pollution regime (defined here as the signing of an international agreement).[2] In the last stage, the regime is implemented and may be successively transformed. Europe and North America are currently in a regime-transformation process in which the very definition of the international air-quality problem under consideration is altering significantly, in both cases from a singular focus on acidic pollutants to a wider concern for multiple air pollutants and multiple effects. East Asia remains in the regime-formation stage.

We turn now to the localization process and the policy trajectories in each region. For simplicity of organization, developments in science and politics are separated for each of the three regions. The interaction of science and

Table 8.1 Chronology of Acid Deposition Histories of Europe, North America, and East Asia

Stage	*Europe*	*North America*	*East Asia*
Birth	1967	1972	1985
Regime formation	1967–1979	1972–1991	1985–
Regime implementation (and transformation)	1979–	1991–	—

politics and their influence on policy is illuminated by considering the relative influence of regional scientific knowledge, how political actors use this knowledge, and the larger political and cultural contexts in which the knowledge is situated (the three hypotheses presented above).

ACID DEPOSITION IN EUROPE

Science

The foundation for acid deposition science was laid primarily in Europe between 1850 and 1967. Since acid deposition is principally a by-product of industrialization, it is not surprising that the emission, transport, and impact of acidic substances was initially recognized in the birthplace of the industrial revolution: England. Robert Angus Smith first created a crude model of the acid deposition problem and, in 1872, first used the term "acid rain."[3] In the late nineteenth century and during the decades leading up to World War II, scientific knowledge on acidification grew slowly as a result of diverse research.[4] However, scientists did not recognize the existence of a clear-cut acid deposition problem.

After World War II the locus of acid deposition science, unnoticed at the time, came to be centered in Scandinavia. In 1948 Hans Egnér established the world's first large-scale precipitation chemistry network in Sweden. It was eventually expanded and transformed into the European Atmospheric Chemistry Network. It allowed for, among other things, the analysis of regional-scale changes in precipitation pH. Meanwhile, other disparate research in the 1950s pointed to acidification problems. After almost a century of random and sporadic accumulation of scientific evidence, in 1967 Svante Odén, a Swedish soil scientist, locked the many pieces of the environmental puzzle together and created the first integrated scientific picture of a long-range, transboundary, multiple-impact acid deposition problem, which he published as a brief government report.[5]

Odén immediately followed this by authoring a provocative article in the October 24, 1967, edition of the Swedish newspaper *Dagens Nyheter,* in which he graphically described an insidious "chemical war" in Europe, a war being conducted with air pollutants. Odén outlined in remarkably accurate detail the mechanisms and impacts of acid deposition, a portrait that has basically stood the test of time. He synthesized a wide range of ideas and data into the following key conclusions:

- Acidified precipitation due to industrial activities was a regional-scale phenomenon in Europe.

- Acidified precipitation in Scandinavia was attributable to long-range transport of sulfur and nitrogen compounds emitted in England, Germany, and central Europe.
- Precipitation and surface waters in Scandinavia were becoming increasingly acidified over time.
- Acid inputs to soils acidified the soils, displacing nutrient cations, reducing biological nitrogen fixation, and releasing heavy metals (especially, mercury) into surface waters.
- Declining fish populations, decreasing forest growth, increasing plant diseases, and accelerated degradation of materials were all in part due to acidified precipitation, and these trends would intensify over time if no measures were taken to control emissions.

In effect, Odén created both the first standard model, or universally applicable scientific knowledge, describing the acid deposition problem and its initial localization to Europe.

A brilliant and unconventional scientist, Odén in one fell swoop both identified a scientific problem and almost single-handedly made it an international political issue. He not only explained previous scientific mysteries (such as the decline in freshwater fish populations in Norway), but also made a series of predictions about future environmental consequences (such as the release of toxic metals into surface waters), some of which were quickly verified. Odén and his scientific supporters postulated that virtually all of Scandinavia's ecosystems were vulnerable to acid deposition, that major commercial interests (especially fishing and forestry) were at risk, and that Scandinavians' cultural heritage (for example, their love of lakes, recreational fishing, and mountain hiking) was under attack. This ominous message stirred serious concern among the mass media, policy makers, and the general publics in Sweden and Norway. To some, national ecological survival seemed at stake. To compound this, the further message was that the source of the problem was out of Scandinavian domestic control. Scandinavia was upwind of the major, extraterritorial pollution sources causing the problem. The scientific picture painted by Odén and his colleagues was sufficient not only to garner support for further research in Scandinavia, but also to ignite a furious debate over acid rain in Europe.

Odén's work spurred the formation of a community of scientists dedicated to studying the acid deposition problem. Sweden's first major research effort by the Swedish Ministry of Agriculture was headed by Bert Bolin of the University of Stockholm. Experts in meteorology, ecology, engineering, soil science, economics, and planning contributed to synthesizing, summarizing, and interpreting what was known about acidification phenomena. This was the basis for a scientific report that Sweden presented at the 1972

UN Conference on the Human Environment, alternatively known as the Stockholm Conference.[6]

Since Odén's synthesis was in part based on previous Norwegian research, Norwegian policy makers did not require a lot of convincing to see the value of investing in further scientific research. In 1972 Norway begin its Acid Precipitation Effects on Forests and Fish Program (SNSF). This massive eight-year endeavor was, and, at the time of this writing, still is, the largest research program in Norway's history. More than 150 scientists from 12 Norwegian research organizations participated. They produced Norway's first national assessment of the problem.[7] This report, the Norwegian localization of the general model set forth by Odén, boiled down to the following key facts: the overwhelming majority of acidic pollutants falling on Norway originated outside its borders; thousands of lakes and rivers located on acidic granite bedrock were being acidified, and, as a result, fish populations and aquatic ecosystems were being destroyed; and some 80 percent of all productive forest land in Norway was dominated by acid-sensitive soils.

Results such as those of SNSF further galvanized Scandinavian scientists. They spearheaded the rapid evolution of acid deposition science. An activist core saw itself as the defender of Scandinavian ecological integrity and the champion of international environmental justice. This group pressed hard to gain scientific and political recognition of the seriousness of the problem. SNSF director, Lars Overrein, called the acid deposition problem Norway's "silent spring."[8]

The first international research program set up to test Scandinavian claims of long-range transport of air pollutants in Europe was an OECD four-year (1972–1976) study. The final report concluded that long-distance transport was indeed occurring and that the area of acidic deposition comprised the majority of northwestern Europe.[9] In 1977 the OECD project was superseded by the establishment of the European Monitoring and Evaluation Programme (EMEP), located in and funded primarily by Scandinavian governments. EMEP was concrete, official acknowledgment of the need to construct a truly Europewide scientific community and research program for the purpose of generating a common scientific understanding of the acid deposition problem (i.e., to collectively construct a European localized version of acid deposition scientific knowledge).

Science evolved rapidly in Europe during the 1980s. The role of scientists and scientific knowledge was almost as prominent as it had been in the 1970s. A scientific infrastructure was established under the 1979 Convention on Long-Range Transboundary Air Pollution (LRTAP; the political process leading to LRTAP is explained below) for creating a continually updated, high-quality, common, and collective scientific understanding of the acid deposition problem by an all-European scientific community. In other words, LRTAP institutionalized a process for creating a (localized) European acid deposition scientific knowledge base. As a by-product of this process, innovative research and new discoveries often added to the universal knowledge base.

The LRTAP executive body established a dense network of subsidiary entities to assist it in its technical (and political) duties—working groups, task forces, international cooperative programs (ICPs), and meetings of acidification research coordinators (MARCs).[10] The working groups and task forces were a mix of bureaucrats and scientists; bureaucrats dominated the working groups, and scientists the task forces, EMEP, and ICPs. These bodies constituted a dense cluster of points of contact between the science-making and policy-making processes by which policy makers were informed of the latest scientific knowledge and scientists were informed of changing political needs.

The most eye-catching scientific event of the 1980s, however, was the emergence of a new acid deposition–related impact—*Waldsterben*, or "forest death" (more properly understood as widespread forest decline).[11] Bernhard Ulrich of Göttingen University was the first to set forth a plausible cause-and-effect explanation for *Waldsterben* around 1980. He traced it to acid-induced changes in forest soils, specifically the leaching of plant nutrients and the release of toxic metals such as aluminum. By 1982 foresters estimated that the complex of symptoms associated with *Waldsterben* affected 8 percent of German forests; by 1983 the estimate had grown to 34 percent; and by 1984 it reached almost 50 percent. *Waldsterben* helped reconfigure the political landscape in Europe (as will be discussed below). Thus, to summarize the 1980s, scientists not only added evidence of a powerful new adverse impact—forest decline—to the European scientific picture, but also argued that virtually all of Europe was touched by acid deposition problems—aquatic impacts in Scandinavia, forest decline in northern and central Europe, and damage to important cultural properties in cities throughout Europe.

In the 1990s the role of science in European politics was still as vital as it was during the 1970s and 1980s. The knowledge-generating structure of the European acid deposition community was institutionally stable and stood at the center of the regime dynamic. Two major scientific developments ensured this role: the coming-of-age of the critical loads concept and the expanding scope of research on multiple air pollutants and multiple effects pointing to the need to move away from a focus solely on acid deposition.

As adopted by the executive body of LRTAP in 1988, a critical load is "a quantitative estimate of an exposure to one or more pollutants below which significantly harmful effects on specified sensitive elements of the environment do not occur according to present knowledge."[12] During the 1980s European scientists pioneered methods for determining the sensitivity of European ecosystems to deposition of sulfur and nitrogen. To aid policy making, beginning in the early 1980s the International Institute for Applied Systems Analysis (IIASA) in Austria developed RAINS, an integrated assessment computer model that applied the concept of critical loads.[13] The intent was to create a model that was codesigned by scientists and users, that was of modular construction, that was simple yet based on more detailed models or data, and that was user friendly. IIASA's effort was successful, and RAINS is now used in LRTAP negotiations.

The second scientific development of the 1990s was intensified research on multiple pollutants and multiple effects. As but one example, EMEP studies documented that nitrogen compounds, primarily nitrogen oxides and ammonia, were a significant contributor to eutrophication (a nonacidification effect) in the Baltic Sea.[14] Scientists estimated that almost one-third of total nitrogen inflow into the Baltic came from direct deposition of airborne nitrogen compounds. Other research during the 1990s addressed pollutants such as carbon monoxide, methane, and heavy metals, and effects such as eutrophication, elevated ground-level ozone, and climate change.[15]

Politics

Odén created an international environmental-policy-making "object"—acid rain. Swedish and Norwegian policy makers vigorously supported acid rain research to flesh out its implications and sought venues to discuss and publicize research results. Indeed, Scandinavia's "pollution victim" status relative to the problem was a major reason Sweden proposed convening the now-famous 1972 Stockholm Conference, the seminal event that firmly placed environmental issues on the international political agenda. Sweden, via its air-pollution report, used the conference as a vehicle to bring the European acid deposition issue to world attention. The report argued that the low rainfall pHs in Sweden and resulting negative impacts were due to transboundary pollutants; that 50 percent of the sulfur deposited on Sweden originated from foreign emission sources, mainly from central Europe and Great Britain; and that if then-current trends continued, roughly 50 percent of all Swedish lakes and rivers would become so acidified in 50 years time that they could no longer support fish. The report recommended increasing European research programs, engaging in international scientific and political cooperation, and pursuing a 50 percent reduction in emissions in the region around Scandinavia. Scandinavian policy makers enthusiastically stumped the report at the conference, but response was cool and skeptical. Scientific uncertainty, particularly concerning Europewide source-receptor relationships, was a primary reason behind the indifferent response.

After the Stockholm Conference, Scandinavian policy makers, increasingly motivated and backed by scientific evidence, labored mightily to get some kind of international discussions off the ground, but were frustrated at every turn. They gained grudging support, though, for expanded research as evidenced by the OECD study. By the mid-1970s, factors other than scientific evidence began swaying political momentum. One factor was the sudden appearance of an appropriate international political forum in which to discuss acid deposition.[16] By sheer good fortune, the 1975 Helsinki Conference on Security and Cooperation in Europe (CSCE), a forum for detente, provided a window of opportunity for the Scandinavians, and they seized it.

As a means of promoting detente, the final act of the CSCE called for cooperation on three major topics, one of which was the environment. The secretary general of the U.S.S.R., Leonid Brezhnev, then proposed holding an "all-European" (including the United States and Canada) conference on the environment. Western governments were cool to the idea. The proposal languished until a 1977 meeting of the United Nations Economic Commission for Europe (UNECE), when it was agreed to hold a ministerial meeting on the environment. Haggling over the specific agenda continued for months until late 1978 when the governments consented to address transboundary air pollution. Sweden and Norway were the prime movers pushing this as a major theme for negotiation. By this time, too, acid rain was a hot political issue (in part because a second regional-scale acid rain problem had been discovered in North America).

Subsequently, what started out as a means of easing Cold War tensions using the UNECE as a forum turned into a battle between northern "acid rain victim" countries (Sweden, Norway, and Canada) and southern "acid rain producer" countries (principally the United Kingdom, West Germany, and the United States). Ironically, the Communist nations became bystanders in this process. After intense negotiations, the acid rain victim states dropped their demand that binding regulation of sulfur emissions be included in an agreement (Sweden and Norway had pressed for a standstill clause—a halt in increases of SO_2 emissions—and a rollback clause—a decrease in emissions), and the acid rain producer states agreed to the need for a statement of intent and for further scientific collaboration. The result was the UNECE LRTAP convention, signed by thirty-three parties, including the United States and Canada, in Geneva in November 1979.

At the beginning of the 1980s, the prime resisters to political action on acid deposition in Europe were West Germany and the United Kingdom. These two nations essentially held veto power over any proposals by the Scandinavians. The discovery of *Waldsterben* and other scientific evidence was influential in changing their policies though. *Waldsterben* provided the spark for reshaping West German acid deposition policy. One of the areas in West Germany hit hard by *Waldsterben* was the Black Forest, an area of great importance in German history. The mass media spotlighted this and it captured the public's imagination. Meanwhile, during the 1970s, the influence of the Green Party had been growing and German industry had developed advanced pollution control systems. This confluence of circumstances (scientific, technological, and political) resulted in widespread support for a policy shift. Helmut Kohl completely switched Germany's position from anti-control to pro-control in 1982.[17] From this point on, Germany joined the Scandinavian nations as a leader in promoting stringent emission controls. The West German turn-about was like an ice dam cracking in a spring thaw and had major international political repercussions. For instance, it helped break the logjam on the adoption of the first pollutant-specific protocol to LRTAP, the Helsinki Sulfur Dioxide Protocol, signed in 1985.

By the middle to late 1980s, the UK was the biggest emitter of SO_2 in western Europe and the fourth largest in the world. Influenced by increasingly bleak reports of forest damage in Europe (including the UK), a rise of NGO activism and a drop in public approval for government policies, Margaret Thatcher announced in 1988, to the astonishment of her rivals, that the environment was "one of the great challenges of the twentieth century." The UK went from laggard status to, if not exactly leader status, becoming a more flexible partner in support of international acid deposition policy.

With this added momentum, by the end of the 1980s, an abundant harvest of control measures had been reaped. LRTAP was fully operational as an institution. In addition to protocols on EMEP financing (1984) and SO_2 (1985), the Sofia Nitrogen Oxides protocol was signed in 1988. As debate over control strategies intensified in the 1980s and early 1990s, both scientific and nonscientific factors vied for policy maker's attention. As already noted, scientific evidence, especially the discovery of *Waldsterben*, heightened concern over acid deposition. Activist scientists and others such as NGOs marketed this information to policy makers. With help from other changes in the political landscape (such as the demand for uniformity of regulations as momentum built for European unification), proponents of acid deposition emission controls succeeded in sufficiently overcoming resistance by segments of industry, certain politicians, many economists, and elements of the general public so as to maintain a strong forward momentum in implementing control measures.

The Oslo Sulfur Dioxide Protocol of 1994, an update of the earlier 1985 protocol, was the first to use the critical loads–based RAINS model. This symbolically marks the beginning of a period of radical transformation in the European acid deposition regime. It was a transformation spearheaded by scientists, but strongly backed by most policy makers. For years scientists had promoted the use of critical loads and the RAINS model because it steered negotiator's attention to the capacity of the environment to absorb anthropogenic acidic inputs. It made nature's ground rules (i.e., the ability of the environment to withstand acidic inputs), rather than political whim, the basis for decision making. This, and the flexibility allowed by the RAINS model, caused it to gain political acceptance. The model linked the impact of pollutant emissions to environmental sensitivity in Europe as the basis for calculating differentiated obligations in emission reductions. In other words, it calculated differential reductions where percentage reductions for each country are based on the amount their emissions contribute to ecosystem damage, not flat-rate emission reductions where each country reduces by the same percentage.

The effectiveness of Europe's acid deposition regime encouraged political support for further control measures. One sign of effectiveness was the conspicuous decrease in SO_2 emissions.[18] Regulation of a wider range of air pollutants began—a volatile organic compounds (VOC) protocol was signed in 1991, and heavy metals and persistent organic pollutants (POPs) protocols in 1998. In 1999 the Protocol to Abate Acidification, Eutrophication and Ground-level Ozone was adopted. It set emission ceilings for 2010 for four pollutants: sulphur, nitrogen oxides, VOCs, and ammonia. This protocol places effects such as eutrophication and regional ozone damage on a par with acidification.

At present, science and policy making maintain a relatively harmonious and intimate relationship in Europe. However, new political factors have appeared. The tighter decision making links between LRTAP and the EU have strengthened the relationship. Issues of technology transfer to eastern Europe and countries of the former Soviet Union and their political involvement in European affairs are challenging the relationship. In conclusion, though, scientists and scientific knowledge continue to be transforming forces in the European transboundary air-pollution regime as it moves toward tackling multiple air pollutants and multiple effects.

ACID DEPOSITION IN NORTH AMERICA

Science

In the prewar period, North American (Canada and the United States) science in general, and acid deposition–related science in particular, lagged behind Europe. The first major interest in acid deposition as an environmental problem emerged from a report by the Natural Resources Council of Canada on damage done by the Trail smelter in Trail, British Columbia, Canada, to agricultural interests on both sides of the border.[19] This report was one input into the ICJ's famous 1939 Trail Smelter Arbitration between the United States and Canada.

A second locus of acid deposition research was the mammoth Sudbury mining complex in Ontario, Canada, which opened in 1886 and which contains a magnificent copper deposit and the world's richest nickel deposit. Sudbury became a virtual laboratory for the negative impacts of acid deposition (due to metal smelting) in the prewar and postwar periods.[20] By 1970 the Inco smelter at Sudbury was the world's single largest SO_2 emission source.

Neither this Canadian research nor any other early research in North America resulted in scientific recognition of a region-scale acid deposition problem. The man who first discerned a pattern of large-scale and increasing acidity in the precipitation of North America was the American ecologist Gene Likens. Though there were others, such as Eville Gorham and Harold Harvey of Canada, who investigated acidification of precipitation as a local problem, it was Likens who first sounded the alarm in 1972 that a regional-scale problem existed in North America.

In 1963 a crucial single monitoring station was established at Hubbard Brook Experimental Forest in the mountains of New Hampshire. Likens and his colleagues were studying ecosystem dynamics at Hubbard Brook. They analyzed the watershed as a whole, tracking everything that went into, was cycled within, and came out of it. One substance that went in was precipitation and the acids it carried. "It was obvious that the rain was acidic, but we just assumed it was a local effect," Likens said years later.[21] It wasn't until he went to Sweden in the late 1960s and talked with Svante Odén and others that he realized that what he had observed at Hubbard Brook was quite likely the tip of an acid rain iceberg; the measurements were really pointing to a regional-scale acid rain problem like that recently discovered in Europe. "I began to wonder if acid rain might be a regional thing. So in 1969 . . . I started making measurements in central New York State. I was surprised to find the same kind of acid rain data that I'd been getting in New Hampshire." He and his colleagues began gathering the records of all previous precipitation chemistry measurements made in the United States, and, upon analyzing them, became convinced they had discovered the world's second large-scale acid rain problem. They published an article in the semi-popular scientific journal *Environment,* warning of an acid rain danger in North America similar to that in Europe.[22]

Likens et al.'s research was the first instance of localization to North America of the universal elements of acid deposition science discovered in Europe. The process was initiated by personal contact with European scientists and well illustrates the importance of a trade in scientific ideas in the localization process. It also illustrates the importance of prior existence of a scientific foundation (e.g., the Hubbard Brook station) that gives meaning to universal scientific ideas related to the environment.

From 1972 on, a small and informal network of U.S. and Canadian scientists began applying the full European acid deposition model to North America. Activist scientists, through newspaper articles, press conferences, interviews, and direct contact with policy makers, soon roused the U.S. and Canadian governments to take the problem seriously. As in Europe, they framed the problem as a potential threat to North American ecosystems. Other scientists, however, especially those associated with the electric power industry, countered that these claims were exaggerated because too little was known. This controversy served to swell political support for acid deposition research, especially after the election of Jimmy Carter in 1976. Acid deposition science matured at an astonishing rate during the later half of the 1970s, and communication between North American and European scientists was active and intense.

In 1976 the Canadian Network for Sampling Precipitation (CANSAP) was authorized. By this time, though, the equivalent European network was almost thirty years old. In the United States more than one hundred scientists from state agricultural experimental stations, universities, and industrial research institutes gathered in 1976 to develop plans for a coordinated and long-term acid deposition research program. This resulted in the National Atmospheric Deposition Program (NADP), which began operation in 1978. By 1980 localization of acid deposition knowledge had progressed such that the outlines of the North American acid deposition problem were clear. In both countries scientists concluded that there were high potential impacts due to acid deposition. They did not find direct evidence of widespread damage to ecosystems, such as in Scandinavia, but they did demonstrate the potential for such damage.

The key elements of the Canadian acid deposition scientific knowledge base were as follows. First, Canada was a victim of transboundary pollution from the United States. Second, vast areas of Canada had a granitic shield bedrock and a glacial history very similar to that of Scandinavia. This meant that inland water bodies were highly vulnerable to acidic inputs. Also, the beautiful Muskoka Lakes–Haliburton Highland region in Ontario, famed as a resort and home of artists, showed evidence of aquatic damage. This touched a Canadian cultural nerve, much as damage to lakes in Scandinavia and the Black Forest in Germany touched cultural nerves in Scandinavia and Germany, respectively. Third, vast tracts of forest were deemed vulnerable to acidic inputs. Canada was more dependent on forests as a mainstay of its economy than most any other nation in the world. Not only were commercial uses of forests assessed to be under threat, but uses related to recreation, tourism, and native people's traditional forest-dwelling lifestyles were also considered to be in danger. The adverse impact message coming from Canadian scientists was ominous.

The localized U.S. acid deposition scientific picture in 1980 was not nearly so ominous, if only because the areas of potential damage were not nearly as extensive as in Canada. Also, the United States, it was quickly discovered, was a transboundary pollution aggressor, not a victim. Rather than transboundary problems per se, the biggest long-range transport problem seemed to be between pollution-importing New England states and pollution-exporting lower Midwest states. The only areas that appeared to be threatened were the New England states and upper New York in the Northeast, the upper Midwest, and certain mountainous areas throughout the country. Aquatic impacts had only been documented in the Adirondacks Mountains of New York. In sum, as compared with Canada, America's ecosystems, economic activities, and cultural heritage seemed less threatened.

However, because of significant scientific uncertainty, in 1980 the U.S. National Acid Precipitation Assessment Program (NAPAP) began operation. A ten-year, $600 million research effort, it was the world's largest integrated research program to date on an environmental issue. It was expressly set up as a policy-oriented scientific research program whose mission was to reduce scientific uncertainty, assess the state of the problem, and define policy options. In other words, NAPAP's purpose was to establish a solid, high-quality (localized) scientific picture of the acid deposition problem in the United States that could be used by policy makers. More than three thousand scientists eventually took part in the program. However, NAPAP was fraught with problems almost from day one because of a shift in presidential administrations from Jimmy Carter to Ronald Reagan. When Ann Gorsuch stepped in as the new EPA administrator in 1980, she set out to undercut the NAPAP process. She, for example, abruptly terminated studies on aquatic effects and excluded the upper Midwest from consideration, even though its ecosystems were considered highly susceptible to acidification.

A much-delayed 1987 interim report argued that acid deposition was not a problem.[23] Its conclusions drew a firestorm of protest. The final writing had been done in close proximity to the White House by politically influenced individuals who were not researchers. Thus, Gene Likens termed it "highly politicized," and the Canadian minister of environment labeled it "voodoo science."

The culmination of NAPAP was its 1990 final integrated assessment.[24] Supported by twenty-seven state-of-the-art volumes, it totaled more than six thousand pages. But criticism continued. The science of NAPAP was to have played a major role in crafting the 1990 Clean Air Act, but it played only a minor role. Some complained it was "totally irrelevant." NAPAP found that there were negative consequences from acid deposition that merited concern, but that the United States did "not confront an acid deposition problem of a size or of an urgency that puts substantial natural resources at major near-term risk or that threaten human health, at least in a major way."[25] In short, no spectacular, headline-capturing negative impacts were uncovered.

NAPAP was a pioneer in attempting a large-scale, multiagency, multidisciplinary environmental assessment for policy input. However, in addition to Reagan Administration tampering that left the public confused and policy makers distrustful, there were problems with the science, particularly the greater emphasis on pure scientific research to the detriment of timely assessment for policy input.

Canada's equivalent to NAPAP, the Canadian National Acid Rain Research Program, did not get off the ground until 1985 and did not compare in scope or complexity to NAPAP. Its final assessment was published in the same year as NAPAP's integrated assessment.[26]

After the passage of the 1990 Clean Air Act and completion of the ten-year NAPAP research mandate, interest in acid deposition in the United States plummeted. The scientific community associated with the problem shrank to a shell of its former self. In Canada interest and funding also declined. The lack of support for research is most directly attributable to the lack of well-documented evidence of large-scale damage of the sort found in Europe. However, some scientists warn the problem may yet reassert itself.[27]

Like Europe, acid deposition science in North America became more complex and comprehensive in the 1990s as more pollutants and impacts were investigated. However, unlike Europe, North America possesses no scientific-political focal point like the critical loads approach (despite the fact that critical loads was first pioneered in Canada). The United States resists use of the critical loads approach, claiming that it overly simplifies the dynamic complexity of acidification processes.[28]

Politics

The announcement in 1972 by Likens and his associates, in provocative language reminiscent of Odén's, warning of an acid rain danger drew instant media attention. But the science was as yet highly speculative and its scope and depth did not begin to compare to that in Europe. Considerable uncertainty remained as to the true nature of the acid deposition problem in North America. In contrast to Europe, policy makers did not immediately act on the scientific information.

However, scientific uncertainty in key areas, such as aquatic impacts, and general public interest in environmental issues propelled political support for research in North America in the late 1970s. U.S. industry, particularly the coal and power companies, reacted strongly against any call for control measures. After a U.S. Council on Environmental Quality assessment of the acid deposition problem, President Carter announced an Initiative on Acid Precipitation in 1979.[29] The initiative called for a ten-year research program, which was realized when NAPAP was established in 1980.

Internationally, a Bilateral Research Consultation Group on the Long-

Range Transport of Air Pollutants (BRCG) was created under the auspices of the International Joint Commission. Its reports were the first official and joint statement concerning the North American acid deposition problem.[30] This group documented the fact of transboundary transport and showed that the United States exported far more acidic pollution to Canada than Canada to the United States. In particular, it asserted that two to four times more SO_2 and eleven times more NO_x crossed the border from the United States to Canada than crossed in the opposite direction. Based on these reports, the two countries signed the 1980 Memorandum of Intent (MOI) Concerning Transboundary Air Pollution, pledging them to develop a bilateral agreement on "the already serious problem of acid rain." The Canadian minister of environment called the transboundary acid deposition problem in 1980 "the most serious environmental threat ever to face the North American Continent."

Most actions in the 1970s in North America related to establishing programs to scientifically clarify the nature and scope of the acid deposition problem. And not surprisingly, scientists were front and center in pushing for increased funding. With a strong start on science and the formation of a "soft" regime (the 1980 MOI), North America seemed headed down the same treaty road as Europe. LRTAP was signed in 1979, and a North American treaty was expected to be signed in the early 1980s. But this did not happen. After the election of Ronald Reagan in 1980, otherwise friendly relations between the United States and Canada were seriously strained over the single issue of transboundary air pollution.

In 1980 the Canadian Parliament amended their 1971 Clean Air Act to improve the government's ability to control transboundary air pollution. In 1982 Canada offered to reduce its SO_2 emissions by 50 percent if a comparable reduction program was mandated in the United States. This was formally rejected by the United States. In 1984 Canada announced it would unilaterally cut in half its 1980 SO_2 emissions by 1994. In that same year the Canadian government called the environment "the longest-standing irritant" in relations between the two countries and expressed its "deep disappointment" at U.S. failure to take action. The mid-1980s were the nadir of Canadian–U.S. relations over the issue because of the noncooperative attitude of the United States.

The Reagan administration brought a very different political agenda to the presidency than the previous Carter administration. As an advocate of private enterprise unfettered by government regulation, Reagan justified inaction on acid deposition on the grounds of lack of scientific certainty and scarcity of conclusive signs of damage. The Reagan strategy was to limit action to research, which Canada labeled a "serious step backward." "We cannot wait for a perfect understanding of the acid rain phenomenon before moving to control it," Canadian officials responded.[31] In the early 1980s at least four scientific reports were published that undermined, but did not change, the Reagan administration's position.[32]

Limiting action to research was not the administration's greatest vice, though. Tampering with the scientific process was. When the Reagan administration took office, it not only threw monkey wrenches into the NAPAP process, but also deliberately set about undercutting the assessment process set up by the MOI by, for instance, replacing all university-based members on four of the working groups, initiating new studies when old ones were almost complete, withholding data, and rewriting draft reports.[33] When the BRCG's final report was published in 1983, it was published only in Canada by an angry Canadian government. This brought MOI assessment activity to an unsatisfactory halt. However, the Canadian government proceeded forcefully and unilaterally to contain its own acid precursor emissions. In 1985 the Eastern Canada Acid Rain Control Program, centering on the seven eastern provinces, was launched. It was at the time the single largest environmental protection program undertaken in Canada.

The U.S. government's attitude toward acid deposition did not change until Reagan left office in 1988. Although the science remained the same, the politics changed with the new Bush administration. Wafted by the winds of environment-as-sensitive-issue during the election campaign (stirred up in part by increasing public interest in international environmental issues such as stratospheric ozone depletion and climate change), President-elect George Bush brought a new presidential attitude to the White House. He proclaimed "a new attitude about the environment" and gave priority to the passage of a new Clean Air Act. The resulting 1990 U.S. Clean Air Act was a massive update and overhaul of the twenty-year old, and no longer well-functioning, 1970 Clean Air Act. Title IV of the new act (Acid Rain Title) was the first control legislation in the United States to directly address the problem of acid deposition. Its most innovative feature was its market-based environmental economics, especially an SO_2 emissions trading system. In a radical departure from the traditional command-and-control approach, the new law simply set a national ceiling on SO_2 emissions from electric power plants (the major emission source) and allowed utilities to figure out the most cost-effective way to achieve compliance. The national ceiling was only indirectly based on the capacity of the environment to absorb acidic inputs. One method of achieving compliance was to trade in SO_2 emission credits. The market (versus scientific) emphasis of the Acid Rain Title was in part intended to satisfy political opponents to the Clean Air Act, who sided with Reagan's orientation to environmental regulation.

Passage of the Clean Air Act finally paved the way for establishing a transboundary air-pollution treaty in North America. A United States–Canada Air Quality Accord was signed in March 1991 by President Bush and Prime Minister Mulroney. Although its initial focus was acid deposition, it was designed to be a "living" framework for dealing with other bilateral transboundary air-pollution problems. Specific emission targets for SO_2 and NO_x for both countries were contained in an annex (the targets were achieved by 2000). The U.S. requirements were the same as those contained in the 1990 Clean Air Act, and the Canadian requirements were the same as those contained in the Eastern Canada Acid Rain Program. To help implement the accord, a bilateral Air Quality Committee (AQC) was established. As of this writing the AQC has issued five progress reports.[34]

The paths to transboundary air-pollution regime formation in North America and Europe were more or less similar during the 1970s, but diverged drastically in the 1980s. Europe operationalized its transboundary air-pollution regime during the 1980s; North America struggled to convert their soft regime to a hard one. There was no *Waldsterben* equivalent in North America to galvanize policy action, and tampering with scientific research by the Reagan administration had been blatant and impaired the credibility of acid deposition science. In Europe a truly international scientific community developed; in North America such development was thwarted. In Europe communication and coordination between scientists and policy makers was outstanding; in North America (at least in the United States) it was poor at best. The science-policy infrastructure set up under the Air Quality Accord was a poor mirror of that under LRTAP. In the end, the forces binding science to policy making in North America proved far weaker than those in Europe, and they resulted in a less dynamic regime.

North America operationalized its regime as soon as the ink was dry on the Air Quality Accord, for it basically rubber-stamped domestic policies already in place. And, the regime began transforming almost immediately. The character and dating of this transformation are more ambiguous than is the case in Europe. Currently, Canada and the United States are working toward new annexes on ground-level ozone, air toxins, and inhalable particles. Also, under the influence of NAFTA, Mexico is slowly being brought into the arena. Driven by economics, not science, the innovative contribution of the North American transboundary air-pollution regime to environmental politics is the development of emission-trading programs. The resounding success with SO_2 led to the establishment of an NO_x trading program. Canada does not currently utilize any market-based instruments.

The North American transboundary air-pollution regime is, like Europe, undergoing transformation toward incorporating multiple pollutants and multiple effects; however, in the absence of a critical loads approach, the integration of science with policy making is feebler and the momentum toward tougher control measures weaker. Currently, business interests and market economics hold greater sway than science and environmental activism.

ACID DEPOSITION IN EAST ASIA

Since 1990, East Asia has emerged as the world's third regional-scale acid deposition hot spot. East Asia is defined to include Japan, China, North and South Korea, Taiwan, far-eastern Russia, and Mongolia. What binds most of this area environmentally is a shared temperate monsoon climate, and what binds most of it culturally is a shared historical connection to ancient China. At present, the dominant states in the issue are Japan, China, and South Korea.

Science

The first step in the birth of acid deposition science in East Asia was the absorption in the nineteenth and early twentieth centuries of the Western intellectual tradition known as "science." Absorption took place differently in each East Asian country. The first country to fully assimilate modern scientific techniques was Japan. Japan's first precipitation chemistry measurements, for instance, took place in the 1880s. Although there were early studies of acidification phenomena in Japan and even a precipitation chemistry network established in the 1930s, scientists did not recognize an acid deposition problem.[35]

Japan was the first country in East Asia to adopt Western acid deposition scientific knowledge. However, the process was slow and sporadic because of the lack of a scientific context with which to justify the full importation and application of the ideas. If I am to single out one individual in Japan responsible for initiating the localization effort, it would be Okita Toshiichi (I am using the Japanese convention of last name first), who was with the Ministry of Health and Welfare's Institute of Public Health in Tokyo when he started his work in the 1960s. Okita became aware of research on the newly discovered acid rain problem in the West through the scientific literature (for instance, through the journal *Ambio* published by the Swedish Royal Academy of Sciences starting in 1972) and later through conferences. He thereafter worked diligently to promote such research in Japan.

Consistently low pH readings recorded throughout Japan eventually provided the scientific context for initiating a full-scale importation effort. These pH measurements were initially an incidental aspect of air-quality monitoring during the heyday of Japan's horrendous air-pollution problems in the 1960s and 1970s. Scientists, Okita among them, finally convinced the Environment Agency (which became the Ministry of Environment in 2001) to establish the Acid Rain Survey in 1983, arguing that the low pH readings, the hubbub the acid deposition problem was causing in the West, and the potential for adverse impacts in Japan justified it. The survey was Japan's first nationwide investigation into acid deposition. To set it up, Japan essentially adopted the West's acid deposition problem framework lock, stock, and barrel (methodologies, knowledge base, and orientation to the problem). Publication of the final report of the first phase of the survey represented the first official picture of a Japan-localized acid deposition problem.[36] Though no serious domestic acid deposition problem was uncovered, over the course of the survey, the first evidence of long-range transport of acidic pollutants from mainland Asia was discovered around 1985. It came from monitoring data on the Sea of Japan coast. Soon after, research by the Central Research Institute of the Electric Power Industry (CRIEPI) confirmed this. Imported pollutants, it turned out, were a "side-effect" of the market-oriented economic reforms initiated in China in 1979.

China lagged behind Japan in all areas related to acid deposition science. Its early absorption of the Western scientific tradition was far less even and complete than Japan's. The first widespread precipitation chemistry measurements were not made until 1979 and were prompted by reports of serious acid deposition damage in southern China, especially around the cities of Chongqing and Guiyang. In 1982 a two-year survey on acid precipitation was conducted. By 1984 it was clear that the source of the problem was coal combustion not only in power plants, but also in small units used for residential heating and cooking, and that southwestern China was the nation's acid deposition trouble spot.

The birth of acid deposition science in East Asia differed fundamentally from that in Europe and North America as transmission of foreign methodologies and knowledge (versus indigenously developed concepts and tools) played a definitive role. It required great effort for East Asian nations to localize the standard acid deposition model of the West. Since the historical base of scientific practice was shallower, scientists struggled to build scientific capacity at the same time that they were locally applying the Western model to the very different natural environments found in East Asia.

Acid deposition science matured much more slowly in East Asia than in the West. Verification of the existence of a transboundary problem was drawn out over many years. Even in 1989, Japan's Acid Rain Survey final report commented only that "meteorological conditions and emissions sources, including those on the continent, need further investigation."[37] Thus, in complete contrast to the explosive appearance of the issue in Europe and North America, its appearance in East Asia was muted.

Aircraft measurements begun in 1991 showed the existence of huge plumes of SO_2 emanating from the continent during the winter months. Their existence became known to the press at a 1992 scientific meeting. A leading newspaper carried the following headline: "Polluted Winter Air Mass 500 Kilometers Wide from Continent Hits Western Japan."[38] Computer models readily verified that long-range transport of significant quantities of air pollutants from the continent, especially during the winter, could occur, but none were capable of reliable estimates of actual quantities transported.

Reports by Japanese scientists at conferences and in the mass media helped launch other programs in East Asia. China expanded its research to cover the whole of the country. South Korea, which like Japan saw itself as a major importer of Chinese emissions, established large projects in 1992 and 1995. In Taiwan, the Environmental Protection Administration established the Taiwan Acid Deposition Study, the first phase of which lasted from 1990 to 1995. The buildup of science had a multiplier effect. It accelerated research programs in all countries except North Korea; it led to international cooperative projects[39]; it triggered the formation of an informal East Asian scientific community and scientific assessment process; it helped bring into existence forums to discuss environmental issues (e.g., the Northeast Asian Conference on Environmental Cooperation); and it eventually persuaded China to acknowledge its uncomfortable position as the leading exporter of air pollutants.[40]

By the mid-1990s, a nascent, localized East Asian scientific picture began to emerge. It was a loose collection of knowledge, scattered and unsystematic as far as East Asia as a whole was concerned. Its clearest, yet motley, expression was to be found in the proceedings of expert meetings on a proposed monitoring network.[41] Scientists basically agreed that there were no dramatic present impacts, but that future impacts might be dire. Areas in Japan, South Korea, and Taiwan recorded low pH values. The only location sustaining clear-cut damage in the region due to acid deposition was southern China, the result of local, not transboundary emissions. However, because of projected massive increases in China's acidic emissions, the economic dynamism of East Asia, and the appearance of circumstantial evidence of forest and lake impacts in Japan and South Korea, the future looked potentially grim.

The major scientific achievement in East Asia of the 1990s was the long and arduous task of establishing the East Asian Acid Deposition Monitoring Network (EANET), a collaborative scientific effort that involves most countries of Northeast and Southeast Asia (as described in greater detail below). Despite the lack of significant, regionwide ecological impacts, such as lake acidification or forest decline, research activity is intense and booming in East Asia today, similar to research efforts in Europe and North America in the 1980s. If national knowledge bases are thought of as pilings for a bridge, then each country in East Asia is energetically sinking its own piling to support the EANET bridge. EANET provides the foundation for institutionalizing the process of locally generating acid deposition scientific knowledge in East Asia. Current EANET efforts, however, do not cover the full range of research activities (monitoring, emission inventories, atmospheric transport computer models, and impacts studies) necessary for this task. At present it focuses on monitoring and some impact studies, but is slowly expanding into East Asian–wide compilation of emission inventories and construction of computer models.

Politics

Japanese science dominated the first stage of East Asia's acid deposition history. The primary locus of political activity was Japanese scientists pressuring bureaucrats in the Environment Agency and the Ministry of International Trade and Industry (MITI) to support acid deposition–related research, which the bureaucracies did at modest levels. The power of the scientists' message lay in Japan's victim status, which goes a long way to explaining its proactive approach to transboundary air pollution in East Asia after its discovery. And Japan's sensitivity to China's reaction goes a long way in explaining why the approach, while proactive, was also very cautious. China, as the overwhelmingly dominant emission source in the region, has unequivocal veto power over international action. In the late 1980s China was preoccupied with its domestic acid deposition problems and vigorously denied that a transboundary problem existed.

In contrast to the acid deposition scientific communities in the West, at no time did "policy-activist" groups emerge in East Asia. There was only a quiet Confucian pressing of the importance of the problem to scientists' home bureaucracies. Japanese scientists argued that even though no major ecological impacts in Japan due to transboundary pollutants may yet be evident, China's economic growth and environmental pollution statistics were more than enough to cause worry about the future. This argument influenced acid deposition's rapid climb on Japan's international political agenda.[42] In a way, Japanese policy makers had all the information they needed to take action. They knew that a transboundary air-pollution problem existed and that the number of coal-fired power plants in China (the main source of the transboundary problem) was projected to skyrocket. Action came on two fronts, one technological and the other scientific. On the technological front, in 1991 MITI announced its so-called Green Aid Plan to aid developing countries in dealing with environmental problems. A major focus was China's coal-fired power plants. On the scientific front, in 1992 the Environment Agency announced that it would seek to establish a regional monitoring network. The first step was the series of previously mentioned expert meetings of scientists and administrators from the region.

By the mid-1990s, nonscientific factors joined scientific factors to keep Japan in the driver's seat on the East Asian acid deposition issue. One nonscientific factor was the keen interest shown by the Japanese public on the issue. Japan has long and deep cultural associations with rain and forests (the most seriously threatened ecosystem in Japan). Public concern over acidified rain prompted in the 1990s a nationwide pH monitoring craze among the general public that helped foster a highly acid deposition–literate citizenry.[43] Although there were few avenues for environmental activism by citizens and NGOs, a combination of citizen attentiveness, enthusiastic mass media coverage of the issue, and pressure by scientists for expanded international research deepened Japanese political commitment to the issue in the absence of evidence of significant ecological impacts. In addition, the Japanese government desired to demonstrate greater leadership on regional issues such as the environment, and the acid deposition issue provided them with a ready opportunity.

In June 1997 at the United Nations General Assembly Special Session on Environment and Development, held to honor the fifth anniversary of UNCED, or the Rio Earth Summit, Prime Minister Hashimoto proposed, as part of Japan's so-called Initiatives for Sustainable Development, the establishment of EANET. Hashimoto signaled Japan's intent to make EANET a central pillar in its efforts to promote sustainability in the East Asian region:

> [EANET] will be dedicated to creating common understanding of the status of the acid deposition problem among the countries in the region through implementation of acid deposition monitoring by each country, central compilation, analysis and evaluation of such monitoring data, and periodic publication of reports. Such common understanding will become the scientific basis for taking further steps to tackle the problem, such as cooperative measures to control precursor emissions.[44]

After long and delicate negotiations, the Japanese government succeeded in gaining the cooperation of other East Asian nations, including China, and EANET began operation in January 2001. By this time some fifteen years had elapsed since the discovery of the transboundary problem in East Asia. In Europe an international cooperative study using uniform methods, the OECD study in 1972, was established only four years after the discovery, and a monitoring network, EMEP, was set up five years later. In North America, a cooperative monitoring network was never established, but Canada and the U.S. quickly coordinated their monitoring activities. The comparatively slow evolution of EANET is indicative of the uphill battle in East Asia to develop regional-scale, cooperative institutions, environmental or otherwise. Development is inhibited by factors such as the legacy of World War II. EANET is therefore an institutional trailblazer.

Acid deposition politics in East Asia currently has two main prongs—establishing a regional scientific structure (via EANET) and cajoling China to deal with its increasing domestic emissions and worsening domestic impacts. Both prongs are fundamentally driven by an essential feature of East Asia's localized acid deposition picture: the mammoth size of China's emissions. This fact was first established by scientists, but is now virtually common sense, such that scientific revelations of impacts are not necessary to fuel current policy-making efforts. If scientific cooperation is successfully instituted through EANET, and if China is willing to give greater priority to environmental, as opposed to economic, development goals, then wider political cooperation may eventually result in a transboundary air-pollution regime like those in Europe and North America.

ANALYZING THE DIFFERENCES

The above comparative analysis demonstrates clearly that universal science does not produce universal policies. There is significant variation in regional policies, despite the fact that all regions used basically the same scientific methodologies and had access to more or less the same universal acid deposition scientific knowledge base during any given time period. This leads us to ask questions based on the three hypotheses introduced at the beginning of the chapter.

Do variations in regional science alone explain differences in regional environmental policy? There is some support for our first hypothesis. Differences in regional scientific knowledge do correlate with regional policy differences to some extent. For instance, verification of a regional-scale problem (with roughly the same size of area affected and roughly the same pH levels in each region) led to establishment of regional-scale scientific programs in each region (EMEP in Europe, NAPAP and the Canadian National Acid Rain Research Program in North America, and EANET in East Asia). Also, verification of greater damage seems to correlate with stronger policies. Confirmation of damage in Europe (aquatic, forests, cultural properties) is greatest in Europe, and the policies are the strongest (LRTAP and its multiple protocols). The damage is less in North America, and the policies there are arguably weaker. And little damage due to transboundary transport of pollutants is evident in East Asia, and the international policies there are weakest. In addition, recognition of victim status based on pinpointing who is upwind and who is downwind correlates with proactive pursuit of emission controls: Scandinavia in Europe, Canada in North America, and Japan in East Asia.

But there are cases in which the scientific knowledge was the same, but the policies were very different. For instance, it is indeed true that large-scale research (monitoring) programs were launched in each region, but there is significant difference in how long it took to get the programs off the ground. In Europe and North America, monitoring networks were established fairly quickly; in East Asia it took over fifteen years. As another example, basically the same scientific knowledge was available to Presidents Reagan and Bush, but whereas Reagan took no action, Bush signed the 1990 U.S. Clean Air Act and the 1991 United States–Canada Air Quality Accord. Therefore, the statement that variations in regional science explain differences in regional policy is not the full truth. Other factors beyond stand-alone scientific knowledge influence policy.

Do variations in how political actors use regional scientific knowledge explain differences in regional environmental policy? As our second hypothesis predicted, policy variation indeed is partly attributable to variations in the way political actors (scientists and nonscientists) use regional knowledge. For example, activist scientists in the West seem to have sparked more rapid evolution of policy than is the case in East Asia. In Europe and North America there was significant political activism by some members of the scientific community; for instance, Odén in Europe and Likens in North America. Scientists are less politically active in East Asia (in large part for cultural reasons), and policy has been slower to develop. As another example, political actors in Canada and the U.S. Reagan administration, using the same scientific knowledge base, interpreted scientific uncertainty in totally opposite ways during the early and middle 1980s, which led to very different domestic policies during this period and an impasse in international policy. As a final example, European and North American policy makers "use" scientific knowledge in almost diametrically opposite ways: the Europeans have put their trust in the critical loads concept and the RAINS model, whereas North American policy makers have no equivalent common scientific touchstone (the North American touchstone is emissions-trading programs). Therefore, there is truth in the statement that variations in the utilization of scientific knowledge by political actors help explain differences in regional environmental policy. But even taken in conjunction with variations in regional scientific knowledge, this does not explain all the variations between regional responses.

Do variations in the political and cultural contexts in which scientific knowledge is situated explain differences in regional environmental policy? The political and cultural context makes a difference, as predicted with hypothesis three. As one example, the state of scientific knowledge and its utilization by policy makers was roughly equivalent in Europe and North America in the late 1970s. However, Reagan's defeat of Carter totally altered the political context of scientific knowledge in North America. Reagan brought in an anti-environmental, pro-business political orientation that virtually precluded international policy action. Given this context, it is understandable that during the 1980s North American and European policies diverged dramatically. Additional examples of differing political contexts include the following: in Europe, the fortuitous appearance of the 1975 Helsinki Conference on Security and Cooperation, which provided a window of opportunity that eventually led to LRTAP; in East Asia, the historical legacy of World War II, which still impairs scientific and political cooperation. Examples of differing cultural contexts include the following: in Europe, damage to the historically important Black Forest helps explain West Germany's policy reversal in the early 1980s; in North America, the threat of damage to its heritage-rich lakes helps explain Canada's strong stance on emission controls; and in East Asia, the specter of injury to its forests, from which it derives its "culture of wood," helps explain Japan's active promotion of cooperation in East Asia. Therefore, besides variations in regional acid deposition knowledge and variations in how this knowledge is utilized by political actors, differing political and cultural contexts must be included to explain nonuniformity of regional policies.

In conclusion, to fully understand the development of regional-scale, acid deposition–related international policies in Europe, North America, and East Asia, requires (at minimum) assessment of variations in regional scientific knowledge, its use by political actors, and the political and cultural context in which scientific knowledge is situated. Indeed, a full complement of scientific and nonscientific factors must be examined to account for regional policy nonuniformity. The alert reader may even discern factors not covered by the three hypotheses that shaped the development of regional-scale, acid deposition–related international policies in the three regions. For example, what is the role of individuals? Does there exist a dominant norm shared among policy makers? If cultural or other ideational factors are important, does that mean that science is irrelevant, beyond outlining the nature of the problem?

NOTES

1. However, growing evidence of intercontinental transport of pollutants (for example, trans-Atlantic and trans-Pacific transport of acidic substances and their flow to the Arctic polar region) is beginning to change this view. See, for instance, Leonard A. Barrie, "Arctic Air Pollution: An Overview of Current Knowledge," *Atmospheric Environment* 20(4) (1986): 643–63; D. M. Whelpdale, A. Eliassen, J. N. Galloway, H. Dovland, and J. M. Miller, "The Transatlantic Transport of Sulfur," *Tellus* 40B (1988): 1–15; and Kenneth E. Wilkening, Leonard A. Barrie, and Marilyn Engle, "Trans-Pacific Air Pollution," *Science* 290 (2000): 65, 67.

2. Regime definitions range from "soft" to "hard." The most famous and commonly cited (soft) definition is: "[Regimes are] sets of implicit or explicit principles, norms, rules, and decision-making procedures around which actors' expectations converge in a given area of international relations" (see Stephen D. Krasner, "Struc-

tural Causes and Regime Consequences: Regimes As Intervening Variables," in *International Regimes*, ed. Stephen D. Krasner, pp. 1–22 (Ithaca, N.Y.: Cornell University Press, 1983). Hard definitions generally equate regimes with international treaties.

3. See Robert Angus Smith, "On the Air and Rain of Manchester," *Memoirs of the Literary and Philosophical Society of Manchester,* Series 2, 10 (1852): 207–17; and *Air and Rain: The Beginnings of a Chemical Climatology* (London: Longmans-Green, 1872) for his model of the acid rain problem. His work is extraordinary for its prescience. Smith's efforts, unfortunately, went unrecognized and lay in obscurity for almost a century.

4. Accounts of the historical buildup of scientific knowledge in the West are contained in Eville Gorham, "Scientific Understanding of Atmosphere-Biosphere Interactions: A Historical Overview," in *Atmosphere-Biosphere Interactions: Towards a Better Assessment of the Ecological Consequences of Fossil Fuel Combustion,* ed., Committee on the Atmosphere and the Biosphere, U.S. National Research Council, 9–21 (Washington, D.C.: National Academy Press, 1981); and "Scientific Understanding of Ecosystem Acidification: A Historical Review," *Ambio* 18(3) (1989): 150–54; and Ellis B. Cowling, "Acid Precipitation in Historical Perspective," *Environmental Science & Technology* 16(2) (1982): 110A–23A.

5. Svante Odén, "The Acidification of Air and Precipitation and Its Consequences in the Natural Environment," Bulletin 1 of the Swedish Natural Science Research Council, Ecology Committee, 1968 (1967 paper translated from the Swedish, Tr-1172, Translation Consultants, Ltd., Arlington, Virginia).

6. Bert Bolin, ed., *Air Pollution across National Boundaries: The Impact on the Environment of Sulfur in Air and Precipitation* (Stockholm: Norstedt & Söner, 1971).

7. Lars N. Overrein, Hans M. Seip, and Arne Tollan, *Acid Precipitation—Effects on Forest and Fish. Final Report of the SNSF Project 1972–1980* (Oslo: SNSF Project, 1980).

8. Lars N. Overrein, "A Presentation of the Norwegian Project 'Acid Precipitation—Effects on Forest and Fish'," *Journal of Water, Air, and Soil Pollution* 6(2–4) (1976): 167–72.

9. Organization for Economic Co-operation and Development, *The OECD Programme on Long Range Transport of Air Pollutants: Summary Report* (Paris: OECD, 1977).

10. A major function of LRTAP is to coordinate scientific research, standardize research methods, disseminate new information to researchers and policy makers, and foster research efforts in countries that might not otherwise have undertaken them.

11. Several early accounts in English of *Waldsterben* include a West German government report: Ministry of the Interior–West Germany, Report of the Causes and Prevention of Damage to Forests, Waters, and Buildings by Air Pollution in the Federal Republic of Germany (Bonn: Ministry of the Interior, 1984); P. Schutt, and E. Cowling, "Waldsterben, a General Decline of Forests in Europe: Symptoms, Development and Possible Causes," *Plant Disease* 69(7) (1985): 548–58; and F. Pearce, "The Strange Death of Europe's Trees," *New Scientist,* 4 December 1986, 41–5.

12. For further information, see Swedish NGO Secretariat on Acid Rain, "Environmental Factsheet No. 6—Critical Loads," *Acid News* 2 (April 1995): 2, 4.

13. Joseph Alcamo, R. W. Shaw, and Leen Hordijk, *The RAINS Model of Acidification: Science and Strategies in Europe* (Dordrecht, the Netherlands: Kluwer Academic Publishers, 1990); Leen Hordijk, "Integrated Assessment Models As a Basis for Air Pollution Negotiations," *Water, Air and Soil Pollution* 85 (1995): 249–60.

14. Christer Agren, "Hit by Airborne Nitrogen," *Acid News*, 4 October 1996, 13–14.

15. Peringe Grennfelt, Øysten Hov, and Dick Derwent, "Second Generation Abatement Strategies for NO_x, NH_3, SO_2 and VOCs," *Ambio* 23(7) (1994): 425–33.

16. C. Ian Jackson, "A Tenth Anniversary Review of the ECE Convention on Long-Range Transboundary Air Pollution," *International Environmental Affairs* 2(3) (1990): 217–26.

17. Informative sources on West German acid deposition science and policy include Sonja Boehmer-Christiansen and Jim Skea, *Acid Politics: Environmental and Energy Policies in Britain and Germany* (London: Belhaven, 1991); and Sonja Boehmer-Christiansen, "Anglo-German Contrasts in Environmental Policy-Making and Their Impacts in the Case of Acid Rain Abatement," *International Environmental Affairs* 4(4) (1992): 295–322. The latter warns that "[b]laming forest die-back is far too simple a view of the German policy response," and that "Waldsterben represented the marketing phase of a policy that had its origins in the late 1970s."

18. European Monitoring and Evaluation Programme, *Transboundary Acidifying Pollution in Europe: Calculated Fields and Budgets 1985–93* (Oslo: Norwegian Meteorological Institute, 1994).

19. National Research Council of Canada, *Effect of Sulfur Dioxide on Vegetation (NRC Report No. 815)* (Ottawa: National Research Council of Canada, 1939).

20. See Richard J. Beamish, "Acidification of Lakes in Canada by Acid Precipitation and the Resulting Effects on Fishes," *Journal of Water, Air, and Soil Pollution* 6(2–4) (1976): 501–14; T. G. Brydges and R. B. Wilson, "Acid Rain Since 1985—Times are Changing," in *Acidic Deposition: Its Nature and Impacts*, eds. F. T. Last and R. Watling (Edinburgh: Royal Society of Edinburgh, 1991): 1–16; and chapter 6 in this volume for discussions of this early work.

21. This and all other quotes from Likens are taken from a newspaper interview: Robert Ostmann Jr., "His Rain Study Caused a Storm," *Minneapolis Star*, 6 December 1979, 1A, 6A.

22. Gene E. Likens, F. Herbert Bormann, and Noye M. Johnson, "Acid Rain," *Environment* 14(2) (1972): 33–40.

23. National Acid Precipitation Assessment Program, *Interim Assessment: The Causes and Effects of Acidic Deposition* (Washington, D.C.: NAPAP, 1987).

24. National Acid Precipitation Assessment Program: (1) 1990 Integrated Assessment Report; (2) Acidic Deposition: State of Science and Technology Summary Report; (3) Acidic Deposition: State of Science and Technology (Washington, D.C.: NAPAP, 1990).

25. Oversight Review Board of the National Acid Precipitation Assessment Program, *The Experience and Legacy of NAPAP* (Washington, D.C.: NAPAP, 1991): 13.

26. Federal/Provincial Research and Monitoring Coordinating Committee for the National Acid Rain Research Program, *1990 Canadian Long-Range Transport of Air Pollutants and Acid Deposition Assessment Report* (Ottawa: Environment Canada, 1990).

27. See James Dao, "Acid Rain Law Found to Fail in Adirondacks," *New York Times,* 27 March 2000, 1, A23; and G. E. Likens, C. T. Driscoll, and D. C. Buso, "Long-Term Effects of Acid Rain: Response and Recovery of a Forest Ecosystem," *Science* 272 (1996): 244–46.

28. U.S. Environmental Protection Agency, *Acid Deposition Standard Feasibility Study Report to Congress (EPA430-R-95–110a)* (Washington, D.C.: EPA, 1995) did not recommend setting an "acid deposition standard," as the concept of critical loads is called in the United States.

29. National Atmospheric Deposition Program, *A National Program for Assessing the Problem of Atmospheric Deposition (Acid Rain): A Report to the Council on Environmental Quality* (Washington, D.C.: NADP, 1978).

30. Bilateral Research Consultation Group on the Long-Range Transport of Air Pollutants, *The Long Range Transport of Air Pollutants Problem in North America: A Preliminary Overview* (Downsview, Ontario: Atmospheric Environment Service, 1979); and *Second Report of the U.S., Canada Research Consultation Group on the Long Range Transport of Air Pollutants* (Downsview, Ontario: Atmospheric Environment Service, 1980).

31. As quoted in Gregory S. Wetstone and Armin Rosencranz, *Acid Rain in Europe and North America: National Responses to an International Problem* (Washington, D.C.: Environmental Law Institute, 1983): 114.

32. National Academy of Sciences, *Atmosphere-Biosphere Interactions: Toward a Better Understanding of the Ecological Consequences of Fossil Fuel Combustion* (Washington, D.C.: National Academy Press, 1981); *Acid Deposition: Atmospheric Processes in Eastern North America* (Washington, D.C.: National Academy Press, 1983); W. A. Nierenbery, *Report of the Acid Rain Peer Review Panel* (Washington, D.C.: Office of Science and Technology Policy, 1983); and Office of Technology Assessment, *Acid Rain and Transported Air Pollutants: Implications for Public Policy* (Washington, D.C.: OTA, 1984).

33. As the executive chairman of Canada's Federal Environmental Assessment Review Office stated in 1982:

> Despite substantial agreement among the scientists within the groups in the production of draft reports, we were treated to the sight of nonexperts rewriting the work group conclusions and unhappy scientists (formally part of the U.S. team), being quietly reassigned. . . . A pattern of external interference or inadequate support of the work has continued over the past year and a half" (quoted in Wetstone and Armin Rosencranz, *Acid Rain in Europe and North America*, 127).

34. The most recent, as of this writing, is AQC, *United States–Canada Air Quality Agreement: Progress Report 2000* (Washington, D.C.: U.S. EPA, 2000).

35. For a detailed history of acid deposition science and politics in Japan, see Kenneth E. Wilkening, *Acid Rain Science and Politics in Japan* (Cambridge, Mass.: MIT Press, forthcoming).

36. Environment Agency of Japan, *Dai Ichi Ji Sanseiu Taisaku Chōsa Kekka (Results of Acid Precipitation Survey—Phase 1)* (Tokyo: EAJ, 1989).

37. EAJ 1989, 2—my translation.

38. "Fuyu no Tairiku kara Nishi ni Hirai, 500 kiro, Sanseiu ni Eikyō," *Nikei Keizai Shimbun*, 8 October 1992, 19 (my translation of article title).

39. For example, the World Bank's RAINS-Asia project started in 1993: Wes Foell, Markus Amann, Greg Carmichael, Michael Chadwick, Jean-Paul Hettelingh, Leen Hordijk, and Zhao Dianwu, eds., *RAINS-ASIA: An Assessment Model for Air Pollution in Asia* (Washington, D.C.: World Bank, 1995).

40. For instance, Qu Geping, then head of China's environment agency, admitted in an interview in 1993 with a Japanese newspaper the possibility of long-range transport of pollutants from China to Japan. The interview was reported in front-page headlines as "Nihon e Ekkyō Osen no Osore" (my translation: "Fear of Transboundary Pollution to Japan"), *Asahi Shimbun,* 24 January 1993: 1.

41. Environment Agency of Japan, *The Expert Meeting on Acid Precipitation Monitoring Network in East Asia (26–28 October), Toyama, Japan* (Tokyo: OECC, 1993); *The Second Expert Meeting on Acid Precipitation Monitoring Network in East Asia (22–23 March), Tokyo, Japan* (Tokyo: OECC, 1995); and *Final Report of the Third Expert Meeting on Acid Deposition Monitoring Network in East Asia (14–16 November) Niigata, Japan* (Tokyo: OECC, 1995).

42. Japan established a Council of Ministers for Global Environmental Conservation in 1989 and soon after approved a Global Environmental Protection Research Plan (GEPRP). The importance Japan accorded the acid deposition issue is demonstrated by the fact that it was included as one of a handful of "global environmental problems" receiving concentrated research attention in GEPRP.

43. Kenneth E. Wilkening, "Culture and Japanese Citizen Influence on the Transboundary Air Pollution Issue in Northeast Asia," *Political Psychology* 20(4) (1999): 701–23.

44. Interim Secretariat of EANET, *Acid Deposition Monitoring Network in East Asia* (Tokyo: Environment Agency of Japan, 1998): 118.

9

The Effectiveness of the UN Convention on the Law of the Sea in Resolving International Fisheries Disputes

The Southern Bluefin Tuna Case

Jeremy Firestone and Tom Polacheck

This case study of a scientific and international legal dispute over the management of the Southern Bluefin Tuna (SBT) considers three important dilemmas in international environmental policy. First, how can one assure the integrity of scientific experiments that can potentially improve the management of a common-pool natural resource, but that also may entail increased environmental risk and substantial economic gain? Second, what role should international institutions play in resolving scientific disputes of shared international resources? Third, how does the interaction of global and regional international institutions affect the governance of common-pool resources? In the case of SBT, a unilateral experimental fishing program (EFP) requiring substantial additional catches from a depleted stock became the subject of a dispute under UNCLOS. The dispute led to questioning the appropriate role of UNCLOS in resolving international disagreements over the science and management of fish stocks.

Under UNCLOS (1982)—a comprehensive convention that governs ocean uses and jurisdiction—the ocean is divided into several zones. For the present purposes, the two relevant zones are each country's exclusive economic zone (EEZ), which extends two hundred nautical miles from the shore, and the high seas, which lie outside the EEZs. While a nation can determine by itself the catch for a fishery that exists wholly within its EEZ, it does not have exclusive jurisdiction over species that migrate beyond an EEZ to the high seas or into another nation's EEZ. In the present context, SBTs traverse several EEZs and the high seas and, consequently, are available to vessels from any nation that wishes to harvest them. Historically, Australia, Japan, and New Zealand were the principal nations harvesting SBT. Without any governance arrangements, the common-pool nature of the resource led to the classic "tragedy of the commons" outcome of overexploitation.[1] These three nations established international governance arrangements, initially informally in 1982. Although these arrangements reduced the rate of exploitation, they did not prevent the resource from declining further or prevent other nations from entering the fishery.

Stock assessments (the scientific process of estimating the current status of a resource and the consequences of future management actions) inherently contain substantial uncertainty. This uncertainty creates dilemmas for management and governance and, in the international arena, can create substantive disagreement about appropriate catch levels, reflecting differing views on appropriate risks to take with stocks and fisheries. In the present context, purportedly to reduce uncertainty in the scientific management advice, Japan began unilateral "experimental fishing" that required substantial catches, over the objections of Australia and New Zealand. In response, Australia and New Zealand took international legal action under UNCLOS to resolve the dispute.

Scientific information is needed for the rational use and management of natural resources. Yet, substantial uncertainty exists in the available information and our understanding of the process that determines resource productivity. Although well-designed experiments and data-collection programs have the potential to reduce uncertainty and, thus, provide an improved basis for informed decision making, collection of such information may require substantial exploitation. This may in turn entail increased risks for the sustainability of the resource and large profits for those conducting the experiments. Disputes about the efficacy of such experiments, as occurred with SBT, are not surprising, given differing scientific perspectives and sociopolitical objectives. Thus, in addition to the dilemmas posed in the opening paragraph, the SBT cases raise general questions about the interplay between science and politics. Is any dispute only about "science," as Japan claimed? Who should be responsible for resolving disputes that entail substantial scientific understanding? Where and how does law fit on the science-political continuum?

THE SCIENCE CONTEXT

Fisheries assessment is an imprecise science, and scientific predictions about the biological consequences of proposed management actions contain much uncertainty. Yet, concrete decisions are required about appropriate catch levels. These decisions must balance the tension between the economic risk of foregoing catches if quotas are set too low and the risk of stock depletion from overfishing.

Fishing reduces the reproductive potential of a stock (the ability to generate recruitment). Scientists use parental biomass (the total weight of all sexually mature individuals) as a measure of a stock's reproductive potential because the number of eggs produced by a mature female fish is proportional to its weight. The relationship between the size of the parental stock and the average number of subsequent recruitments has a density-dependent component: the number of recruits per unit of parental biomass generally increases as a stock declines. The mechanisms underlying this increase are not well understood, but the resulting increased production provides the basis for sustainable exploitation. However, annual recruitment (the number of new fish produced each year) can vary greatly even when parental biomass levels remain basically unchanged. Most marine fish spawn a very large number of eggs, but a large number of interacting and highly variable biological and physical factors (such as currents, water temperatures, food availability, predations, and disease) determine the ultimate number of eggs that survive to become juvenile fish.[2]

Estimating the current status of a fishery resource is inherently uncertain (how does one count the number of fish in the sea?). Fishery scientists conduct stock assessments to provide advice on the current status of the stock relative to the past and on the consequences of potential future actions. Stock assessment is a multistep procedure, and within each step, there are uncertainties about data to be included in the model and the processes controlling the underlying dynamics. For SBT, the magnitude of historical catches, the size distribution of the catches measured by sampling, and the growth rates used to estimate the age of fish from their size are all uncertain and open to differing interpretations. Other critical input parameters, such as natural mortality rates by age and age of maturity, also are uncertain.

Scientists use a variety of statistical techniques and mathematical population models to estimate current and historical abundances. For SBT, virtual population analysis (VPA) has been the primary assessment model. VPA combines information on the size of a cohort contained in the age structure of catches over time with information on relative abundance and fishing-mortality rates from catch rates, relative abundance surveys, and tagging to produce estimates of the historical number of fish by age.[3] Large structural or model uncertainties come from assumptions about age-specific natural mortality rates, selectivity in different fishery components, the stock/recruitment function, and catch rates as a measure of relative abundance.[4] The estimation of the plus group (the older component of the population for which size does not provide a basis for estimating age) has been one of the key problems in SBT assessments because of changes in the SBT fisheries and because these age-classes constitute a large proportion of the spawning biomass.[5]

To advise on the consequence of proposed management actions, scientists calculate probability estimates of future stock sizes given different levels of future fishing intensity based on estimates of current stock size and assumptions about future recruitments from estimated stock/recruitment relationships. Uncertainties in the estimates of current stock sizes and in the process controlling future dynamics make such predictions problematic. In particular, recruitment processes are often highly variable, which means that the longer the time frame, the more uncertain projections are likely to be.

Recruitment overfishing—when fishing reduces a stock below the threshold where recruitment notably declines—has been the mechanism associated with many fisheries' collapses.[6] For depleted stocks, the main concern is that environmental variability will combine with the vulnerable state of the resource to cause an abrupt recruitment decline and a subsequent further decline in the parental biomass. It is difficult to quantify the probability of this occurring. A stock can remain at low levels without collapsing, even occasionally producing large recruitments. However, the lower the level of the parental stock and the longer that it remains at low levels, the higher the probability of an abrupt recruitment decline. Detection of stock and recruitment collapse at the time they are actually occurring is nevertheless difficult. It generally takes several years of observations to reliably confirm the strength of recently recruited cohorts, and it is a feature of stock assessments that the most recent estimates are generally the most uncertain.

SCIENCE AND GOVERNANCE

The uncertainties associated with stock assessment make governance of a depleted resource particularly difficult. The inability to quantify the risk of collapse and the fact that a stock has not collapsed at current catch levels can make it difficult to justify drastic or urgent actions (and increasingly so, the longer a stock remains overfished without collapsing). Yet, stock collapses do occur with devastating consequences. When they do, commercial fisheries become economically unviable and biological recovery of the stock (and thus the fishery) is unlikely to occur for a long time. On the other hand, preventive measures like large reductions in catch designed to rebuild the fishery within a reasonable time frame are likely to have large economic and social consequences. Capital investment is often large and highly specialized, and overcapitalization (and the need to service the debt underlying these investments) is commonly one of the major factors contributing to overfishing in common-pool resources. In addition, fisheries can be a major source of employment, particularly for small coastal communities. All of the above can result in intense political pressure not to reduce catches, although short-term reductions could lead to substantially higher future catches.

The short- and long-term risks and costs are very asymmetric, creating a large dilemma for fishery management. The biological risk of not reducing catches in any given year can be small, but over a longer time horizon, the cumulative risk keeps increasing (how long can one flip a coin without expecting at least one tail?). Similarly, while the short-term cost of reducing catches may appear high from the industry perspective, these costs would be more than compensated for in the long term by the increased catches that could be taken from a recovered stock. Moreover, because of the likely long recovery time, if a stock does collapse, the long-term cost to industry would be enormous compared to the short-term reduction that might be needed to ensure recovery.

THE SBT RESOURCE

SBT is a long-lived, late-maturing, highly migratory fish that is among the most prized species for the Japanese sashimi market. Current catches are around 16,000 metric tons, with a total annual value estimated at at least US$250 million. SBT is distributed widely throughout most of the southern temperate oceans, except in the more easterly regions of the South Pacific. All information suggests that it is a single stock with only one known spawning ground in tropical waters south of Indonesia. Juveniles (ages one to five) tend to spend their summers in the coastal waters off Australia, where they are harvested by Australian surface fisheries, and their winters in deeper oceanic waters, where they are harvested by various longline fisheries. After about age five, SBT are seldom found in near shore surface waters, at which point they are only vulnerable to longline gear. Many aspects of their reproductive biology are not well understood, but the most recent information suggests that the mean age of maturity is between ten and twelve years.[7] Once mature, the animals return to the spawning grounds between September and April. Individuals can live in excess of forty years.[8]

Australian surface and Japanese longline fisheries for SBT developed in the 1950s. The Japanese longline fishery, which was initially concentrated on the spawning ground, expanded rapidly. The largest catches of SBT ever taken occurred in 1960 and 1961 (79,371 tons and 81,605 tons, respectively), with subsequent substantial declines in both catch and catch rates (see figure 9.1). In the mid-1960s, the Japanese longline fishery moved away from the spawning ground to southern feeding grounds because the quality and price of those fish on the Japanese sashimi market was much higher. The Australian surface fisheries expanded through the 1970s and early 1980s. A major component of the surface fishery (that off the southeast coast of Australia) collapsed in the early 1980s. Very high exploitation rates of juveniles continued into the 1980s.[9] In response, beginning in 1984, Australia dramatically reduced its catch of juveniles in the surface fishery.

Stock assessments have consistently found that SBT parental biomass is at historically low levels and that it steadily declined through the 1980s and most of the 1990s.[10] Assessments in the 1980s also indicated that recruitment declined substantially when parental biomass decreased below the 1980 level: the stock was "recruitment overfished." For over a decade, parental biomass has been estimated to be substantially below that level, and in 1998, it was estimated to be on the order of 25 to 53 percent of the 1980 level.[11] Based on these results, the Scientific Committee of the Commission for the Conservation of the Southern Bluefin Tuna (CCSBT) found that the SBT parental biomass was below commonly used thresholds for a biologically safe population (i.e., recruitment overfished). It repeatedly has concluded that "the continued low abundance of the SBT parental biomass is cause for serious biological concern."[12]

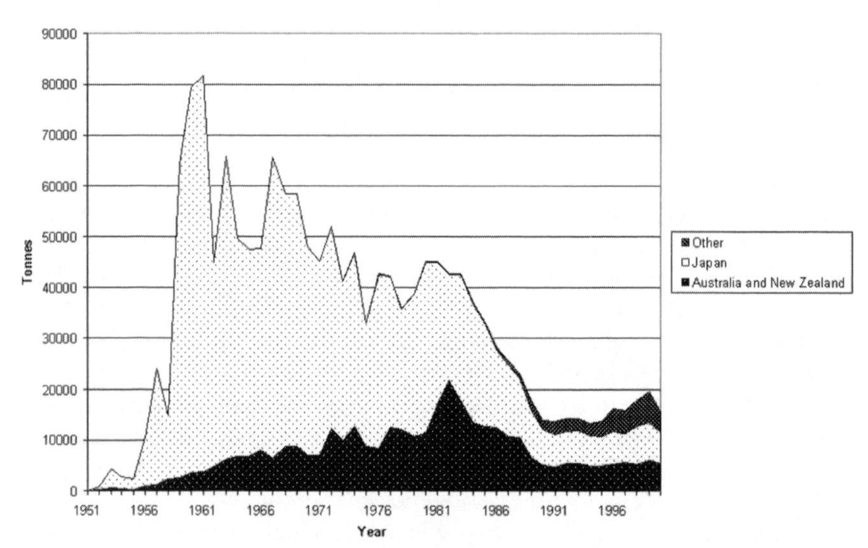

Figure 9.1 Estimates of SBT Catches in Weight

THE POLITICS CONTEXT

Informal international governance arrangements involving Australia, Japan, and New Zealand were initiated in the early 1980s and were subsequently formalized in 1993 when these countries adopted the Convention for the Conservation of the Southern Bluefin Tuna (SBT convention) and established the CCSBT, as well as the CCSBT Scientific Committee, an advisory body for conducting stock assessments and related scientific activities and providing management recommendations (by consensus). Parties to the commission are free to select whomever they wish to attend the meeting. The core participants have been working scientists from marine laboratories in the member countries and invited external scientists. Government officials with some scientific training from national management agencies also have been included within scientific delegations (almost exclusively by Japan), and in some cases industry representatives have been included as well. On occasion, both of these groups have been quite active and vocal within the meetings.

As envisioned under the convention, a primary role of the CCSBT is to set total allowable catch (TAC) limits and allocate the TAC among the nations of the convention. This task has been complicated by the fact that the SBT convention only defines general objectives (conservation and optimal utilization of the resource) and the commission has failed to develop an agreed-on decision-making framework for turning scientific assessment results into explicit TAC levels. In addition, all decisions are made by consensus, and all parties must resort to alternative dispute mechanisms.[13]

TAC limits were introduced in 1985 and were progressively lowered, with a major reduction of around 50 percent for the 1989 fishing year. The early limits caused reductions in catches and fishing mortality rates from the surface fisheries. It was not until the 1989 fishing year that the catch limits restricted the Japanese longline fleet; prior to that year, the Japanese longline fishery had not been able to catch its limit. Although the TAC has remained fixed at 11,750 tons since 1989, total catches have increased in recent years, as a result of longline vessels from countries not a party to the CCSBT having increased their efforts and of unilateral experimental fishing by Japan in 1998 and 1999 (figure 9.1).[14]

Based on the scientific committee's conclusion, the CCSBT established the rebuilding of the parental biomass to at least the 1980 level as a primary management objective for reasons discussed above. The initial time frame for achieving this objective was 2010, but it was subsequently extended to 2020. Because the CCSBT has no defined framework or decision rules for setting quotas, the national commissioners to the CCSBT set quotas through negotiations. The commissioners are high-level government officials from departments responsible for international fishery management (e.g., assistant secretaries, councilors, and policy managers). In the negotiating process, results from the scientific committee reports are used as the formal basis for supporting differing national positions on the TAC. For example, at the Fourth Annual Commission Meeting, Japan based its argument for a TAC increase on the recovery estimates calculated by Japanese scientists.[15] New Zealand argued for a TAC reduction focusing on the agreed-on depleted status of the stock, low recovery probabilities as estimated by New Zealand and Australian scientists, and increasing nonparty catches. Using similar arguments, Australia "favored a precautionary approach" without stating a preference for an increase, decrease, or maintainance of the status quo. At the core, there is disagreement on what information is critical and an absence of well-defined procedures for turning scientific estimates on stock status or recovery probabilities into specific TAC levels. Without a technical decision-making framework and with differing perspectives among the parties, the CCSBT has had difficulty reaching a consensus on TACs and has failed to agree on one since 1997.

DEVELOPMENT OF SCIENCE: EXPERIMENTAL FISHING

On the surface, the principle problem has been that, despite agreement about the depleted state of the stock, projections under constant catches have yielded very different estimates about the future status of fish stocks and the probability of achieving the recovery target if catches are maintained at their current levels (table 9.1).[16] Since 1995, Japanese projections have yielded high (>75 percent) estimates of the probability of recovery of stocks, while those of Australia and New Zealand have yielded low and progressively lower estimates over time.[17] Australia and New Zealand also have estimated a greater than 50 percent probability that parental biomass would continue to decline.[18] Given these disparities, Japan has consistently argued within the CCSBT for substantial increases in the TAC (25–50 percent), while Australia and New Zealand have variously argued that the TAC should be maintained at its current level or decreased.

A number of sources of uncertainty are considered within the SBT assessment. For each source, results are generated for several alternative hypotheses, which are intended to cover the range of uncertainty. Overall assessment and projection results are obtained by assigning weights (i.e., relative probabilities) to each hypothesis and then calculating a weighted average. The differences in probabilities of recovery among the national scientific delegations stemmed mainly from different weights that each delegation assigned to the various hypotheses.[19]

Table 9.1 Estimates of the Probability of Recovery Assuming Constant Catches at the 1997 Level Taken from the Results Presented in the 1998 CCSBT Scientific Committee Report

Method	*Australia Input Weightings*	*Japan Input Weightings*	*New Zealand Input Weightings*
Australia–1997 priority uncertainty set (convergence output weighting only)	0.14	0.65	0.07
Australia–1997 priority uncertainty set (all output weightings, including lack of fit)	0.06	0.26	0.04
Australia–preferred[a] uncertainty set (convergence output weightings only)	0.09	NA	NA
Australia–preferred[a] uncertainty set (all output weightings, including lack of fit)	0.11	NA	NA
Japan–1997 priority uncertainty set (convergence output weightings only)	0.24	0.76	*
Japan–preferred[a] uncertainty set (convergence output weightings only)	NA	0.87	NA

Source: Anon, "Report of the 1998 Scientific Committee Meeting."
Note: Under the method column, Australia and Japan refer to the national delegation that performed the calculation. Uncertainty sets refer to alternative sets of hypotheses included in the analyses. Output weighting schemes refer to different methods for integrating the results to estimate an overall probability. National input weightings refer to the set of relative weights each national scientific delegation assigned to the different hypotheses within an uncertainty set. Only the fact that a wide divergence of results existed and that the differences were primarily due to difference in the input weights given by the different national delegations is important here.

[a] Preferred was used to indicate a delegation's "best" estimate.

* Results did not converge.

One important source of uncertainty was how best to interpret catch rates. Catch rates (usually referred to as CPUE) provide indicators of trends in the stock (e.g., high catch rates should indicate high densities) and form an important component of the SBT stock assessment, although a number of factors beside changes in fish abundance can affect catch rates. Large changes in the location and total area fished have occurred. This has necessitated consideration of different hypotheses about the abundance of fish in areas where there has been no fishing (and thus no data on the density of fish in those areas). The various hypotheses yield similar results about long-term trends, but differences in the most recent trends (figure 9.2), which in

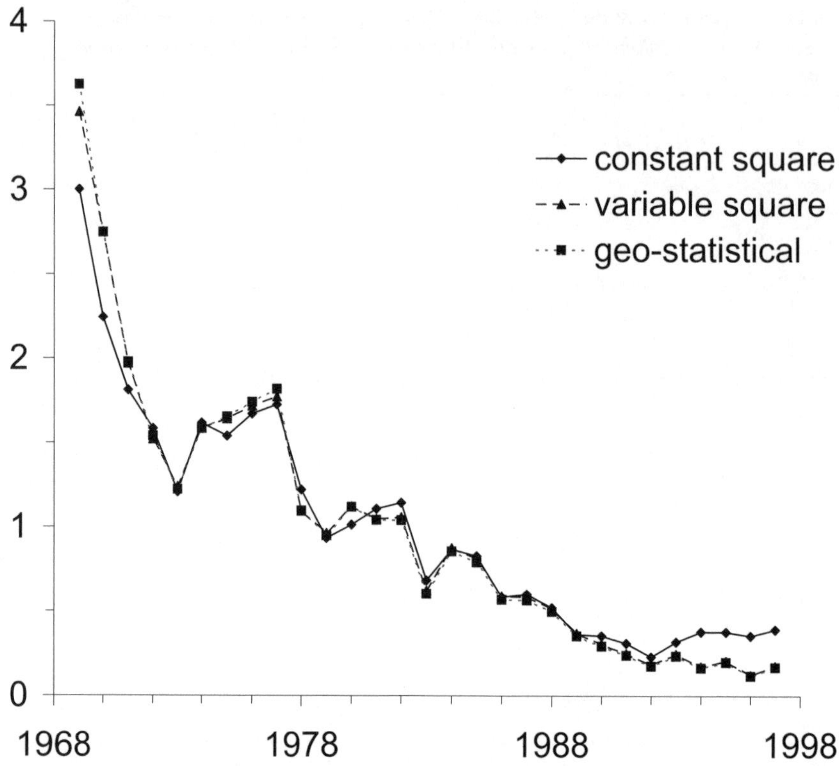

Figure 9.2 Comparison of the Three Primary CPUE Indices of Abundance Used in the 1998 Stock Assessments for Fishes Age Eight to Eleven

Source: Indices are taken from Tom Polacheck and A. Preece, "Documentation of the Virtual Population Analysis Methods and Model Inputs Used for Estimating Current and Historical Stock Sizes of Southern Bluefin Tuna: Updates and Modifications," 1998, CCSBT/SC/9807/37.

Note: The degree of similarity among the indices was similar for other age groups. The indices have been standardized by their mean since it is the relative trends that are important in the stock assessment context.

turn yield different projection results. Thus, collecting data from those areas where fishing has not been occurring would resolve at least some of the uncertainty in the interpretations.[20]

In 1998, there were no major substantive differences in the weights assigned to the principle CPUE interpretations by Australian and Japanese scientists (figure 9.3), although there were major differences in the weights assigned to hypotheses for other sources of uncertainty. The differences in the weights assigned to different CPUE interpretations had only a minor effect on their respective estimates of the probability of stock recovery (figure 9.4). Even if the most optimistic CPUE interpretation could have been

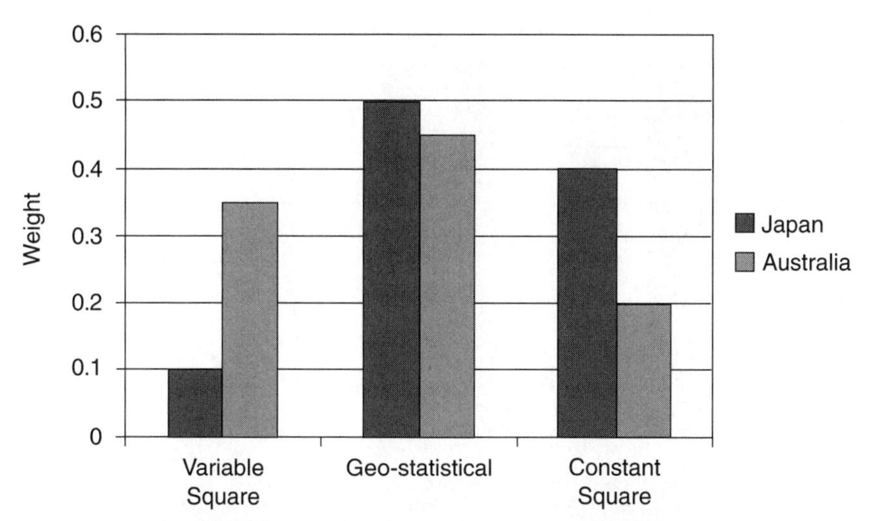

Figure 9.3 Relative Weights Assigned to the Primary CPUE Indices Used in the 1997 and 1998 SBT Stock Assessments for the Priority Set of Uncertainties

Source: Anon, "Report of the Third Scientific Committee Meeting of the CCSBT," 28 July–8 August 1997, Canberra.

shown to be correct with no uncertainty, the best estimate of the recovery probability by Australian scientists would have been 36 percent and 40 percent for New Zealand.

Fishery scientists have recognized that an EFP can be an effective tool for improving the governance of a fishery resource.[21] In the context of the CCSBT discussions, an EFP is a designed experiment that allows for short-term additional catches, taken in a controlled manner so as to provide specific information to improve the management of the stock. In other words, an EFP is not simply a scientific exercise for improving understanding and knowledge, but is meant to be a program directly linked to future governance decisions. The reason for considering an EFP is that, in some cases, controlled additional catches through a well-designed experiment may provide sufficient information to reduce uncertainty and divergence of views in stock assessments. However, a poorly designed EFP can result in increased uncertainties (e.g., about the magnitude, size, composition, and spatial distribution of catches because of inadequate monitoring of the experiment), inappropriate conclusions, and increased risk.

An EFP requires agreement on the design of the experiment (e.g., where, when, and the extent of fishing to take place) and the methods to be used for analyzing the data in order to assess whether the EFP provides informative

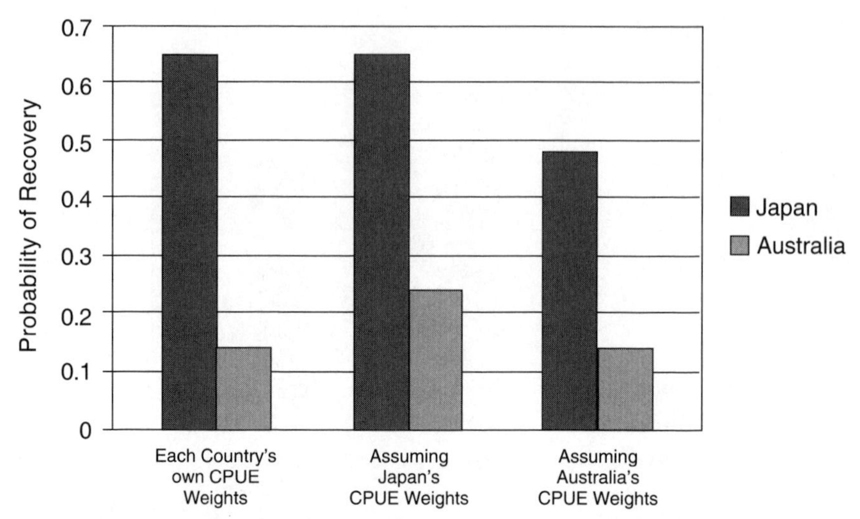

Figure 9.4 Comparison of Australian and Japanese Estimates of Probability of Recovery from the 1998 Stock Assessment for Different Weights Assigned to the CPUE Interpretations

Source: Anon, "Report of the 1998 Scientific Committee Meeting"; Polacheck and Preece, "Virtual Population Analysis for SBT."

results. In theory, this should be a "scientific" process. However, in practice, it is essential to engage industry and managers in the design process because of logistical and resource constraints and the need for industry cooperation to carry out the experiment. The differing perspectives and objectives of the different participants can make designing an EFP difficult, particularly when the data collection can entail substantial quantities of commercially valuable catches.

Fundamental to the concept of an EFP is prior specification of decision rules that reflect how the scientific results will be interpreted and used in future governance decisions. Scientists can provide technical advice on the expected effect of alternate decision rules, but their actual specification is a policy decision that balances competing governance objectives (e.g., stock recovery versus short-term catches). Without the establishment of decision rules prior to experimentation, it is impossible to objectively compare the potential benefit from the additional information with the increased risk to stocks from the additional catches. In a depleted stock, any short-term increase in catch inherent in an EFP increases the immediate risk of stock collapse. Thus, an EFP needs to provide information that decreases the probability that future catches will exceed levels that would prevent the stock from recovering: an EFP must reduce uncertainty about the appropriate levels of catch.

The idea of using experimental fishing within the CCSBT stemmed from a suggestion by an invited expert at the 1994 scientific committee meeting that the interpretation of CPUE could be addressed by the division of the TAC between currently fished areas and areas where fishing had been absent in recent years.[22] At the following scientific committee meeting, Japanese scientists presented a three-page proposal for extensive experimental fishing.[23] This proposal was not pursued at the subsequent CCSBT meeting. Instead, Japan sought a six-thousand-ton increase in the TAC. Because no agreement resulted, a follow-up special commission meeting was held one month later. There, Japan continued to seek a six-thousand-tons increase, or alternatively, agreement on a three-year six-thousand-ton additional annual experimental fishing quota. This pattern of seeking substantial increases in the TAC or, alternatively, a large experimental fishing quota continued through 1999 (see table 9.2).

Japan's 1995 request for an experimental fishing quota was accompanied by a formal written proposal, but the proposal provided little detail on the experimental design, its analysis, its links to future management decisions, or the basis for the magnitude of the catch sought. Nevertheless, the CCSBT eventually agreed that "work to evaluate a possible implementation of an experimental fishing program was warranted," at least in part as a way to overcome the current impasse in setting the TAC.[24] It further agreed on "Objectives and Principles for the design and implementation of an experimental fishing program," as well as a process both to evaluate the impact of additional removals and for designing and evaluating proposals for experi-

Table 9.2 Japan's Proposals for TAC Increases and Additional Quota for Experimental Fishing (in tons)

Year	*Proposed Annual TAC Increase*	*Proposed Annual Increase in EFP Quota*	*EFP Increase over the TAC (%)*	*Increase over Japan's National Allocation (%)*
1995	6,000	6,000	51.1	98.9
1996	6,000	6,000	51.1	98.9
	3,000	1,500	12.8	24.7
1997	3,000	2,010	17.1	33.2
1998	3,000	2,010	17.1	33.2
		Japan's EFP take: 1,464	12.5	24.1
1999	3,000	Japan's EFP take: 2,198	18.7	36.2

Source: Adapted from United Nations, *SBT Case*, Australia's and New Zealand's Reply on Jurisdiction, vol. 1.

mental fishing.[25] A fundamental part of the Objectives and Principles was agreement on the need for decision rules linking the results from any EFP to future management actions.

The CCSBT held numerous workshops and meetings between 1996 through mid-1998 aimed at the development and evaluation of experimental fishing proposals based on the agreed-on objectives, principles, and evaluation process. These meetings illustrate the difficulties of implementing a process as complex as an international EFP that requires a shared understanding and consensus among the scientists, managers, and industries from the participating nations. A detailed history of these meetings is beyond the scope of this chapter. Suffice it to say, the process failed to reach consensus.

Although the parties had not completed the agreed-on evaluation process, over Australia and New Zealand's strong objections, Japan began unilateral experimental fishing in July 1998. The issues underlying Australia and New Zealand's objections to Japan's experimental fishing included the fact that no direct link (i.e., decision rules) existed between the results of the EFP and subsequent management actions; estimated probabilities from the 1998 assessment were basically insensitive to which delegation's CPUE weights were used and, thus, the program would not be capable of resolving the differences in the projection of future fish stocks, and the experimental design (particularly the method chosen by Japan for distribution of fishing effort) was not statistically and scientifically appropriate and robust. Indeed, small changes in the location fished by the vessels could significantly change the results, and substantial concerns existed about the increased short-term biological risk to the stock without a demonstrated long-term benefit.

Japan argued that the increased risk was minimal and that its 1998 experimental fishing had demonstrated the value and need for experimental fishing to continue.[26] In particular, Japan claimed that the 1998 results had already yielded dramatic increases in the probability of recovery to over 65 percent for all scientific delegations (based on Japanese calculations).[27] Japan also considered its experimental design, analysis, and interpretation of the results from the 1998 experimental fishing appropriate. It further considered that its commitment to pay back its experimental fishing catches if harm to the stock was demonstrated was a sufficient definition of a management response and decision rule. Japan argued that its experimental fishing was essential because "research on the SBT stock must be done in the hope that it will show that there is room for a larger TAC."[28]

FROM A SCIENTIFIC TO A LEGAL DISPUTE

In response to Japan's decision to commence the EFP, on August 31, 1998, Australia and New Zealand formally notified Japan of the existence of a dispute over the EFP and alleged that Japan had breached the SBT convention, UNCLOS, and customary international law.[29] Citing a breach under UNCLOS, Australia and New Zealand initially requested that urgent consultations, followed by negotiations, be convened under Article 16 (the dispute settlement provision) of the SBT convention. Negotiations commenced shortly thereafter and resulted in an agreement to establish an Experimental Fishing Program Working Group (EFPWG) to once again attempt to reach agreement on a joint EFP.

The EFPWG contained managers, national scientists, industry representatives, and a range of other government officials, as well as three invited external scientists. The terms of reference for the working group required that an account be taken of the 1996 Objectives and Principles and listed eight specific tasks, which were similar and consistent with basic development and evaluation steps agreed on previously.[30] In particular, one of the specific terms of reference for the group was to "decide on appropriate decision rules governing the interpretation of the EFP results for the management and conservation of SBT . . . and any necessary response to the impact of the EFP catch."[31] The EFPWG failed to reach agreement in four months of meetings, though that is not surprising given the complexity of the tasks required and the short time frame.

Although Japan agreed to both the Objectives and Principles and the terms of reference for the EFPWG, it failed to embrace a critical component—the need for reaching agreement on explicit management responses. The history and record of the discussion on this matter is too extensive to describe in detail here. Japan's earliest proposals contained no reference to decision rules or management responses. Although some versions of its later proposals included a section on decision rules, these were not clearly specified and never contained specific predetermined management responses. Moreover, while the proposed decision rules allowed for increases in TAC if the results were positive, they did not contain allowance for decreases in the TAC (beyond a possible pay back of EFP catches) if the results were negative. The parties' different perspectives on the linkage between the results of an EFP and an agreed-on management response was a key element preventing agreement throughout the entire discussions.

In May 1999, Japan indicated that it intended to recommence its unilateral EFP on June 1, 1999, unless the parties accepted Japan's proposal for a joint EFP. The parties thereafter exchanged a series of diplomatic notes. Because they recognized weaknesses in the SBT convention's nonbinding dispute settlement provisions, in the course of the diplomatic exchange Australia and New Zealand reiterated their view that the dispute arose under UNCLOS as well and stated that a full exchange of views for the purposes of UNCLOS's informal dispute settlement provisions had occurred.[32] Japan recommenced its unilateral experimental fishing in July 1999.

TESTING THE LAW OF THE SEA

UNCLOS, to which Australia, New Zealand, and Japan are parties, requires disputes regarding its interpretation or application to be settled by peaceful means. Parties to a dispute are required in the first instance to attempt to settle their dispute informally. Under international law, dispute-resolution procedures have traditionally not bound disputants. However, early on in the negotiations over UNCLOS, it was recognized that compulsory dispute-settlement procedures were necessary to ensure that rights and obligations created under UNCLOS were meaningful. Thus, in certain disputes, including high-seas fishery disputes, UNCLOS prescribes a compulsory binding-decision procedure when informal attempts are unsuccessful.[33] Pursuant to Article 287 of UNCLOS, states are free to select one or more dispute settlement institutions: the International Tribunal for the Law of the Sea (ITLOS), the ICJ, an arbitral tribunal, or a special arbitral tribunal composed of experts. If a party does not make a declaration or the parties to a dispute have not selected the same method, an arbitral tribunal is employed.

On July 15, 1999, Australia and New Zealand initiated formal dispute proceedings with Japan under UNCLOS before an arbitral tribunal. The applicants (Australia and New Zealand) alleged that Japan had failed to cooperate and to conserve SBT and that Japan's EFP threatened serious and irreversible damage to the SBT population, thus breaching Japan's obligations under Articles 64 and 116–19 of UNCLOS.[34] Those provisions require nation-states to cooperate to ensure the conservation and optimum utilization of highly migratory fish species and other living resources.

On July 30, Australia and New Zealand sought interim relief from ITLOS. Although ITLOS is only one of several institutions that nations may select for dispute settlement, UNCLOS designates ITLOS as the institution to consider and order provisional measures (essentially equivalent to preliminary injunctions in U.S. national law), pending the establishment of an arbitral tribunal in all disputes not involving the seabed.

Article 290 of UNCLOS provides that "when the urgency of the situation so requires," ITLOS may prescribe "provisional measures" (measures that cannot await the establishment of the arbitral tribunal). In a departure from international precedent, provisional measures under UNCLOS can be ordered not only to "preserve the respective rights of the parties," but also to prevent "serious harm to the marine environment," pending a final decision (Article 290).

The applicants invoked Article 290 and filed papers with ITLOS requesting that it enter an interim injunction against Japan that would prohibit certain actions prior to a final decision by the arbitral tribunal on the merits. More specifically, they requested that ITLOS order Japan to immediately cease the unilateral EFP and restrict Japan's catch to its national yearly allocation as last agreed on by the SBT commission, subject to being reduced by the unilateral EFP fishing catch in 1998 and 1999, as the EFP had effectively increased Japan's catch over its quota by more than 30 percent. Although not invoking the precautionary principle by name, they also asked that the parties be ordered to act with precaution to ensure the sustainability of the stock and refrain from prejudicing a decision on the merits by the arbitral tribunal.

On August 9, Japan filed a response to the applicants' requests for interim relief. Japan maintained that the SBT convention and not UNCLOS governed the dispute. Because in Japan's view, once constituted, the arbitral tribunal would not have jurisdiction to resolve the dispute, Japan claimed that ITLOS did not have the power to order provisional measures. Japan desired to continue dispute settlement under the SBT convention. Under that convention, settlement of a dispute is binding only if each party consents to the referral of the dispute to the ICJ or an arbitral tribunal. In other words, each party essentially has a right of veto. The SBT convention's consent requirement thus effectively rendered its dispute-resolution provisions not binding on Japan.

Assuming for the sake of argument that ITLOS decided that the matter was properly before it, Japan claimed that ITLOS should not order provisional measures because there was no "urgency" as required by Article 290. Japan also filed a counter-request for provisional measures that, if ordered by ITLOS, would direct Australia and New Zealand to resume good-faith negotiations. In the event those negotiations were unsuccessful, the matter would be referred to independent scientists.[35] This last request was consistent with Japan's view that the dispute "is and always has been scientific," rather than legal in nature.[36]

The hearing commenced on August 16, 1999, and lasted three days. The oral presentations addressed both jurisdictional and scientific issues, and involved witness testimony and oral argument. Underscoring the historic nature of the case, an expert witness testified on behalf of Australia and New Zealand only after first being examined by Japanese legal counsel as to his competence and potential bias in a process referred to as a voir dire. Although voir dire of a witness is a common occurrence in U.S. courts, it was one of the few instances of the use of this procedure in an international tribunal. Upon conclusion of the applicants' examination of the scientific expert, Japan had an opportunity to cross-examine the witness.[37]

In its decision, ITLOS first addressed the question of whether it had prima facie jurisdiction (whether there was sufficient evidence to support a contention that the arbitral tribunal *might* have jurisdiction) to order provisional measures.[38] To the extent ITLOS might play an important role in international fishery disputes, the decision on prima facie jurisdiction was critical. Under an expansive view of jurisdiction, which was put forward by Australia and New Zealand, UNCLOS's "obligations remain and . . . are fundamental," even though UNCLOS envisions that important substantive and procedural obligations "should be discharged through appropriate subsidiary organizations" such as the SBT commission.[39] UNCLOS's "obligations are not excluded, diluted, or modified—let alone eliminated by the creation of such organizations."[40] The applicants noted that if the contrary were true, a party could evade UNCLOS's compulsory dispute-resolution provisions by entering into a regional compact such as the SBT convention.[41] This would dilute or eliminate the governance powers of global institutions like UNCLOS in favor of potentially weaker regional institutions in which individual countries could have a right of veto.

ITLOS ultimately rejected Japan's view that the dispute was solely scientific and that the existence of the SBT convention precluded resort to UNCLOS's dispute-resolution provisions.[42] Having concluded that it had prima facie jurisdiction, ITLOS turned its attention to the merits of the request for provisional measures. In proceedings before a legal tribunal such as ITLOS, scientific judgments have to be tied to legal standards—in this case to the urgency requirement set forth in Article 290, and perhaps to the notion of irreparable harm advanced by Japan.[43]

The applicants considered the situation to be urgent from both a procedural and a substantive standpoint. Procedurally, it might take three or more months until the arbitral tribunal was constituted and capable of functioning.[44] Substantively, they called for the institution of provisional measures to prevent further declines in the parental biomass and deterioration of the SBT stock as a whole.[45] Japan, for its part, argued that the situation was not urgent because its 1999 EFP would end in two weeks and was insignificant when compared to the annual global catch.[46] In any event, Japan claimed it could reduce its catch in the future should the arbitral tribunal find that the EFP had a substantial adverse impact on the stock.[47] Japan also noted that to the extent the SBT stock was at risk, that risk was attributable to catches of nations that were not parties to the SBT convention. Those nations in the aggregate accounted for at least one-third of the global catch, an amount that far exceeded Japan's EFP catch.[48] On the question of the relationship between urgency and irreparable harm, Japan argued that "irreparable harm" is the "core concept" needed to prove urgency.[49] According to Japan, it "has been firmly established throughout the long history of international jurisprudence" that a complaining party has the burden of demonstrating irreparable harm prior to the prescription of provisional measures.[50] Japan also noted that urgency requires that the irreparable damage be imminent.[51] In response, the applicants claimed that the "drafters of UNCLOS deliberately chose to give [ITLOS] a broader as well as a more effective provisional measures jurisdiction" than that possessed by the ICJ.[52] They noted that Article 290 does not require a showing of irreparable harm before provisional measures may be imposed and that such measures may be instituted not only to maintain the status quo, but also to prevent environmental harm.[53]

After listening to the arguments of the parties, ITLOS found in favor of Australia and New Zealand. ITLOS noted that while the parties disagreed on the effect of the EFP on the SBT stock, the parties agreed that the SBT stock was "severely depleted" and at its "historically lowest level," raising "biological concern."[54] ITLOS then expressed its view that the parties should "act with prudence and caution to ensure that effective conservation measures are taken to prevent serious harm to the stock."[55] Although ITLOS acknowledged that it could not "conclusively assess the scientific evidence presented by the parties," it found that "measures should be taken as a matter of urgency."[56] Finally, by equating the need to avoid harm to the SBT stock (the "conservation of living resources of the sea") with the "protection and preservation of the marine environment" and noting that provisional measures should be undertaken to "preserve the rights of the parties and to avert further deterioration" of the SBT stocks, ITLOS relied on both the traditional and novel bases set forth in Article 290 for the prescription of provisional measures.[57]

Although the ITLOS orders did not explicitly address the relationship between urgency and irreparable harm, Judge Laing, writing separately, stated his belief that the tribunal had "not chosen to base its decision on the criterion of 'irreparability,' which is an established aspect of the jurisprudence of some other institutions. . . . [T]hat grave standard is inapt for application in the wide and varied range of cases that, pursuant to UNCLOS, are likely to come before this Tribunal."[58]

By a large majority, ITLOS ordered the parties to refrain from any action that "might aggravate or extend the disputes" or "prejudice the carrying out of any decision on the merits" and directed them not to exceed their last agreed-on national allocations and to count any country's experimental catch against its national allocation.[59] ITLOS also recommended that the parties resume negotiations and make further efforts to reach agreement with nonparties to the SBT convention.[60] In sum, the decision showed UNCLOS and ITLOS to be powerful instruments for resolving international fishery disputes, but this interpretation was later called into question by the arbitral tribunal, the first such tribunal constituted under UNCLOS.

Following the constitution of the five-member arbitral tribunal and the exchange of legal papers setting forth the parties' respective positions, the arbitral tribunal convened a four-day hearing in May 2000. On August 7, 2000, the arbitral tribunal rendered its decision. While the arbitral tribunal agreed that the dispute arose under both the SBT convention and UNCLOS, it ruled in a four-to-one decision that the nonbinding dispute-resolution provisions of the former governed the dispute.

More specifically, the arbitral tribunal noted that the main elements of the dispute had been addressed within the SBT commission and were related to obligations under the SBT convention.[61] It stated that there was no disagreement between the parties over whether the dispute fell within the ambit of the SBT convention. Rather, the issue was whether the dispute also arose under UNCLOS.[62] Although Japan maintained that UNCLOS was merely a "framework or umbrella convention that looks to implementing conventions to give it effect" and that the SBT convention has "subsumed, discharged and eclipsed" those provisions of UNCLOS that bear on the conservation of SBT, the arbitral tribunal observed that often more than one treaty governs a particular dispute and that, in any event, UNCLOS creates obligations beyond the SBT convention, such as barring discrimination and requiring nation-states to conserve living resources on the high seas.[63] Consequently, the arbitral tribunal held that while the dispute over the EFP was "centered" on the SBT convention, it also arose under UNCLOS.[64]

Although the arbitral tribunal essentially upheld ITLOS's conclusions that led it to find prima facie jurisdiction, the arbitral tribunal nevertheless found jurisdiction wanting based on an issue not discussed by ITLOS. Specifically, the arbitral tribunal noted that Article 281(1) of UNCLOS permitted parties to resolve disputes by peaceful means "of their own choice." Such a choice could include placing limits on the applicability of UNCLOS's compulsory binding dispute-resolution provisions. Although the SBT convention did not explicitly exclude binding dispute resolution, the majority found that the SBT convention implicitly placed just such limits on UNCLOS.[65] As a consequence, the arbitral tribunal held that it lacked jurisdiction to consider the merits of the claim. Accordingly, it unanimously revoked provisional measures put into force by ITLOS. At the same time, it stated that revocation of the order prescribing provisional measures did "not mean that the Parties may disregard the effects of that Order or their own decisions made in conformity with it."[66] It emphasized that if the parties abstained from unilateral acts that might aggravate the dispute, the prospects for a successful settlement would be enhanced.[67] One might interpret the panel as saying, we do not have authority as a matter of law to maintain and enforce the provisional measures, but we think that the measures that ITLOS had put into place are sound as a matter of policy.

LESSONS LEARNED

Science, Prediction, Law, and Politics

This chapter raises many questions about the political uses of scientific methods, the relationship between science and governance of a common-pool resource, and the uses of international institutions. At the outset of this chapter, we asked how to ensure the integrity of scientific experiments that may entail increased environmental risk and substantial economic gain. This question may be more easily approached if one considers three subsidiary questions: what standards should govern whether a proposed scientific experiment should be undertaken, where should the burden of proof lie, and who should judge the appropriateness of such experiments? From a legal perspective, the burden of proof and the governing standard are not abstract ideas, but concepts that often determine the outcome of disputes, particularly in the face of uncertainty. Although there are no right or wrong answers to the questions posed, one can gain insight into the answers by looking at how the law treats science and uncertainty in other contexts.

First, consider prescription drugs. The marketing and use of a new prescription drug can be thought of as an experiment that a manufacturer wishes to conduct. To obtain approval to market a drug, a manufacturer must demonstrate (it has the burden of proof) to the FDA (the decision maker) that the drug is "safe and effective" (the governing standard).

On the other hand, consider the use of a chemical in society—essentially an experiment in the marketplace that may expose human populations, wildlife, and the environment to the risk of harm. Although U.S. law requires a firm to submit test data ninety days prior to manufacturing a new chemical (essentially the burden of coming forward with the evidence) that it believes shows that the chemical will not present an unreasonable risk, government regulators bear the burden of proving that the chemical "may present an unreasonable risk or injury to health or the environment" prior to regulating the chemical's use. The U.S. legal system likewise typically places the burden of proof on a plaintiff—the individual or entity seeking redress of injuries alleged to have been caused by chemical use. These rules go a long way toward shifting the uncertainty burden from chemical manufacturers to the public.

Perhaps a closer analogy to the SBT dispute is how the United States treats species that are threatened with extinction. The Endangered Species Act protects any species that is endangered or threatened with extinction throughout all or a significant portion of its range. However, species receive no protection until the government has acted—often at the legal prompting of NGOs. Thus, initially, the burden of proof is on the proponent of protection rather than those entities engaged in activities that threaten or endanger a species. Once an endangered species is added to the list, however, a person may not take (hunt, capture, kill) a member of that species unless he has obtained either an exemption or an incidental take permit. Both mechanisms effectively place the burden of coming forward with evidence and the burden of proof on the applicant, and contain exacting standards. In a somewhat similar vein, the Marine Mammal Protection Act seeks to maintain marine mammal populations above their "optimum sustainable population" and prohibits taking with certain enumerated exceptions.

It is worth noting that the governing standards in all of the above examples are legal standards without equivalent or easily translatable scientific standards. Thus, concepts such as "safe," "effective," "optimum sustainable population," and "unreasonable risk" do not have comparable scientific definitions. They contain implicit value judgments in terms of desirable outcomes. Indeed, environmental laws and the institutions created to enforce them (such as UNCLOS) are humanly devised constraints that shape human interaction by introducing an agreed-on and structured approach to decision making and governance. Laws are developed in a sociopolitical context where social norms and values, bargaining strength, existing property rights, economics, and politics shape the outcome.[68] Yet, the results of scientific investigations are an important component of this sociopolitical context and must be evaluated to determine whether legal standards are in fact being met. To facilitate such evaluations, legal standards need to be translated into operational, scientifically appropriate, and objective performance measures.

Moreover, legal standards are generally not framed in words that recognize the inherent uncertainty and limitation of scientific inquiry and predictions. Scientific results are reported in terms of probabilities and rarely provide absolute certainty—particularly in natural resource and environmental problems. The translation of legal standards into scientific practice and vice versa requires distinguishing clearly between the role of governance (decision making) and science (provision of information and advising on consequences). Further complicating matters is the need to judge the adequacy of scientific evidence being brought to bear in a given dispute.

It is also worth thinking about a fundamental difference in the above legal examples and the SBT case: the SBT case raises the issues of defining the legal limitations on scientific activity and the burden of proof on what constitutes valid science. Science tends to be seen as its own justification in modern society, and Japan consistently maintained in its legal arguments that the "case involves nothing more than a disagreement about a matter of science."[69] It further contended that such issues are "measures that lie within the margin of appreciation allowed to States."[70] Should science be considered a sufficient justification? This is not a simple question. Academic freedom and lack of political interference in universities and related institutions are generally considered fundamental principles. However, at the national level, states clearly recognize that the pursuit of science is not an unbounded right and impose legal restrictions on scientific experiments (e.g., ethical standards on the use of animals, human cloning, embryonic stem cell research). At the international level, Japan's contentions demonstrate that there is no consensus that international treaties can override a state's right on such issues.

The failure of the SBT case to proceed to the merits phase before the arbitral tribunal left unresolved the appropriate international standards for large-scale experiments in the governance of natural resources, such as Japan's EFP. Such experiments, if properly designed and executed, are recognized as powerful tools that can result in improved management and reduction in the overall biological risk of stock collapse. However, the catch in such experiments can be highly valuable—the value of Japan's EFP catches for a single year was on the order of US$24 to US$48 million. Without rules governing the collection of data and well-defined standards and international mechanisms for judging the validity of proposed experiments, no burden of proof is required of a state that chooses to unilaterally undertake an experiment. In such a circumstance, other states have no power to prevent the experiment, even if from their perspective, the scientific process is being used as a pretext for increased catches or as a mechanism to assuage domestic political concerns or further political, economic, or social interests. To minimize political factors, states need to develop and agree on rules governing the collection and interpretation of data prior to any party undertaking an experiment. Current controversies over Japan's ongoing scientific whaling in the Antarctic and North Pacific and previous scientific whaling by Norway in the early 1990s reinforce the need for a mechanisms for resolving conflicts over scientific catches.[71]

The 1995 Straddling Stocks and Highly Migratory Fish Stocks Agreement (FSA), which is now in force, however, is likely to alter the landscape fundamentally. The FSA adopts a precautionary approach (Article 6) as a fundamental standard for managing internationally shared fishery resources.[72] It, thus, effectively shifts the burden of proof from those nations that have concerns regarding species depletion to the proponent of an activity. The precautionary approach might have particular force in those instances where uncertainty can lead to irreversible changes, such as the collapse of a fishery. Moreover, because the FSA adopts UNCLOS's dispute-resolution procedures, it also provides a binding mechanism for setting standards if disagreements arise.

Global and Regional Institutions

At this point it is worth considering what role international institutions might play in managing shared natural resources. While common pool resource problems are perhaps intractable if individuals are viewed as solely self-interested profit maximizers, international institutions provide states (hence, individual citizens, as states are merely agents of those individuals) with forums where they can act as members of a community. It is this community-based decision making—or what Hardin refers to as mutual coercion—that provides an answer and an alternative to the tragedy of the commons. An alternative solution to common-pool resources is the issuance of property rights (e.g., land titles in the case of grazing commons), thereby eliminating the basic source of the problem. Although rare within the United States, property rights systems, generally referred to as individual transferable quotas (ITQs), are becoming increasingly common and are often seen as highly effective in fisheries management.[73] In such cases, government and management institutions still play a critical role in establishing and enforcing the rules of the market.

From a policy standpoint, the SBT case also raises the following questions: Can global institutions like UNCLOS be effective if their rules and policies can be superceded by regional arrangements? As with federal–states' rights issues under the U.S. Constitution, when should regional and more specialized institutions have the final authority for governing common-pool resources? Can global governance (or at least oversight) be practical and effective? In the case of SBT, the arbitral tribunal left Australia and New Zealand at the mercy of a less effective SBT convention and suggested that a nation can condition its participation in a regional organization on terms that allow it to avoid UNCLOS's compulsory dispute-resolution provisions, taking much of the bite out of that far-reaching convention. Whether or not this will be the case will depend on a state's view of international law and sovereignty and the power it is willing to cede to a regional institution. For those states that have been reluctant in the past to agree to compulsory dispute-resolution procedures, a consequence of the SBT case might be that they could become even more intransigent in the drafting of new regional agreements, knowing that they could prevent UNCLOS from being invoked if a dispute arose. Thus, the effect of the SBT decision may be to reduce collaboration among states on resolving specific international common-pool resource issues. Yet, nations need not, and all will not, structure their regional agreements in such a manner. Indeed, the unsatisfactory outcome (in the sense that the merits of the dispute were never heard by the arbitral tribunal) in the SBT case may leave nations wary of constructing dispute procedures that do not ultimately mandate resolution. But again, binding procedures are difficult to negotiate because states are often reluctant to place greater limits on their sovereignty.

Additionally, as the arbitral tribunal noted, the FSA, which all three parties had signed and which is now in force, may go a long way toward resolving many of the substantive and procedural problems that parties encountered.[74] Article 30 of the FSA requires nations to "agree on efficient and expeditious decision-making procedures" within regional organizations and to "strengthen existing decision-making procedures as necessary" in order to prevent disputes. The FSA also applies UNCLOS's binding dispute-settlement provisions (with appropriate adjustments) to disputes regarding the interpretation and application of regional fisheries agreements such as the SBT convention. Finally, a novel provision strengthens the hand of regional fishery organization against nonmembers by allowing states that are parties to the FSA and to a regional agreement to take compliance measures against another state party to the FSA, regardless of whether or not such state is a party to the regional agreement.

More generally, what did dispute-resolution procedures through a global institution accomplish in the present case and what might they accomplish in other cases? First, the filing of legal proceedings forces government officials at the highest levels to examine the dispute in a larger political context. For example, a state might not want to be so publicly at odds with an important strategic or trade partner, or, alternatively, it might not wish to exacerbate other lingering disputes between it and the other state. Second, the submission of a dispute to a neutral body requires parties to measure their position against legal standards rather than against their relative political power. Third, preparing and putting a legal and scientific case before a tribunal such as ITLOS forces each party to reexamine the issues in dispute and to consider the arguments that favor the positions of its adversaries. And fourth, formal proceedings bring the dispute out into the glare of the public spotlight.[75] In that regard, the SBT dispute is noteworthy in that not only were the orders and awards by ITLOS and the arbitral tribunal placed on the Internet, but the transcripts of the proceedings were as well.[76] In the case of SBT, the combination of provisional measures and the heightened public and political attention that the case brought appears to have been the catalyst that allowed the parties to move forward.[77]

Finally, the SBT case demonstrates aptly that the shape and character of institutions and the legal arrangements supporting those institutions matter. Indeed, they can have a significant effect on the manner in which international policies are operationalized and implemented and in which states govern common-pool natural resources (and thus ultimately on the actual conservation and sustainable utilization of those resources), as well as on the recourse states have when disputes arise over how those resources should be managed. Given the uncertainty associated with natural resource systems such as fish stocks, scientists, lawyers, resource managers, and policy makers need to work closely during the drafting stage of international agreements to ensure that those agreements incorporate effective governance mechanisms and procedures with realistic expectations of the role and ability of science to provide advice on and predictions about the consequences of management action in complex international problems. This is essential so that scientific principles and results are not lost in a legal soup, but rather, can be brought to bear on international environmental problems.

NOTES

1. Garret Hardin, "The Tragedy of the Commons," *Science* 162 (1968): 1243–48. While common-pool problems are not unique to the sea, when a resource such as

SBT migrates across sovereign boundaries and international commons, it poses additional management challenges. R. Bratspies, "Finessing King Neptune: Fisheries Management and the Limits of International Law," *Harvard Environmental Law Review* 25 (2001) 213–58. Straddling stocks and so-called highly migratory fish are a human-made, rather than a biological, phenomenon.

2. A single SBT female will spawn several million or more eggs a day for an unknown period of at least many days. The number produced is greatly in excess of the number needed to replace the current population.

3. For more detail, see R. Hilborn and C. J. Walters, *Quantitative Fisheries Stock Assessment* (New York: Chapman and Hall, 1992).

4. Selectivity is the differential in the relative rate of harvest for different age or size classes in a particular fishery. The stock/recruitment function is a mathematical expression of the average relationship between the size of the spawning stock and the subsequent number of recruits produced.

5. T. Polacheck, A. Preece, N. Klaer, and A. Betlehem, "Treatment of Data and Model Uncertainties in the Assessment of Southern Bluefin Tuna Stocks," in *Fishery Stock Assessment Models*, eds. T. Funk, J. Quinn II, J. Heifetz, J. N. Ianellis, J. E. Powers, J. F. Schweigert, P. J. Sullivan, and C.-I. Zhang (University of Alaska Sea Grant, AK-SG-98–01, 1999).

6. Hilborn and Walters, *Quantitative Fisheries.*

7. T. Davis, J. Farley and J. Gunn, "Size and Age at 50 Percent Maturity in SBT: An Integrated View from Published Information and New Data from the Spawning Ground." CCSBT-SC/0108/16, 2001; J. Gunn, J. Farley and N. Clear, "The Age Distribution and Relative Strength of Cohorts of SBT on the Spawning Grounds," CCSBT/9807/39, 1998.

8. J. M. Kalish, J. M. Johnston, J. S. Gunn, and N. P. Clear, "Use of Bomb Radiocarbon Chronometer to Determine Age of Southern Bluefin Tuna Thunnus Maccoyii. Mar." *Ecol. Prog. Ser.* 143 (1996): 1–8.

9. A. E. Caton, "Review of Aspects of Southern Bluefin Tuna Biology, Population and Fisheries," in *World Meeting on Stock Assessment of Bluefin Tunas: Strengths and Weaknesses. Special Report,* eds. R. B. Deriso and W. H. Bayliff, (La Jolla, Calif.: InterAmerican Tropical Tuna Commission, 1991): 181-357.

10. The most recent assessments are unable to determine if the parental biomass is still slowly declining, increasing, or remaining unchanged. Anon, "Report of the Sixth Meeting of the Scientific Committee," CCSBT, Tokyo, Japan, 28–31 August 2001.

11. Anon, "Report of the 1998 Scientific Committee Meeting," CCSBT, Tokyo, Japan, 3–6 August 1998.

12. Anon, "Report of the Thirteenth Meeting of Australian, Japanese and New Zealand Scientists on Southern Bluefin Tuna," Wellington, New Zealand, 19–29 April 1994; Anon, "Report of the First Meeting of the Scientific Committee of the CCSBT," Shimizu-Shi, Japan, 10–19 July 1995; Anon, "Report of the Second Meeting of the Scientific Committee of the CCSBT," Hobart, Australia, 26 August–5 September 1996; Anon, 1998 scientific committee meeting report.

13. This means that all decisions of the commission are binding on the parties, which is generally not the case in international organizations.

14. Officially reported catches by Australia, Japan, and New Zealand have generally been within their respective national allocation of the TAC, with relatively small overcatches reported in some years. However, concern exists that actual fishing-related mortalities since 1989 may be substantially greater due to unreported catches and discarding of poorer quality or juvenile fish (small fish are worth substantially less per kilogram).

15. Anon, "Report of the Fourth Annual Meeting," First Part, CCSBT, Canberra, Australia, 8–13 September 1997.

16. One can question, as has the CCSBT Scientific Committee, whether long-term projections under constant catches are an appropriate basis for setting TACs (Anon, 1998 scientific committee meeting report).

17. Anon, 1995, 1996 and 1998 scientific committee meeting reports. Note that the first stock assessments since 1998 were conducted in 2001 because the dispute disrupted the normal CCSBT scientific process. The results of the 2001 assessments all indicated a low probability of recovery, and consensus existed at the 2001 scientific committee meeting that "at current catch levels there is little chance that the SBT spawning stock will be rebuilt to the 1980 levels by 2020, and substantial quota reductions would be required to achieve that goal." (Anon 2001, scientific committee meeting report).

18. Anon, 1998 scientific committee meeting report.

19. Over seventeen different factors with multiple hypotheses have been considered within different assessments. For example, uncertainty about the age of maturity has resulted in scientists considering a range of values from age eight to age twelve. Estimates of current stock size and projection results differ depending on which age is used within the stock assessment model. A weighted average is used to get an overall estimate of the results for differing ages of maturity (and all other factors considered). The weights used in averaging the results reflect different scientific delegations' judgments that a particular hypothesis was true. Thus, for the age of maturity, the Japanese delegation gave 100 percent weight to age eight (i.e., absolute certainty to this value), while Australian and New Zealand delegations gave their greatest weight to age twelve, but considered that there was a nonnegligible possibility that the age of maturity may be lower. Anon, 1998 scientific committee report.

20. This is not as simple as it might appear because it still does not provide data on how abundance has changed over time in the varying areas not fished. Moreover, uncertainty in the interpretation of CPUE is only one of a number of substantive factors contributing to the difference in the projection results.

21. In the fishery literature, the concept of an EFP is more commonly referred to as adaptive management. C. J. Walters, *Adaptive Management of Renewable Recourse* (New York: Macmillian, 1986); Hilborn and Walters, *Quantitative Fisheries*.

22. Anon, 1994 thirteenth meeting report.

23. NRIFSF, "Research Proposal for the Southern Bluefin Tuna (SBT)," revised, SBFWS/95/23, 1995. The written proposal indicated that it would require "far higher catches than the Japanese quota alone," but provided little detail.

24. Anon, "Report of the Special Meeting of the Commission for the Conservation of Southern Bluefin Tuna," Canberra, Australia, 17–19 January 1996.

25. Anon, "Report of the Second Special Meeting of the Commission for the Conservation of Southern Bluefin Tuna," Canberra, Australia, 23 April–3 May 1996.

26. ITLOS, *Southern Bluefin Tuna Cases (New Zealand v. Japan; Australia v. Japan),* Japan Response, Case Nos. 3 and 4, Order issued on 27 August 1999; Proceedings available at www.itlos.org/start2_en.html, accessed 21 July 2003. References to other documents in the ITLOS proceedings are denominated "ITLOS, SBT Cases," and then indicate the document at issue such as "Statement of Claim."

27. NRIFSF, "Evaluation of CPUE Interpretation through the Analysis of Results from the 1998 EFP"; Attachment E, First Joint Experimental Fishing Program Working Group Meeting (EFPWG(1)); Record of Discussions and Work Program. Commission for the Conservation of Southern Bluefin Tuna. Tokyo, 1–3 February 1999. Australian scientists were not able to reproduce these calculations and stated that the Japanese claim was inconsistent with previous projection results presented at the 1998 CCSBT Scientific Committee meeting.

28. ITLOS, *SBT Cases*, Japan Response.

29. ITLOS, *SBT Cases*, Statement of Claim, ¶¶ 18–19.

30. Anon, "First Joint Experimental Fishing Program Working Group Meeting (EFPWG(1))." Record of Discussions and Work Program. Commission for the Conservation of Southern Bluefin Tuna. Tokyo, 1–3 February 1999.

31. Anon, 1999 First EFPWG Working Group Meeting.

32. ITLOS, *SBT Cases*, Statement of Claim, ¶¶ 20–35.

33. A. O. Adede, *The System for Settlement of Disputes under the United Nations Convention on the Law of the Sea* (Dordrecht, the Netherlands: Martinus Nijhoff, 1987).

34. ITLOS, *SBT Cases*, Statement of Claim, ¶¶ 37, 45, 51–62, at http://worldbank.org/csid/bluefintuna/SBT-Statement-Claim.pdf, accessed 21 July 2003.

35. This contention assumes for the sake of argument that ITLOS has jurisdiction and irreparable harm is not required.

36. ITLOS, *SBT Cases*, Transcript 19 August 1999, Morning, 30.

37. Wigmore heralded cross-examination as the greatest legal tool ever employed to discover the truth. Cross-examination thus may be seen as the legal counterpoint to the view of science as a search for knowledge and truth that was first explored in the introductory chapter of this casebook. J. Wigmore, *Evidence*, vol. 5, § 1367, rev. ed. (Boston: J. Chadbourn, 1974): 32.

38. A showing of only prima facie jurisdiction, rather than jurisdiction in fact, at this stage, is consistent with the concept of urgency embodied in provisional measures.

39. ITLOS, *SBT Cases*, Transcript 18 August 1999, Morning, 24.

40. ITLOS, *SBT Cases*, Transcript 18 August 1999, Morning, 24.

41. ITLOS, *SBT Cases*, Proceedings, 16 and 18–20 August 1999.

42. ITLOS, *SBT Cases*, Order, ¶¶ 43–4, 55.

43. Irreparable harm is an injury that cannot be adequately compensated monetarily.

44. ITLOS, *SBT Cases*, Proceedings.

45. ITLOS, *SBT Cases*, Proceedings.

46. ITLOS, *SBT Cases*, Proceedings.

47. ITLOS, *SBT Cases*, Proceedings.

48. ITLOS, *SBT Cases*, Proceedings.

49. ITLOS, *SBT Cases*, Proceedings.
50. ITLOS, *SBT Cases*, Transcript 19 August 1999, Morning, 11. The necessity of showing irreparable harm prior to an injunction being issued is the standard applied in Anglo-American jurisprudence as well.
51. ITLOS, *SBT Cases*, Proceedings.
52. ITLOS, *SBT Cases*, Transcript 18 August 1999, Afternoon, 25. The ICJ, which was established in 1946, is the primary judicial body of the UN.
53. ITLOS, *SBT Cases*, Proceedings.
54. ITLOS, *SBT Cases*, Order, ¶¶ 71–4 and ¶ 77.
55. ITLOS, *SBT Cases*, Order, ¶ 77.
56. ITLOS, *SBT Cases*, Order, ¶ 80.
57. ITLOS, *SBT Cases*, Order, ¶¶ 70, 80.
58. ITLOS, *SBT Cases*, Opinion by J. Laing, ¶ 3.
59. ITLOS, *SBT Cases*, Order, ¶ 90.
60. ITLOS, *SBT Cases*, Order, ¶ 90.
61. United Nations, *Southern Bluefin Tuna Case—Australia and New Zealand v. Japan*; "Award on Jurisdiction and Admissibility," rendered by the arbitral tribunal constituted under Annex VII of the United Nations Convention on the Law of the Sea (8 April 2000), ¶ 49, available at www.worldbank.org/icsid/bluefintuna/main.htm, accessed 5 August 2002.
62. United Nations, *SBT Case*, Award, ¶ 50.
63. United Nations, *SBT Case*, Award, ¶¶ 51–2.
64. United Nations, *SBT Case*, Award, ¶ 52 and ¶¶ 53–60.
65. United Nations, *SBT Case*, Award, ¶¶ 56–59.
66. United Nations, *SBT Case*, Award, ¶ 67.
67. United Nations, *SBT Case*, Award, ¶¶ 65–8.
68. Douglass C. North, *Institutions, Institutional Change and Economic Performance* (Cambridge, UK: Cambridge University Press, 1990): 3, 46–7.
69. ITLOS, *SBT Cases*, Japan Response.
70. United Nations, *SBT Case*, Japan Memorial on Jurisdiction, 84.
71. Many of the legal and environmental issues surrounding these experimental whaling catches are quite different from those involved in the SBT case. For example, the animals have not been taken from highly depleted stocks. Moreover, the International Whaling Convention has a specific provision that allows states to unilaterally take whales for scientific purposes and the commission under the convention has established standards and guidelines for such takes. However, as in the SBT case, there is no institutional arrangement for resolving the ongoing controversy about the efficacy of the actual programs and whether they meet the commission's guidelines. See www.iwcoffice.org (accessed 5 August 2002) and the annual reports of the IWC and its scientific committee meetings.
72. The FSA uses the term "approach" rather than "principle." Approach is often used in fisheries management where management errors can have irreversible consequences, yet socioeconomic considerations are prominent.
73. R. Shotton, "Use of Property Rights in Fisheries," *Proceeding of the FishRight99 Conference*, Freemantle, Western Australia, 11–19 November 2000, FAO Tech. Pap 404/1 (342) and 404/2. The CCSBT convention requirement to establish national allocation within the TAC represents a form of "national" property rights.

74. The FSA came into force on 11 December 2001. At the time of this writing, Japan has not ratified the FSA; Australia and New Zealand have.

75. Bill Mansfield, "The Southern Bluefin Tuna Arbitration: Comments on Professor Barbara Kwiatkowska's Article," *International Journal of Marine and Coastal Law* 16(2) (2001): 361–6.

76. Barbara Kwiatkowska, "The Australia and New Zealand v. Japan Southern Bluefin Tuna (Jurisdiction and Admissibility) Award of the First Law of the Sea Convention Annex VII Arbitral Tribunal," *International Journal of Marine and Coastal Law* 16(2) (2001): 239–93.

77. At the formal level, the parties have negotiated an end to the dispute. Moreover, significant progress has been made at the functioning level of the CCSBT in developing agreed-on processes and programs for management and science within the commission. See e.g., Anon, Report of the Eighth Annual Meeting, CCSBT, Miyako, Japan, 15–19 October 2001. Thus, the commission has implemented as a high priority a process to develop a management procedure (a decision rule for future TACs). An agreed-on CCSBT Scientific Research Program that will improve data for stock assessment, but not require additional catches, also has been adopted. The 2001 scientific committee meeting provided a consensus report in which it was agreed that the probability of meeting the CCSBT recovery objective under current catches was low. Despite all of this progress, the parties remain unable to reach an agreement on the TAC, with Japan still arguing for an increase and with Australia and New Zealand arguing for the status quo or a decrease (Anon, eighth annual meeting report, 2001).

10

No Fence Can Stop It

Debating Dioxin Drift from a Small U.S. Town to Arctic Canada

Michael S. Carolan and Michael M. Bell

The growth of environmental concern in recent decades has presented politicians, scientists, academics, and citizens alike with a significant challenge: how to protect both human life and the environment from the risks associated with modern technology, while simultaneously reaping the benefits associated with such technology. The significance of this challenge resides in the invisibility and uncertainty associated with many of today's risks, making them increasingly difficult to comprehend, which leads to great contention. The hole in the ozone layer is beyond direct human perception. We can be exposed to dangerous levels of radiation or radon without even knowing it. Nor can we see, taste, or feel dioxin. Consequently, we need to trust others, and even the machines of others, to do this seeing, hearing, smelling, and tasting for us. The inability to perceive these potential risks on our own leads to the contentious nature of the politics of scientific uncertainty.

As early as 1972, Alvin Weinberg observed that policy makers were increasingly being called upon to construct public policy based on uncertain science.[1] Concerns over the impact of modern production technologies on the environment now routinely compel decision makers to act before the scientific community reaches a consensus definition of the problem and appropriate solutions. The possible consequences (for example, global warming, human suffering, potential economic impacts) of waiting are often too great to justify delaying action any further.[2] Consequently, since Weinberg's article, increasing attention has been paid to the (often tenuous) relationship between science, environment, and health policy making.[3]

Yet the tension between science and politics does not simply reside in a need to make a decision before a scientific consensus is reached, but in the fear of the consequences of inaction. What are decision makers to do when, as is increasingly the case, "science" contradicts itself? What are they to do when group X has an "expert," who, using scientific evidence, states global warming is human induced, and group Y has an "expert," who uses similar scientific evidence to refute that claim? Knowledge thus becomes as much a matter of social relations and interpretations as of objective empirical facts.[4]

As philosophers of science (from Kuhn to Feyerabend to Latour) have argued, science always needs to be placed within an interpretative context.[5] Science is a product of social networks that shapes people's impressions of what the facts are. When a fact is brought to our attention, we immediately ask, "Whom did you hear that from?" In doing this we are identifying knowledge. You can see examples of this throughout this chapter (and indeed throughout this book). Whenever we reference an idea, concept, or fact, we are linking it to a particular expert, giving it its own history and its own identity. We are therefore saying, "You can trust us as speaking the truth because others (who themselves can be trusted through processes of peer review) have made similar statements." A similar process goes on in daily life. Whenever a friend or family member or even a stranger on the street informs us of something, we are often eager to know the origin of that knowledge. Was it something they had learned from the newspaper or the TV? Perhaps they heard it from another friend, family member, or stranger. Thus, when we encounter knowledge, we want to know not just what and which knowledge it is, but whose knowledge it is, so we can determine how it fits with our personal sense of trustworthiness.[6]

This is not to suggest, however, that knowledge cannot be public, that it is possessed and judged only by its proponents. But much scientific knowledge of botany and physics, of Linnaeus's categorizations and Mendelian genetics, and of Boyle's Law and Brownian motion is broadly accepted and, thus, more secure and perceived to be "better." As most of us are not botanists or physicists, we trust the specialists if accepted scientific knowledge is to have any meaning for us. These examples only further illustrate the importance of identifying knowledge, for here it is named, literally. And by naming knowledge, we establish its social relations and our own relation to it.

The importance of whose knowledge—of identifying knowledge—can be witnessed within the political arena as well. This is exemplified by policy makers' use of Nobel Prize scientists, scholars from esteemed Ivy League universities, and intellectuals from world-renowned think tanks and organizations. The implicit statement is that their knowledge is better in some sense than that of an unknown scientist working, for example, for the radical environmental group "Earth First!" Knowledge means nothing to us and has no political value for us, if we cannot first place it within a certain social network, thereby giving it an identity. And if we believe that social network to be trustworthy, we will likely feel the same about the knowledge to come from that social network.[7]

"Truth" is also a social, as opposed to a purely scientific, concept; the "laws" of science are only hypotheses, tentative and open to further testing and change (for example, Newtonian mechanics gave way to relativity and quantum mechanics, which may be giving way to string theory). Fundamentally, science is an orderly investigation of reality using defined methods that include deductive theorizing, rational and systematic investigation, peer review, and replication of empirical evidence, on the basis of which we form hypotheses on how to act. Ultimately, the "laws" and "truth" produced by scientific investigation achieve legitimacy by social acceptance (and the social acceptance of the scientific method) in modern political economies.

Scholars of international environmental policy tend to focus on explaining what has been done to mitigate or remediate international environmental problems and how these efforts can be improved. It is unusual to find a scholarly case study of an international environmental problem that has not been addressed by the international community (a case study of the dog that did not bark, as Sherlock Holmes would say). This and chapter 7 by Radoslav Dimitrov in this volume may be the first.

In October of 2000, a report by Barry Commoner, the well-known biologist and environmentalist, alleged that dioxin produced by the municipal power plant in Ames, Iowa, was a major source of dioxin pollution in Arctic Canada. However, this report received little attention: the alleged victims as well as the international community were either uninformed or uninterested. Why was this? Did dispassionate scientific analysis show the errors in Commoner's work? Or were powerful social groups able to use other knowledge to persuade decision makers that Commoner was wrong?

The story begins by introducing the reader to dioxin (what is it? what are the human health effects of dioxin exposure?) and some of the science and politics associated with it. This is followed by a detailed chronological description of the debate and the particular "science" around which this debate revolves. We then continue with an analysis of the social relations of trust and truth. In the final section, we take a step back and frame this discussion within a global and international context, presenting the reader with some of the ethical, political, and social questions that debates of this kind evoke.

INTRODUCING DIOXIN

The chemical loosely referred to as "dioxin" is in fact 210 compounds—75 dioxins and 135 furans—with similar structures and properties, of which only seventeen are toxic, with considerable deviation among these seventeen in their toxicity.[8] Dioxins are technically known as polychlorinated dibenzo-p-dioxins (PCDD), while furans are polychlorinated dibenzofurans (PCDF). Of these 210 compounds, one in particular—2, 3, 7, 8 tetrachorodibenzo-p-dioxin (2, 3, 7, 8-TCDD)—has been reported as being the most toxic synthetic chemical ever tested in the laboratory.[9] It is this compound, and a few other highly toxic relatives, that are generally referred to when scientists and politicians speak of dioxin.

Dioxins are the unintended byproducts of such industrial processes as bleaching wood pulp, manufacturing certain chlorinated chemicals, and incineration of municipal or industrial wastes (specifically the burning of plastics). Lower levels of dioxins and furans can also be produced by such sources as forest fires and backyard trash burning. The basic ingredients for dioxin are hydrocarbons (like wood or paper or coal), a chlorine source (like salt or plastic), and heat in the range of 700°C to 1,000°C.

One clearly demonstrated effect of exposure to dioxins is a disfiguring skin condition, chloracne, that can persist for years after exposure.[10] Other reported symptoms associated with exposure to dioxin include nausea, headaches, depression, and sexual dysfunction.[11] Likewise, a number of varieties of cancer have been associated to dioxin exposure; however, this remains a subject of controversy. Early epidemiological studies focusing on the incidence of cancer and birth defects among humans exposed to 2, 3, 7, 8-TCDD yield conflicting results.[12] The most significant human exposure to dioxin has occurred during such circumstances as industrial catastrophes and warfare—situations that are not conducive to scientific scrutiny because of the difficulty in estimating who was exposed and to what degree.[13] More recent studies, however, have concluded that dioxin is in fact highly carcinogenic, causing the EPA recently to upgrade dioxin's risk factor to one of the highest of any substance that it regulates.[14]

In 2000, by analyzing samples of people's body fat, the EPA determined that the population of the United States ingests seven billionths of a gram of dioxin per day through eating. From this it was concluded that the average person has a lifetime risk of acquiring cancer from dioxin of between one in one hundred and one in one thousand.[15] Human exposure to dioxin occurs almost entirely through animal foods (approximately 98 percent), specifically through the consumption of animals high in fat.[16] This is primarily because dioxin is highly soluble, causing it to accumulate in fatty tissue. Larger animals have a higher likelihood of being contaminated with dioxin because of their longer life span. Thus, a cow would very likely have higher levels of dioxin than would a chicken in a contaminated environment.

Dioxin enters the food chain through the air. In temperate climates, it is believed to be deposited on crops and subsequently consumed by farm animals in their feed. Consequently, in contaminated areas, dioxin is frequently found in milk and beef (again due to the relatively long life span of cows), which account for approximately two-thirds of the dietary protein in the American diet. In the Arctic, however, risks increase greatly because indigenous peoples have a diet high in marine mammals, such as seals and walruses and Arctic terrestrial mammals (e.g., caribou), which are long-lived and have a high fat content and, thus, a high concentration of dioxin.

No Fence Can Stop It

The geographic proximity of the United States and Canada and the earlier transboundary problems like acid rain (see the chapters by Don Munton and Kenneth Wilkening in this volume) have caused both countries to look at what the other is releasing into the atmosphere. But one toxic chemical in particular has received significant attention by both countries in recent years: dioxin.[17] And it appears that Canadians may have more reason to worry about dioxin from the United States entering their food chain than vice versa.

When emitted into the atmosphere, dioxin has a tendency to move northward in summer atmospheric conditions (for example, there is a northward shift in the jet stream, and warm winds blow up from the Gulf of Mexico).[18] As it moves northward and reaches the colder atmospheric temperatures of arctic Canada, it begins to cling onto other particles within the atmosphere, which ultimately causes it to fall to the ground in the form of either toxic rain or snow, contaminating the soil, plant life, animals, and eventually humans.[19]

A report released by the "International Air Quality Advisory Board" in 1998 is quite clear about the transboundary dioxin pollution that occurs between the United States and Canada (with Canada overwhelmingly on the receiving end of this exchange). The report notes that pollutants released from the Midwestern United States and the Ohio Valley travel across southern Ontario, southern Quebec, and into the northeastern United States, while pollutants released from the southeastern United States travel up the northeastern corridor of the United States into the Atlantic provinces of Canada.[20] Dioxin levels in Arctic Canada, where there are no significant sources of dioxin and few people, are consequently far greater than one would expect. The high-fat diet of the Arctic people leads to their increased exposure to dioxin. Taken within this context, the emission of dioxin is an international problem; dioxin does not stop at a country's border. But it also is an international problem because the sources of dioxin production, modern industrial techniques, can be found in most countries.

Other examples of transboundary dioxin pollution can be easily found. Dioxin in the Great Lakes has been reported to come from as far away as Florida, California, and Mexico.[21] Dioxin has also been found in wildlife on remote Pacific Islands thousands of miles from its point of origin.[22] The Swiss Environment Ministry reports that while national emissions of dioxin from municipal incineration facilities are just sixteen grams per year, uncontrolled burning of waste by householders emits between twenty-seven and thirty grams of dioxin each year, which then proceeds to be carried through the atmosphere to the colder climates of the arctic. This is so, despite the fact that only 1 to 2 percent of Switzerland's municipal waste is burned illegally while 46 percent is burned in properly managed plants.[23]

Testing for Dioxin

While testing for dioxin in the soil, plant life, and animal life in a given area is a relatively straightforward process, its detection in municipal power plants like Ames's can be much more problematic. In something as complex as a power plant, where should dioxin be tested? In the combustion chamber? What about the stack? Plume studies are designed to test for dioxin in the plant emissions and its distribution in the atmosphere. Yet with plume studies, you again encounter the same question: where do you test? Maybe all of these aforementioned tests should be conducted. Such tests, however, are not cheap—a point that always weighs heavily among both politicians and taxpayers. One estimate, for example, places the cost of a comprehensive plume study as high as a million dollars.[24] Additionally, given the advances made in dioxin testing over the last decade, the tests have become so sensitive that it is very likely that some level of dioxin will be detected. The questions then become as political as they are scientific and have no easy answers: how much dioxin is acceptable, who makes this decision, and how is this decision ultimately made? These questions we shall return to below.

To curtail the high cost of dioxin tests, computer modeling and secondary data are frequently used to assess the dioxin emission levels at municipal incinerators like the power plant located in Ames. Yet such modeling is predicated on numerous assumptions, as we will show later in the chapter. The question then boils down to what model and "whose" numbers most closely approximate the truth? In other words, it boils down to the social relations of knowledge.

The question, how much dioxin is acceptable? is scientific in so far as science measures damage to an individual's health, but it is political with respect to the costs, including the economic opportunity cost of mitigation and the social and personal cost of lives lost. Yet even the assessing of damage to human health involves value judgments, as well as scientific issues. For instance, should the representative body for measuring the incidence of disease from dioxin accumulation be that of a small-framed woman, a child, an elderly person, a pregnant woman, or a very large man? They all have different thresholds to disease from dioxin exposure. What level of exposure risk is appropriate? What risk threshold is acceptable considering the costs of mitigation—a one in a hundred chance over a person's lifetime of acquiring cancer; or perhaps one in a thousand; what about one in a million? Individuals would probably prefer no risk of death from human-made chemicals, but policy decisions are framed in terms of social costs and benefits. Ultimately, even the most apparently scientific questions make implicit statements about reality, thereby making them as much ethical questions as they are political and scientific questions.

The Commoner Report

The dioxin dispute in Ames, Iowa, emerged in October of 2000 in response to the publication of a study by the North American Commission on Environmental Cooperation (NACEC), on which the lead author was the well-known biologist and environmentalist Barry Commoner.[25] The Commoner report claimed that the high levels of dioxin found among the Inuit people of Nunavut, Canada (the new Canadian province), came primarily from just a few dozen sources in the United States. It argued that one of the ten most significant was the waste-burning power plant located in Ames.[26]

Throughout the 1990s, Commoner engaged in numerous research projects to investigate dioxin generation and transmission. (As a result, he is now widely regarded as the United States's leading dioxin expert and activist.) In the mid-1990s, dioxin was found in Inuit hunting localities throughout Nunavut. Although there are no significant sources of dioxin in Nunavut or within five hundred kilometers of its boundaries, dioxin concentrations in Inuit mothers' milk were found to be twice the levels observed in southern Quebec. It was eventually concluded this was due to the elevated dioxin content of the indigenous diet—namely, foods such as caribou, fish, and marine mammals.[27]

As a result of these findings, NACEC commissioned the Center for the Biology of Natural Systems, with which Commoner is affiliated, to conduct a study to understand the origins of the Nunavut dioxin. Commoner thus set out to model, on a continental scale, the rates of deposition of airborne dioxin in the new Canadian arctic territory of Nunavut, with the intention of identifying the major contributing North American sources.

The computer model used in the Commoner report employed two major components: an air transport model and a dioxin emission inventory. The air transportation model—with the cumbersome but scientific-sounding name Hybrid Single-Particle Lagrangian Integrated Trajectory, or HYSPLIT—was originally developed by the National Oceanic and Atmospheric Administration to track the movement of inorganic radionuclides. Air transport models have been widely tested and validated by numerous research groups and can be expected to give reasonable semiquantitative results for well-characterized sources and sinks of transported chemical compounds. As part of an earlier study, Commoner had adapted the model to the transport of dioxin in the atmosphere. The adaptation required the addition of a mechanism for photolytic destruction of organic pollutants in the atmosphere by sunlight, but this does not bring into question the basic soundness of the air transport model. See figure 10.1, which shows the successive monthly air transport coefficient maps for the Ikaluktutiak land receptor. (Notice how nearly half of the total annual dioxin deposition occurs in only two months: September and October).

Commoner used this air transport model to rank North American sources of airborne dioxin deposited in Nunavut. Source emission data was obtained from national Canadian and U.S. inventories prepared by Environment Canada and the U.S. EPA, respectively, to which were added other data on back-

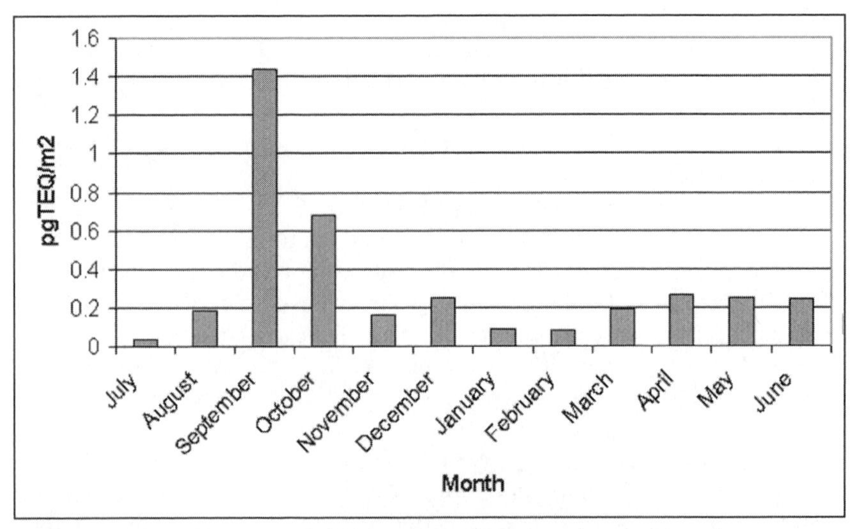

Figure 10.1 Monthly Dioxin Air-transport Coefficients and Deposition Flux to Ikaluktutiak Land Receptor, July 1996–June 1997

Source: Commoner et al., 2000.

Note: The scientific notation "pgTEQ/m²" indicates the number of picograms of Toxic Equivalent Quotient per square meter, a measure of the rate of toxic deposition.

yard trash burning and several other point source classes. However, these inventories of sources were not actual samples of emissions from each of the 44,091 registered North American sources. Using a method developed by the U.S. EPA, emissions from sources were estimated by allocating each source into one of four standardized emissions categories by source type. The Commoner report assigned the Ames plant to the category for incinerators and calculated its emissions based on that assignment. The model then calculated the amount of dioxin emitted by each source at its geographical location that is deposited at each of a series of receptor sites in Nunavut over a one-year period: July 1, 1996, to June 30, 1997. The model assumes that the dioxin is emitted as four-gram "puffs" at four-hour intervals from each source and tracks its predicted location and content at one-hour intervals. When a hypothetical puff is plotted as passing over one of the sixteen Nunavut receptor sites, the model records the calculated amount of dioxin that is deposited there.

The Commoner report cited the municipal power plant in Ames as being one of the largest contributors of dioxin in Nunavut. The EPA limit on dioxin emission for municipal incinerators for the size of the one located in Ames is 3.9g TEQ/yr., where the toxic equivalent quotient (TEQ) is scaled to the most toxic compound among the group of 210 related compounds. The Commoner report, however, cited the Ames plant as having a predicted dioxin emission of 58g TEQ/yr., well over the EPA limit. See figure 10.2, which lists the top nineteen individual dioxin sources (notice that Ames's municipal waste incinerator is denoted as the largest number-one-ranked source).

Scientific Reaction to the Commoner Report

The power plant in Ames is a publicly owned facility, designed and built in the late 1970s with the expertise of the engineering department of Iowa State University (also located in Ames). The charge leveled by the report resulted in an immediate response by public officials and university scientists, responses both vigorous, and perhaps predictably, defensive. The report was deemed "bad science" in newspaper stories, in discussions with the local city council, and in personal conversations between officials and the public. As a member of the Ames City Council stated to one of the authors (Bell) on the day after the story first appeared in the *Ames Tribune*, "our science is better than theirs." A faculty member from the industrial engineering department at Iowa State University used stronger language: "[T]he Barry Commoner report that our power plant is causing dioxin problems for the Inuit is flaky, discredited, and invalid."[28]

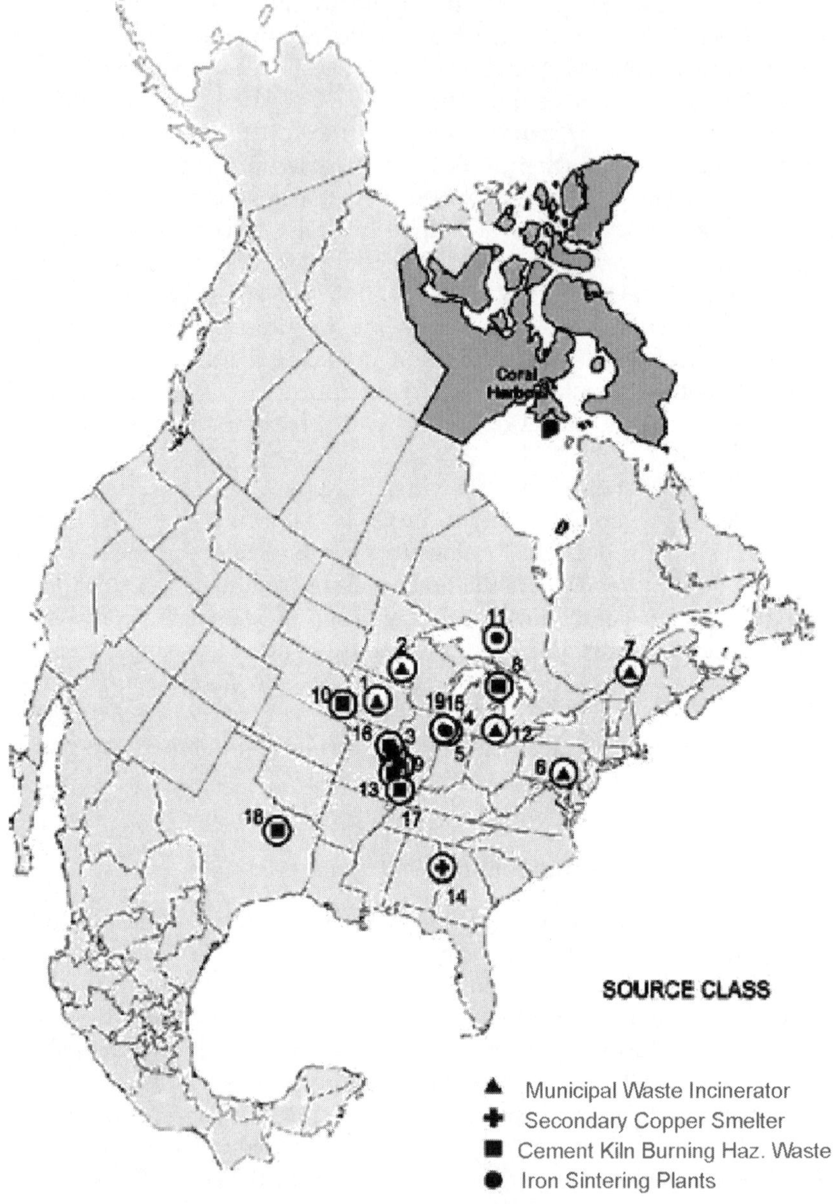

Figure 10.2 Highest-ranked Individual Sources that Contribute to 35 Percent of Total Dioxin Deposition at Coral Harbour Land Receptor

Source: Commoner et al., 2000.

The contention by the city and local scientists that the Commoner report represented "bad" and "flaky" science was based on two main arguments. First, they argued that the report was based on computer modeling of climate patterns, not actual samples from the Ames power plant. Second, they argued that although incinerators are frequently criticized as dioxin sources, the Ames plant uses an unusual method of incineration, a process called "cofuel" incineration. (City officials venomously reject the label "incinerator" in reference to the plant, preferring instead the designation of "cofiring plant" or "coal-fired electric generating plant.") Only 10 percent of the fuel source for the plant is waste; the remainder is coal, which raises the temperature of the combustion process to approximately 1,600°C. Supporters of the power plant argued that since dioxins only form between 700°C and 1,000°C, dioxins could not possibly be produced in the Ames plant.

The supporters also argued that the sulfur in the coal mitigates dioxin formation and that there is relatively little plastic in the waste burned at the plant and, thus, little chlorine, an essential ingredient of dioxin. They were quick to argue that a "cofuel" power plant also greatly lessens the need for landfills, often describing the plant as using "cutting-edge" technology that represents a potentially broader environmental solution. Most importantly, power plant supporters cited a 1981 test of the plant that detected no dioxins.

The Commoner report presented a different view. Commoner himself made a number of local public statements about the report's findings, and he even visited Ames twice following the release of the report. With regard to the argument by plant supporters concerning computer modeling, Commoner agreed that the report did not directly implicate the Ames power plant. However, the report was based on standardized models of power plant dioxin emissions, developed over many years by the EPA. Commoner also raised the findings of an earlier study he had led in the mid-1990s, which, again based on modeling, suggested that the Ames power plant was a leading contributor to dioxin found in milk from dairy farms in Iowa's neighboring state of Wisconsin. In this study, a research team led by Commoner examined dairy farms in Wisconsin, measuring dioxin levels in the air, pastures, mixed feed, and the milk.[29] According to computer models used in that report, 60 percent of the dioxin measured at one site located on a southeastern Wisconsin farm could be traced to the Ames incinerator. In a public speech made in Ames in the fall of 2000, Commoner claimed that this finding should not be surprising in light of this recent dioxin debate, stating, "of course you're going to find dioxin from the Ames incinerator in states like Wisconsin. More comes down in Wisconsin than Nunavut because it's closer."[30]

With regard to the second argument of supporters, Commoner argued in numerous public statements that the high temperature of combustion in Ames did not matter. Dioxin, Commoner claimed, can easily form after combustion in the power plant stack during cool-down of the plant's effluent. Commoner also argued that dioxin can form as a result of the electrostatic precipitator in use at the Ames plant to reduce ash emissions, due to the electrostatic sparks that create mini-environments with optimal temperature range for dioxin formation. In fact, based upon his "expert" opinion, Commoner stated at a public appearance that "the Ames incinerator has got to be producing dioxin," citing an EPA report released in 1978 (which did not directly test for dioxin) that listed all the organic and chlorine compounds detected at the Ames plant, including furans, to support his claim.[31] Based on this report, Commoner stated, "it's clear in 1978 the Ames incinerator was generating chlorinated organic material, much of which was far less stable than dioxin."

Finally, Commoner was also very critical of the dioxin emissions test conducted in 1981, arguing it was based upon "primitive, inadequate, and antiquated methods," citing recent technological advances in dioxin testing.[32] Again, Commoner agreed that there was no direct evidence that the Ames power plant was producing dioxins. But he did call on city officials to conduct a thorough dioxin test of the plant, using the latest methods, and to check for dioxin fallout in nearby fields. If dioxin were found, Commoner argued, the only reasonable response would be to shut down the plant altogether and use alternative sources of energy, such as the new wind power farms that have recently been established in northern Iowa (including one which is reputed to be the world's largest).

THE POLITICS OF TESTING

For eight months after the report was published, a debate ensued in Ames over whether or not to test the plant. On the side against testing were the city officials and the local university's engineering professors; on the other side was a loose and poorly organized group of local citizens, including a few professors from other departments in the university. Probably the strongest local public voice in favor of the testing was the Ames Quality of Life Network (AQLN), a local environmental group that has often been at odds with the city government. The AQLN and others attempted to advance the argument of Commoner, receiving a significant boost when Commoner came to speak at the Iowa State University campus.

Eventually, on April 24, 2001, the Ames City Council decided to hold off on a comprehensive dioxin test of the plant, citing its potential high cost. Instead, the council decided to commission an "engineering study" of the plant to determine if the test is necessary. Rather then choosing an independent third-party to conduct the preliminary engineering study, however, a local university engineering professor was asked to investigate.

The critics of the plant were hardly satisfied by this decision, which to them amounted to asking the fox to count the chickens to see if any had been taken. But in the view of the city, engineering professors, and other plant supporters, the critics would never be satisfied. Dioxin detection had become so sensitive, the supporters argued at the April 24 council meeting (in something of a change of position) that some dioxin is sure to be detected, albeit likely at 1/1,000 of the levels Commoner had suggested. If the city commissions a test, "the number will not be zero," testified Robert Brown, professor of mechanical engineering at Iowa State University.[33] The *Ames Tribune* story further reported, "Brown cited an additional unresolved question if comprehensive testing were adopted—it would be expensive, if not impossible, to thoroughly satisfy the plant's critics."

Then there's the question, according to Brown, of where to test? The city could test for dioxins in the furnace, the plant's cooler, or the stack and even perform plume studies to see if dioxin is formed in the atmosphere after smoke leaves the plant. "Plume studies could cost $1 million," Brown said. "And where do you test?"[34]

The assistant city manager of Ames went on to testify, "Our uniqueness was not appreciated by this [the Commoner] study," pointing again to Ames's unusual cofueling process designed by Iowa State's engineers from Brown's department. But, as the paper went on to report, the assistant city manager of Ames "added that he wanted to reassure the community and said the city's 'only objective was finding out the truth.'"[35]

The "truth" that the city council and Iowa State University engineering professors ultimately chose was that presented by the report of the Iowa State University engineering professor they had selected, Robert Brown, who recommended not to test for dioxin pollutants at the Ames municipal power plant, thereby officially removing the dioxin debate from the city council's agenda.

The Brown Report

Although the Brown report offers some support for aspects of the Commoner report, specifically the air transport modeling, it is most critical towards Commoner's "suspect" emission calculation.[36] First, Brown questions the Commoner report's assumption that the Ames power plant burns 210 metric tons of waste per hour (tph), which led to the calculation of a dioxin emission of 58g TEQ/yr. Brown reported that the Ames facility was designed to burn only 45 tph of fuel (coal and municipal solid waste [MSW]), nearly one-fifth the amount (210 tph) assumed in the Commoner report. Furthermore, the plant typically operates at only a fraction of its design capacity. For instance, in 2000 the plant operated at 60 percent capacity with an average throughput of 27 tph, a difference, Brown cites, as immediately reducing the upper bound on average dioxin emissions to 7.5g TEQ/yr.

Brown also notes that the Commoner report states that emission factors for conventional sources, such as incinerators, are based on fuel throughput, not the amount of MSW combusted. In fact, less than 20 percent of the total tonnage usually burned in Ames is MSW, the balance being coal, which Brown cites as being an insignificant source of dioxin. In 2000, just 10 percent of the fuel blend burned in the Ames power plant was MSW. Assuming that dioxin emission is proportional to the amount of highly chlorinated fuel burned, the upper bound on average dioxin emissions from the power plant becomes further reduced to 0.75g TSQ/yr. According to Brown, this "reduction in the estimated dioxin emission from Ames arises solely from errors Commoner made in estimating the throughput of highly chlorinated MSW at Ames" (emphasis in original).[37]

The Brown report then continues to demonstrate how additional factors overlooked by Commoner (namely, overestimates in the dioxin emission factor for the Ames facility) further reduce one's calculation for the amount of dioxin emitted from the plant. The Commoner report assumes an average dioxin emissions factor (DEF) of 31.57 ng TEQ/kg of combusted fuel. However, according to the Brown report, a more accurate estimate for an incinerator like the one located in Ames would be closer to a DEF of 0.13 ng TEQ/kg, citing other studies to support this argument.[38] Ultimately, by taking these miscalculations into consideration, Brown arrives at an annual emission of 0.14g TEQ/yr., well below the 3.9g TEQ/yr. limit set by the EPA on incinerators of comparable size.[39]

In its conclusion, the Brown report counters the Commoner report's claim that dioxin could be forming in the gas plume leaving the exhaust stack downwind of the plant. This argument by Commoner is based on the fact that dioxin can synthesize at much lower temperatures than was previously thought (namely, at temperatures lower than 700°C). The Brown report concurs that poststack synthesis of dioxin is a credible possibility in situations where the temperature of exhaust gases are several hundred degrees Centigrade. But Brown argues that in the Ames power plant, flue gas is cooled to less than 250°C before it is emitted from the plant, which is below the temperature range in which dioxin synthesis is thought to occur. The Brown report, however, does not address Commoner's argument that dioxin may form in the plant stack after combustion and before emission during the cool-down phase.

The final paragraph of Brown's findings nevertheless indicates a possibility that dioxins are still being produced at the Ames power plant, however slight. Brown writes:

> Although Commoner's argument for post-stack synthesis of dioxin is flawed, it illustrates the difficulties of attempting to respond to the charges he has made

against the Ames facility. If Ames undertook in-plant testing demanded by Commoner and dioxin emissions were found to be negligible, the controversy would not be at an end. Instead, Commoner could turn to his post-stack theory of dioxin synthesis and suggest plume sampling as the only way to resolve definitively the controversy. It is unlikely that such difficult measurements have ever been undertaken. However, the cost would undoubtedly be several fold more expensive than the $80,000 in-plant testing then being contemplated by the community of Ames.[40]

THE SOCIAL RELATIONS OF KNOWLEDGE AND SCIENCE

The language used in the final paragraph of the Brown report brings us back to the social relations of knowledge and the interrelationship between science and politics.[41] That paragraph suggests that Commoner is not a neutral scientist, but that he is motivated by an agenda and "would turn to his post-stack theory of dioxin synthesis" if the in-plant testing did not give the results he wanted. Yet, Brown too was attempting to shape the community's actions, implying that Commoner would never be satisfied and that, therefore, no testing at all was the best option. The Brown report, like the Commoner report, relied on theories of what is going on in the plant, not direct evidence, and it ignored the direct evidence that dioxin from somewhere is getting to Nunavut. There was substantial uncertainty on both sides—from both incomplete and inaccurate knowledge. And yet, the Brown report argued not only that enough was already known, but something even stronger—that more information would be harmful to the process. Still, we know that the city council accepted Brown's recommendation, not Commoner's, voting unanimously on June 26, 2001, not to test for dioxin.

What were some of the underlying social and political forces at work in this debate? Why was such an esteemed scientist as Commoner (and his conclusions) so quickly distrusted by the city and other plant supporters? Why was a lesser-known professor of engineering with substantially less expertise in dioxin given such credence? Why were some people in Ames in such haste to brush this whole controversy under the carpet, while others were so compelled to make it an issue on everyone's tongue? These questions begin to touch on the hidden social relations of knowledge present in all political disputes.

In this case, two main social networks and their associated knowledges underlay the dispute. On the one hand was the network of leading citizens—the elite of the local business community, city government, and the university, and those who are similarly concerned for the pride and respectability of Ames. On the other hand was the network of those who are often critical of the business community, the city government, and the university—activists of various sorts, generally not well placed in the elite social circles of the town. It would not be fair to say, however, that the activists are uninterested in the pride and respectability of Ames. But their vision of what gives a community a right to feel proud and respectable is a substantially different one, and substantially different from recent and current developments in the city.

It could be argued that the response of city government and the university engineers is simple defensiveness, that they see their reputations being challenged, and, therefore, they are out to protect themselves. We suspect that there is, in fact, a good measure of simple self-interest involved in the response from city government and university scientists. But it is also possible that they envision the measure of their reputations as requiring a rapid decision to test the plant for dioxin. Indeed, this is precisely how the AQLN envisioned the measure of their reputations. The fact that the city government and the university engineers do not see their reputations in this way indicates that they embed themselves in different networks of social honor. In other words, interest is certainly an important factor in the response of all local actors to the dioxin controversy, including the AQLN. But how local actors understand their interest is itself constituted through the same networks that constitute their sense of the truth.

Additionally, one could argue that more than accepting scientific interpretations from people whom they trusted, citizens in Ames accepted scientific interpretations that conveniently matched their own beliefs and interests. We suspect there to be some truth in this argument as well. But beliefs and interests are themselves embedded within social relations, including social relations of trust and distrust. You are, after all, in part, who and what you trust and distrust. Thus, by accepting scientific interpretations based upon personal beliefs and interests, these individuals would still be basing that acceptance, at least in part, upon social relations and, ultimately, upon trust.

Near the end of the dioxin debate, the mayor of Ames remarked, "The report by professor Brown which shows that our plant does not emit dioxin just gives credence to what we've been doing all these years."[42] Here the mayor takes the Brown report as a vindication of the status quo and the honor, reputation, and beliefs of those involved. As we mentioned earlier, the Ames power plant has long been a source of community pride both because of its cutting edge technology of cofuel incineration and because of the town-gown partnership that led to its construction. Dozens of local movers and shakers were involved in bringing the plant into being, and they no doubt felt their reputations threatened by the Commoner report.

On the other hand, the voices of the activists (other than Commoner's) were harder to hear, as they were rarely mentioned in the newspapers. As one member of the AQLN relayed, "the council says it wants to do the responsible thing. But they have a different idea of what responsibility is. They think it's saving money. I think it's protecting the public health. But I don't think they'll listen to us on that. They just listen to their friends."[43] Although the group has recently had some success with an effort to turn a local quarry into a park and a backup water supply reservoir for Ames, AQLN members frequently feel that their views on a wide variety of issues are little appreciated by local decision makers. AQLN members are not without friends, of course, but their friends similarly feel a degree of alienation from the local powers that be. As far as these local powers are concerned, AQLN members are troublemakers, and people with "agendas" (like Commoner) are insensitive to local history and blind to local realities. Ultimately, however, the feeling of alienation goes both ways.

Why, however, was not a compromise made between the AQLN members and the city government? Granted, there is little room for compromise when one side wants to test while the other does not (or at least does not believe it is necessary). But why was a compromise not made earlier in the debate as to, for instance, who would conduct the engineering study; perhaps an individual acceptable to both parties could have been selected to perform this analysis?

To compromise with AQLN would have given the group legitimacy. As French philosopher Michel Foucault argues, social relations are also relations of power.[44] In short, those with access to the dominate social networks also have an avenue through which to express their opinions and to have those opinions heard, all of which has bearing on how others perceive them as speaking the truth. In the case of Ames, if the city council were to arrive at a decision (at least in part) because of AQLN—for instance, reaching a compromise—this act would have given AQLN access to the social networks of power associated with the state, thereby making them part of the "establishment." In making a decision independently of such groups, however, the state was able to retain the social boundaries that exist between itself and these groups.

It is also our impression—although we have not conducted a comparative survey on this point—that the local activists are more likely to be recent migrants to Ames and even to the Midwest. Certainly, this applies to the leadership of the AQLN, which contains only one native Iowan out of five officers, and no native of Ames. By making this observation, we do not mean to besmirch the commitment of AQLN members to the Ames community. Indeed, new residents to a community may have a greater degree of local pride then long-term residents; after all, it was in part their excitement about their new community that caused them to move there. But since they are new, they are less likely to identify with the people who made local decisions long ago and their visions of what constitutes a quality community. At the very least, newcomers are less likely to have had an opportunity to meet these longer-term citizens and to get integrated into their social networks. Moreover, there are cultural differences between Midwesterners and people from elsewhere in the country, and these too may stand in the way of the easy integration of some newcomers into the networks of local respectability.

Therefore, when the Commoner report appeared, these two different networks—the "respectable" citizens versus the local activists—were already prepared to receive it quite differently. This is not only a result of the ideological orientations of these two groups, but also of whom these individuals associate with in the first place. Let us emphasize the "also," for the two reinforce each other. We gather our sense of ideological rightness—our sense of the standards for assessing knowledge claims—in large part from our associates, and we gather our associates in large part from our ideological sense of who is worthy to be associated with. The Commoner report clearly threatened one network and its ideological underpinings and supported another. There was face lost and face gained. Commoner himself does not live in Ames and had no prior social history in the community, so his face was not at issue locally, at least initially. But as the local debate developed, he too was socially placed within the community, at least tacitly.

This was similar for the Brown report. AQLN members expressed private outrage over the selection of an Iowa State engineering professor for the study of whether the city should test the plant. To them this choice seemed extremely biased. However, to the network of respectable citizens, Brown seemed the perfect person to conduct the study. Here was an engineering professor from the local university. Here was someone who knew the plant and the people who built it. Here was someone that could be trusted to understand the local situation and to evaluate the Commoner report appropriately. Here was someone whom local people knew and liked. Here was someone whom local people could trust—at least the local people making the decisions, that is.

But the relationship between trust in social networks and trust in knowledge was not simply a matter of a local social cleavage. In science we look for a source of knowledge free of social bias. Throughout the debate in Ames, there were continued efforts to root knowledge claims in this realm of truth considered to be beyond the social. The Commoner report and the counter-challenges to it were continually represented by their respective adherents as being based in "science." No matter which side in the debate they took, all the participants claimed that their views were objective and scientific.

The asocial status of science is not so clear, though, when experts looking at the same body of evidence come up with such different conclusions. The suspicion that social factors might be behind the different interpretations quickly comes to the fore. Science becomes a matter of "our science" versus "their science."[45] Or even more strongly, science becomes what our side follows, while the other side is deluded by their biases—a claim that both sides of the debate in Ames tried to assert. Thus, science becomes personalized as scientists versus nonscientists. For some in Ames, Barry Commoner is an environmental activist, a radical, a former presidential candidate with the left-wing Citizen's Party, and someone with little regard for Ames. For others, Barry Commoner is a noted scientist, an internationally known professor of biology, and perhaps the country's leading authority on dioxin. For some in Ames, Thomas Brown is a handpicked pawn that could be counted on to give the answer the respectable citizens wanted. For others, Thomas Brown is a top-notch academic, a respected professor of engineering with a close understanding of the town and its power plant.

It is a striking feature of political debates that involve science that those who are quick to see social bias in the science of others are typically just as slow to see social bias in their own science. But as Bruno Latour and other sociologists of science have argued, to see science as asocial science is to misunderstand both how science does work and how it should work.[46] The concerns of people who direct the science necessarily are addressed first. After all, science does not speak for itself—its interpretation is, at least in part, socially constructed. And problems can ensue over the interpretation of this knowledge and the appropriateness of scientific methods, especially when socioeconomic factors are concerned (such as economic costs and environmental effects). This is what appears to have occurred in Ames.

The best science recognizes this necessary human, therefore social, character. It acknowledges that it needs people, with all of their subjective values and biases, for it to speak and have meaning. By considering how others might perceive its evidence and explanations, such a science might be one that unites rather than divides us. While such science may be less pure, it may be more socially effective.

To our mind, one of the most striking social dimensions of the science under debate in the Ames dioxin dispute is that it seems no effort was made by any of the parties involved to engage the people of Nunavut, Canada, in the debate. Ames politicians made no effort. The AQLN made no effort. Barry Commoner made no effort. Robert Brown made no effort. As far as we are aware, no one thought to pick up the phone and call someone at the North American Commission for Environmental Cooperation for a suggestion of someone in the Inuit community who might contribute to the discussion. Commoner had likewise earlier identified rural Wisconsin—a large producer of dairy products—as polluted by dioxin from Ames. But the residents of that region were similarly excluded. Had the debate been designed to include more voices, we suspect that the outcome might have been quite different.

PUZZLES AND QUESTIONS

How, then, are politics and science interrelated? How is all science, in a sense, social science? This case would appear to support a postmodernist view—that all knowledge is contextual and relative—but only because the questions asked were beyond the purview of what "normal science" is capable of answering.[47] The science used was incorrect and inadequate.

On one side, the science was poorly done, using general models to identify point source–effect connections and apparently incorrect information on plant use and construction. On the other side, the science was deliberately foreshortened (for reasons of cost, civic pride, interest, and reputation). Politics usually needs a defined problem—for instance, that dioxin is carcinogenic—in order to pursue a policy remedy. Issues may derive from available solutions, from observation, from scientific evidence, or from events, but they become framed as problems only within a social context. Scientific evidence may identify that a million Americans are homeless. That is an issue. But is it a problem? Only if society decides it is. In Ames, the problem was that the data defining the issue—dioxin production and drift—was incomplete and insufficient to create a consensus. Thus, polarization and reliance on beliefs were almost inevitable. Without sufficient data or with incorrect data, belief in the rectitude of your side's position and trust in leaders becomes even more essential to joining the fray and taking a side.

Not all science is therefore equal; the point of departure between sciences is ultimately determined by what questions are asked. The question, what is the atomic weight of oxygen? is fundamentally different from the question, at what level of exposure is dioxin safe? The latter is a question for what some have called "research science," the former, for "policy science."[48] A similar distinction has also been made between "normal" (also referred to as "applied" science) and "postnormal" science.[49] In short, as questions become exceedingly complex and uncertain and the decision stakes increase, science not only does become more social, but it should.

Ultimately, recognizing that all science is social—that all science is social science—is not without its problems, though. When we are dealing with the uncertain and unclear, it can be difficult to convene a dialogue sufficiently broad to take into account all of the potential ways of understanding a situation. Indeed, with science we are almost always dealing with the uncertain and the unclear, which is why we turn to the special tools of science to resolve difficult questions. Difficult questions are, quite simply, difficult. Even with the best intentions and all of the time in the world, we may still discover that we do not all agree on what is truly going on. Even if we trust each other completely and share the same large network of the entire human community, we might still come to different conclusions about the truth. We are all different, after all. That is why we see things differently and, thus, have something to offer each other in determining what the truth actually is.

But decisions also have to be made. We cannot talk forever, especially if there really is a problem. Science, especially one that recognizes that it is social, will never give us a final answer anyway—nor is it meant to. At its foundation, science is solely an orderly investigation of the world around us—without it we are left to a clash of beliefs much like the one that the medieval Inquisition tried to quell. Indeed, without science, the debate in Ames would not have occurred because dioxin would never have been identified as carcinogenic. Science is, therefore, not about the presumed sanctity of its conclusions, but rather about inquiry. In short, it is about asking more questions.

This is the issue that the debate over the Commoner report raised in Ames and, by extension, in Nunavut, Canada. This is the issue that environmental disputes around the world so often raise. When do we know enough to make a decision? And who decides when we have talked enough and that a decision needs to be made?

But there are additional questions that must also be addressed—especially regarding issues pertaining to environmental and health policy making. For instance, in issues of risk assessment, what represents a "safe" level of risk? Regarding the Ames power plant, given that at least some dioxin is being produced, how much dioxin is too much? And who makes this decision—only the residents of Ames? What about those in surrounding communities or surrounding states (remember the study by Commoner regarding the dairy farms in Wisconsin), or surrounding countries (what about the Inuit in northern Canada)?

Likewise, what is the connection and difference between risk and uncertainty? Does risk presuppose uncertainty (and if so, does risk cease to be risk once it becomes known with some certainty)? Does uncertainty presuppose risk? Uncertainty is almost always with us, as rarely do we truly know anything with certainty. Does this therefore mean that risk must also always be with us? And if so, why should we be so concerned with minimizing risks if risks are always present in some form?

What about science's relationship to knowledge and truth? Is science primarily about the former, and only in moments of overwhelming hubris about truth? Or is science primarily about the latter, and only in moments of overwhelming humility about knowledge? How is one to even differentiate between truth and knowledge (do we not hold our knowledge to be true; otherwise, would we still consider it knowledge)?

Other questions can be asked regarding the motivations of the actors in the debate. For instance, why did the council appoint Brown: to save money by making sure that the expenditure on a plume study was justified, to deny the problem and save face, or to defend civic virtue? And what about Commoner? Why did he implicate Ames when the evidence was only circumstantial, by his own later admission: to gain a little fame (as well as infamy), to highlight a problem, or perhaps to encourage changes in incineration practice and energy production?

One possible solution to the incomplete scientific analyses that occurred in our case study could be what Oran Young has described as a "fair" policy.[50] Young suggests that when there is serious uncertainty about what the state of affairs really is, international negotiators usually choose to be fair. Specifically, they seek to create an abstract formula or rules to divide the costs and benefits on some equitable basis. What would have been fair in this case and to whom?

Another potential solution might reside in what is known as the precautionary principle.[51] When an activity raises threats of harm to human health or the environment, this principle argues that precautionary measures should be taken, even if some cause-and-effect relationships are not fully established scientifically. In this context, the proponent of an activity, rather than the public, should bear the burden of proof. The precautionary principle reverses the long-held belief that an activity is safe until proven (by often the public) otherwise. If such a policy were implemented in Ames, what would be the result? Would the plant be shut down immediately or would it remain online? Who would bear the burden of proof of the plant's safety: the Ames city council, the university engineering department, the plant managers?

Ultimately, the interrelationship between science and politics is more than just interesting intellectual conjecture, something appealing only to philosophers of science. By acknowledging the social relations underlying knowledge and the inherently political nature of science, we can critique science and the claims that it makes. If we ignore the social aspects of science, we relinquish this ability to question and critique scientific claims: a claim then becomes true simply on its merits of being scientific. Yet, such a position can lead to horrific results, as we witnessed in Nazi Germany with the Holocaust, where Hitler used "science" to confirm the Nazi belief of Aryan superiority and Jewish inferiority.

If we are to learn anything from this appalling atrocity, it must be that we can and should critique science. Yet, as we saw in the Ames controversy, whom we include in that critique—and consequently whom we exclude—may turn out to be just as important, and just as contentious, as the critique itself. Do we involve the people of Nunavut? What about residents outside, yet near, the community of Ames. If we choose to include these communities, to what extent should that invitation be extended? Answers to these questions are not just a matter of science or politics, but perhaps of the most social matter of all: ethics.

NOTES

1. Alvin Weinberg, "Science and Trans-Science," *Minerva* 10(2) (March 1972): 209–22.

2. Kathryn Harrison and George Hoberg, *Risk, Science, and Politics* (Montreal: McGill-Queen's University Press, 1994).

3. Some have attempted to label this relationship between science and politics as "trans-science" (Weinberg, "Science and Trans-Science"), "science policy" (Liora Salter, *Mandated Science: Science and Scientists in the Making of Standards* [Boston: Kluwer Academic Publishers, 1988]), and "regulatory science," (for example, Harrison and Hoberg, *Risk, Science, and Politics,* and Sheila Jasanoff, *The Fifth Branch: Science Advisors As Policymakers* [Cambridge, Mass.: Harvard University Press, 1990]).

4. Extensive literature exists on the instrumental use of science, particularly in environmental issues. See, for instance, Ulrich Beck, *Risk Society: Toward a New Modernity* (London: Sage, 1992); Ulrich Beck, *World Risk Society* (Cambridge, UK: Polity Press, 1999); John Dryzek, *Rational Ecology: Environment and Political Economy* (Oxford: Blackwell, 1987); and Eric Darier, ed., *Discourses of the Environment* (Malden, Mass.: Blackwell Publishers, 1999).

5. A vast amount of literature details the social character of knowledge. See, for instance, Bruno Latour, *Science in Action: How to Follow Scientists and Engineers through Society* (Cambridge, Mass.: Harvard University Press, 1987); Bruno Latour, *Pandora's Hope: Essays on the Reality of Science Studies* (Cambridge, Mass.: Harvard University Press, 1999); Thomas Kuhn, *The Structure of Scientific Revolutions* (Chicago: University of Chicago Press, 1962); and Michael Polanyi, *Personal Knowledge: Towards a Post-Critical Philosophy* (Chicago: University of Chicago Press, 1962).

6. See, for instance, Michael S. Carolan and Michael M. Bell, "In Truth We Trust: Discourse, Phenomenology, and the Social Relations of Knowledge in an Environmental Dispute," *Environmental Values* 12: 225–45; Michael Mayerfeld Bell with Gregory Peter, Susan Jarnagin, and Donna Bauer, *Farming for Us All: Practical Agriculture and the Cultivation of Sustainability* (College Station, Penn.: Pennsylvania State University Press, 2003); and Michael S. Carolan, *Trust and Sustainable Agriculture: The Construction and Application of an Integrative Theory,* unpublished dissertation, Iowa State University, Ames, Iowa, 2003.

7. Carolan and Bell, "In Truth We Trust."

8. Robert Brown, "An Evaluation of Expected Emission of Dioxin from the Ames Municipal Power Plant: A Report Prepared for the City of Ames, Iowa," unpublished manuscript, Engineering Department, Iowa State University, Ames, Iowa, 2001.

9. U.S. EPA, "Heath Assessment Document for Polychlorinated Dibenzo-p-dioxins," EPA/600/8–84/014F, 1985.

10. U.S. EPA/600/8–84/014F.

11. See U.S. EPA, "A Cancer Risk-Specific Dose Estimate for 2, 3, 7, 8-TCDD," EPA/600/6–88/007Aa (1988), for a review of the health effects of dioxin.

12. U.S. EPA/600/6–88/007Aa.

13. Kathryn Harrison and George Hoberg, "Setting the Environmental Agenda in

Canada and the United States: The Cases of Dioxin and Radon," *Canadian Journal of Political Science* 14(5) (February 1991): 3–27.

14. For a review of these early studies of the health effects of dioxin, see Michael Gough, *Dioxin, Agent Orange: The Facts* (New York: Plenum Press, 1986).

15. The EPA has been using a factor of one in a million to represent the maximum threshold of an acceptable risk of cancer in evaluating the toxic effects of a substance. Thus, a factor of between one thousand in a million and ten thousand in a million represents a risk of cancer over a person's lifetime, which is one thousand to ten thousand times greater than the "expectable levels" determined by the EPA.

16. Barry Commoner, "Globalization and the Environment," lecture in Iowa State University Lecture Series, Ames, Iowa, November 2000.

17. Harrison and Hoberg, *Risk, Science, and Politics,* and "Setting the Environmental Agenda in Canada and the United States."

18. Barry Commoner, Paul Woods Bartlett, Holger Eisl, and Kimberly Couchot, "Long-Range Air Transport of Dioxin from North American Sources to Ecologically Vulnerable Receptors in Nunavut, Arctic Canada." Final Report, North American Commission for Environmental Cooperation Publication, Montreal, Canada, 2000, at www.cec.org/programs_projects/pollutants_health/312/index.cfm?varlan=english, accessed 12 February 2001.

19. Commoner, "Globalization and the Environment."

20. International Air Quality Advisory Board, "Special Report on Transboundary Air Quality Issues," at www.ijc.org/boards/iaqab/spectrans/chap7.html, accessed 23 May 2002.

21. Kurt Kleiner, "Long-Lived Pollutants Threaten the Great Lakes," *New Scientist* 13 (July 1996): 8.

22. United Nations, "Toxic Chemicals," at http://earthwatch.unep.net/toxicchem/pops.php, accessed 23 May 2002.

23. Our Planet, "Population, Waste and Chemicals," at www.ourplanet.com/aaas/pages/waste01.html, accessed 23 May 2002.

24. Brown, "Evaluation of Expected Emission of Dioxin."

25. NACEC is a three-nation group made up of Canada, the United States, and Mexico, which formed shortly after the signing of NAFTA.

26. Commoner et al., "Long-Range Air Transport of Dioxin."

27. Commoner et al., "Long-Range Air Transport of Dioxin."

28. Personal e-mail communication to Mike Bell on 6 October 2000.

29. Barry Commoner, "Quantitative Estimation of the Entry of Dioxins, Furans, and Hexachlorobenzene into the Great Lakes from Airborne and Waterborne Sources," final report, Center for the Biology of Natural Systems, Queens, New York, 1995.

30. Commoner, "Globalization and the Environment."

31. Commoner, "Globalization and the Environment."

32. Commoner, "Globalization and the Environment."

33. David Grebe, "Council Takes on Dioxin Dispute," *Ames Tribune,* 25 April 2001, A1: 8.

34. Grebe, "Council Takes on Dioxin Dispute."

35. Grebe, "Council Takes on Dioxin Dispute."

36. Brown, "Evaluation of Expected Emission of Dioxin," 3.
37. Brown, "Evaluation of Expected Emission of Dioxin," 4.
38. J. D. Kilgroe, "Control of Dioxin, Furan, and Mercury Emission from Municipal Waste Combustors," *Journal of Hazardous Materials* 47(3) (November 1996): 163–94.
39. Incidentally, the annual emissions of dioxin measured at the Ames municipal power plant by the 1981 test was <0.9g TEQ/yr., according to Brown, "Evaluation of Expected Emission of Dioxin."
40. Brown, "Evaluation of Expected Emission of Dioxin," 13.
41. Highlighting the social relations of knowledge helps us to detail the processes underlying the social construction of knowledge. The social relations of knowledge, therefore, affect not only how knowledge becomes (socially) constructed, but also what becomes constructed as knowledge. We have intentionally chosen not to use the terminology "the social construction of knowledge" within this text, however, because of the nihilistic and postmodernist undertones that are often associated with it.
42. City council meeting, Ames, Iowa, 26 June 2001.
43. Personal communication to Mike Bell on 28 June 2001.
44. See, for example, Michel Foucault, *The Archaeology of Knowledge and the Discourse of Language* (New York: Harper Colophon, 1969); Michel Foucault, *Discipline and Punish: The Mirth of the Prison* (New York: Vintage, 1979); and Michel Foucault, *Power/Knowledge: Selected Interviews and Other Writings 1972–1977* (New York: Random House, 1981).
45. The "our science" and "their science" rendering in the Ames debate reflects an extensive literature in the instrumental use of science in environmental issues. See, for instance, Beck, *Risk Society* and *World Risk Society*, for further discussion.
46. Bruno Latour, *Science in Action* and *Pandora's Hope*.
47. S. O. Funtowicz and J. R. Ravetz, "Three Types of Risk Assessment and the Emergence of Post-Normal Science," in *Social Theories of Risk*, S. Krimsky and D. Golding, eds. (Westport, Conn.: Praeger, 1992): 251–73.
48. Weinberg, "Science and Trans-Science."
49. Funtowicz and Ravetz, "Three Types of Risk Assessment." For a brief overview (and a visual chart) detailing "normal" and "post-normal" science see www.nusap.net/sections/php?op=viewarticle&artid=13, accessed 15 January 2003.
50. Oran R. Young, *International Cooperation: Building Regimes for Natural Resources and the Environment* (Ithaca, N.Y.: Cornell University Press, 1989).
51. Indur M. Goklany, *The Precautionary Principle: A Critical Appraisal of Environment Risk Assessment* (Washington, D.C.: CATO, 2001).

11

International Cooperation in Environmental Politics

Ecosystem Management of the Great Lakes and the Baltic Sea

James Eflin

A maturing of environmental policy and management during the late twentieth century increasingly led scientists and policy makers to acknowledge that adequate measures to manage environmental affairs must begin by replacing the governmental jurisdiction as the primary spatial and institutional scale for implementation. In its place, the biophysical unit of "ecosystem" was elevated to primacy. Through this redefinition of scope, environmental politics and the scientific basis of environmental management were transformed radically. By taking seriously the concept of "system," the geographic boundaries of ecosystems were held to delineate more appropriately the spatial range of both environmental problems for analysis and the scope of their solutions. Out of this, a new interdisciplinary discipline was born, one that called for integrative, holistic thinking and acting. It valued the inclusion of a broadened range of stakeholder's views at several levels of scale and involvement; it stressed adaptive resource management; and it demanded that environmental management be based on a system of cooperative, participatory arrangements (commonly referred to as comanagement). Essentially, ecosystem management provided a strong and unified view for using what visionary geographer Gilbert White called "holistic thinking."

Importantly, the emergence of ecosystem management as a practice began to involve cooperative actions across transpolitical boundaries. Two significant and representative examples of this are seen in the International Joint Commission (IJC) which governs water quality and remediation in the Great Lakes ecosystem of Canada and the United States (see figure 11.1), and the Helsinki Commission (HELCOM), which similarly governs water-quality and remediation efforts in the Baltic Sea ecosystem (see figure 11.2). By looking at these two examples of international cooperation in environmental politics, these case studies provide an account of how science and politics in environmental management jointly emerged as ecosystem management, beginning with its basis in science and moving through its application as a cooperative tool for environmental policy at varying levels of governmental and nongovernmental organizations.

At the outset, we might wonder why society would choose not to take an ecosystems approach to management. After all, everyone knows what an ecosystem is, right? We are taught from our earliest years in school that the natural world is made up of many interlinking parts, and that the world is composed of living and nonliving parts that form a whole. We are given the term "ecosystem" as the rhetorical label that signifies this interlinkage. Yet, if it is so obvious to us as school children, why has the application of ecosys-

Figure 11.1 Great Lakes and St. Lawrence Ecosystem
Source: U.S. Army Corps of Engineers, Detroit District

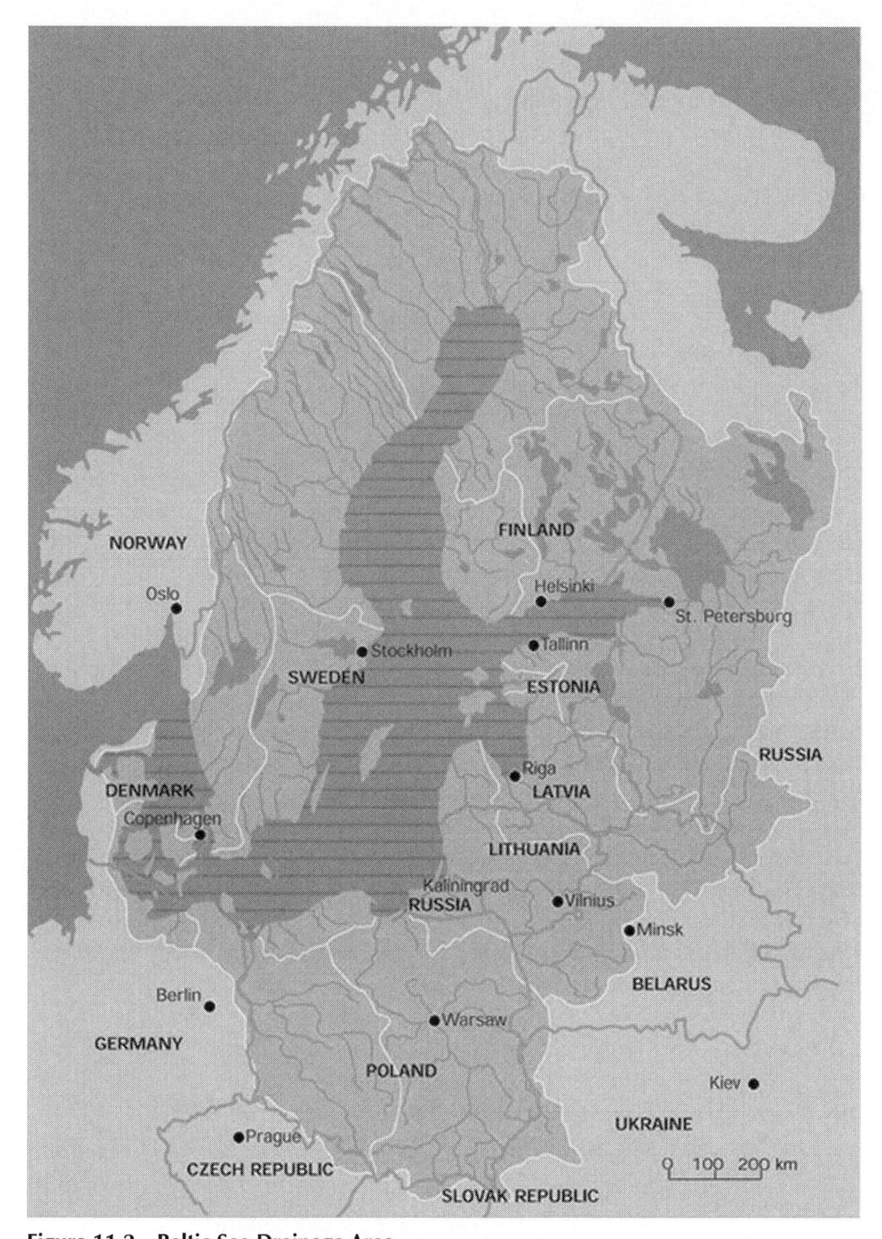

Figure 11.2 Baltic Sea Drainage Area

Source: Baltic Marine Environment Protection Commission [Helsinki Commission], www.helcom.fi/stc/images/docs/drainagearea.jpg, accessed September 25, 2003.

tem management been heralded as almost revolutionary? And why did it take so long to emerge? Answers lie at the heart of this volume, revealing the slow paths by which scientific debates feed ideas into the political arena, whereby policies are crafted and environmental resources are managed. Paradigm shifts within science may occur quickly, but the resonance by which they influence public policy is impeded by the dissonance between these two very different contexts. Further, outcomes of this case study point to conceptual weaknesses in ecosystem management, illustrating that environmental management along ecosystem lines works better when those lines cross certain jurisdictional boundaries, rather than others. In this case study, a pair of international agreements was needed to oversee management of two similar environmental systems. The IJC and HELCOM responded by creating two distinctly different frameworks for action, each of which was viewed, from the onset, as having realistic potential to solve problems that had been revealed. In each case, new ground was being broken for the political regimes involved; eventually, these two regimes took notice of one another, forming yet another pact that led to cooperative learning through the advent of the Great Lakes–Baltic Sea Partnership Program.

Similarities between the cases highlight different problems, approaches, implementations, and outcomes of resource-focused policy. Both the Great Lakes and the Baltic Sea have long and storied pasts of cultural development and commerce that affected present arrangements of social and economic power on their respective continents. Being largely internal to its continent, each water body took the brunt of great industrial development and consequent shipments of raw and finished goods as the economic systems reliant on these resource bases were propelled into the era of late capitalism. The earlier inward focus that developed powerful interlinkages and rivalries between cities, city-states, or city regions within their respective continents was replaced or complemented by more recent globalizing tendencies of which these regions are important parts. This nesting of interlocking systems of the Great Lakes and Baltic Sea regions—through trade routes and resource flows—expands the context by which political entities are interlinked with a larger world. The two regions also may be seen to exist within nested environmental systems. The political-economic and environmental similarities between the cases serve to highlight their very different applications of ecosystem management.

SCIENTIFIC BASIS FOR ECOSYSTEM MANAGEMENT

Rhetoric within environmental studies has long used the word "ecosystem" as a hallowed term meant to represent the complex reality of the living and nonliving worlds. Coined by ecologist Arthur Tansley in the 1930s, "ecosystem" united concepts of ecological components—plants, animals, minerals, and the energy flows and nutrient cycles that link them into a dynamic whole—with the emergent conceptualization of "systems theory."[1] Thus, "ecosystem" conceptualizes an organizational structure to the workings of nature.

In the twentieth century, systems were being viewed everywhere in science, from the organization of microscopic complexes of atoms and molecules to the cosmic relationships between planets and galaxies. In time, systems would dominate the scientific rhetoric as explanatory conceptions of the interactions between social actors[2] to the organization of industrial space[3] to the way in which city "systems" are organized into an economic space of commodity exchanges[4] to the operational principles of mechanical parts comprising modern airplanes. By using the language of systems theory, phenomena of nature and society were conceptualized by an appeal to underlying "wholes," the basis of holism: "Wholes have no stuff, they are arrangement. Science has come round to the view that the world consists of patterns, and I construe that to be that the world consists of wholes."[5]

In this way, the concept of an "ecological system" provides a rich understanding of the complex web of linkages (their arrangement or patterns) between otherwise discrete components of nature (their "stuff"). Ultimately, ecosystems can be treated (and understood) only as wholes, which suggests the controversial claim that looking at parts is irrelevant.

An ecosystem is a collection of living and nonliving components and their interconnections. Living (biotic) components of an ecosystem include plants, animals, and other organisms that are interconnected in a complex food web, transferring chemical energy from the level of producers (green plants) to successive levels of consumers (animals). Energy transfers through this chain undergo progressive entropy, as would the chemical matter of these organisms if it were not for the existence of decomposers and detrivores, which act to recycle matter among living and nonliving components. Thus, the nonliving (abiotic) components are essential parts of ecosystems. They provide chemical building blocks for nutrients and other material cycles, and serve as substrates on which living things build their foundations (essentially, rocks and minerals, and through their decomposition, soils). Abiotic components also include perhaps the most important ecosystem materials, water and air. The unique properties of water enable countless chemicals to circulate both within and between organisms, while its relative abundance or scarcity provides a regulatory medium for the dynamic character of ecosystems. Similarly, air acts as a medium for exchanges of chemicals and energy, both within and between organisms. Finally, there is the driving force of energy, entering the ecosystem from outside in the form of radiant solar energy. Solar energy begins the flow of energy transformations through the photosynthetic processes in green plants; successive transfers of the chemical energy thus liberated follow the food web or are dissipated as low-quality heat energy in the process of entropy. Together, these living and nonliving parts are the totality, or whole, of an ecosystem. Stated succinctly, "the functional ecosystem is the conception where biota are explicitly linked to the abiotic world of their surroundings."[6]

Dynamic properties attributed to ecosystems helped environmental scientists express the functions they saw among component parts of systems. Species were said to be adapted to the niches they occupied in ecosystems. Together, associations of species regularly found together were said to exist as a "community," "an array of interacting species."[7] In some cases, a species' role was defined by its competition with other species for scarce resources; in others, it was a symbiotic and mutually supporting network of relationships that were seen to exist. In either case, long-term tendencies toward ecosystem integrity, or homeostasis of systems conditions, were said to have evolved to the point where the ecosystem was seen as a durable, sustainable enterprise of living and nonliving dynamic systems, codependent on one another and representative of a system of permanence.

Scientific calls for a new resource-management policy based on functional ecosystems may be traced to early work in the 1930s. A comprehensive nature sanctuary system was called for in 1932 by the Ecological Society of America's Committee for the Study of Plant and Animal Communities, while biologists George Wright and Ben Thompson drew focus to the mismatch between park boundaries and functional ecosystems "by virtue of boundary and size limitations."[8] This early attention was not successful in grounding public land management in terms of ecological or landscape units, however. This awaited a renewed focus on functional ecosystems in the work of policy analyst Lynton Caldwell, whose advocacy for ecosystem-management principles in the early 1970s acknowledged that such an approach "would require that the conventional [political] matrix be unraveled and rewoven in a new pattern."[9] By the end of the 1970s, popular attention was attracted to ecosystem management through research involving the natural range of grizzly bear populations in the Yellowstone National Park area. There, biologists Frank and John Craighead concluded that the local grizzly bear population required more than five million acres of habitat, far in excess of the National Park Service boundaries.[10]

Ultimately, specifying the physical delimitation of the ecosystem to be managed has more than just theoretical significance. The arena of public policy—grounded in its own way on conceptions and theories of organization, power, and process within sociospatial systems—sets the parameters for what the policy governs. Water—perhaps the most valuable component of ecological systems—seems a fitting delimiter of the practical boundaries for managing resources, particularly if management is to be based on the ecosystem approach. Water resources, like the atmosphere itself, often do not coincide in any practical way with the spatial limits that jurisdictions set for themselves. Despite being a sink into which much of the unwanted residuals of human activities flow, water has increasingly become too valuable to relegate to an entirely public commons for society's wastes. It did not take the dawning of a resource-management system based on the ecosystem approach for civilizations to recognize the importance of water basins and watersheds. Armies and navies long defended them; nineteenth-century French geographers recognized them as the "natural" landscape element, *pays*, by which to designate natural regions.[11] Nor did civilizations await the 1980s and 1990s for political conflicts over water resources to signal a need for management planning to acknowledge that water systems can transcend the jurisdictional boundaries of political states. Political fragmentation is rapidly being seen as a challenge for resource management, as numerous governing entities with legal authority over an ecosystem make rational policy difficult.

As understanding emerged that environmental problems demand input from many scientific specialties, scientists became used to working collaboratively across disciplines, and integrative disciplines like ecology attracted increasing attention and resources. But is it possible for science to be truly holistic? Can environmental science meet the political demands for holism? How can environmental policy appropriately address challenges created by politically fragmented spaces?

SOCIOPOLITICAL PROBLEMS OF ECOSYSTEM MANAGEMENT

> An ecosystem approach does not depend on any one program or course of action. Rather it assumes a more comprehensive and interdisciplinary attitude that leads to wide interpretation of its practical meaning.[12]

If an ecosystem can be neatly conceived in the mind, it is not so neatly delineated on the ground. Part of the beauty of the conceptual ecosystem is its elegant denial of spatial boundaries—but this metaphysical beauty causes many social, political, and policy problems. "Size is not the critical characteristic, rather the cycles and pathways of energy and matter in aggregate form the entire ecosystem."[13] Hence, the spatial level of analysis for ecosystems must address the cycles and pathways that matter most in the totality—the whole—of the system.

The policy and politics of managing natural resources were not so cleanly conceived. One long-dominant mode of resource management was organized around Western principles of property ownership and, by extension, jurisdictional domain. Another dominant mode was organized around discrete components of the whole, such as individual resources (water, soil, vegetation), wildlife (game or nongame species), or commodities (oil, minerals, land). Arising from Western society's conceptions of property rights, a preeminent mode of resource management is based on the jurisdiction, which can be designated by spatial references to property. Frequently, that is further established on the basis of an individual resource (e.g., a water-conservation district, a forest-management unit). Hence, the conventional approach to resource management is built on the principle of the single, often top-down, mandate: manage the component, not the system. Frequently, this exacerbates poor environmental management arising from the frequent mismatch between ecological boundaries and political jurisdictions, which are often not related to one another.

Against this approach is an upstart conceptualization of how resources should be managed, one that grew during the 1980s to receive critical acceptance—by some. Built on the emergent concepts of sustainable development from the UN WCED,[14] ecosystem management was advanced as holistic and integrative in its approach: "Ecosystem management integrates scientific knowledge of ecological relationships within a complex sociopolitical and value framework toward the general goal of protecting native ecosystem integrity over the long term."[15]

The practice of focusing on the ecosystem for resource management was presaged in the 1973 Endangered Species Act (ESA), although not with the term "ecosystem management." Provisions of the ESA did not call for "saving or maintaining viable populations and individual species. Rather, it was centered on public and scientific concerns with the maintenance of ecosystem functions."[16] Since the ecosystem "concept signifies the study of living species and their physical environment as an integrated whole," the significance of ecosystem management therefore "is understood to lie in a comprehensive, holistic, integrated approach"[17] that acknowledges "that humans are part of, not separate from, the ecosystem."[18] This integrative inclusiveness is a "holistic approach to natural resource management, moving beyond a compartmentalized approach focusing on the individual parts of the [environment] . . . in order to integrate the human, biological and physical dimensions of natural resource management. Its purpose is to achieve sustainability of all resources."[19]

Sustainability of all resources is a tall order, and not the only one that remains poorly understood in the process of ecosystem planning and management. If planning leads where management follows,[20] each needs a spatial focus, which was conventionally provided by jurisdictional boundaries. By contrast, in ecosystem planning the politically contrived administrative boundaries take a back seat to a focus on the scale of resource interactions.[21]

Thus, the ecosystem approach confronts the political process squarely by insisting on a process in which the participation of stakeholders is actively sought in an effort to achieve effective, integrated management of the ecosystem of which humans are acknowledged to be a part and in which they play important roles. Loomis proposed a continuum of ecosystem planning and management scales, from a single federal ecosystem-management plan through successively smaller scales to "a single ecosystem management plan for all landowners."[22]

> A continuum of planning units ranges from rather small isolated areas for some resources, such as a particular plant species, to nearly global planning units for atmospheric resources. The smaller the planning unit, the more likely it is that a single landowner will be able to manage the resource effectively. The larger the ecosystem interaction, the more likely that many landowners and agencies will be involved.[23]

Clearly, as the scale enlarges it can be expected that the number, variety, and potential for conflict among stakeholder interests will increase.

Cooperation between sovereign countries has been problematic for centuries, and holism demands greater participation from an increased number of actors. Is holism in politics possible? Can political decision-making processes meet the scientific objectives created by holism? Once more, how can environmental policy appropriately address challenges created by politically fragmented spaces?

CASE STUDIES IN INTERNATIONAL ECOSYSTEM MANAGEMENT

On the basis of this short history of ecosystem management, both in terms of its scientific roots and its emergence within political processes, a series of questions might arise regarding the practice of ecosystem management in the Great Lakes and Baltic Sea regions. How has the interaction of science and policy that led to ecosystem management unfolded? In comparing this emergent practice, have different visions of ecosystem-management principles and practices led to different outcomes? In the case studies to be presented, we might ask the specific question, how did different ecosystem-management principles and practices develop in two scientifically comparable cases?

Environmental Comparisons

Great Lakes. Cooperative international management of water resources has been underway for most of a century in the Great Lakes of North America. The five Great Lakes—Superior, Michigan, Huron, Erie, and Ontario—form a connected water system that holds 18 percent of the world's freshwater. They represent the largest system of freshwater on Earth (exceeded only by the polar ice caps), covering over 244,000 square kilometers and draining nearly 522,000 square kilometers.[24] With the exception of Lake Michigan, each of the lakes and its interconnecting waterways is shared by two nation-states, and together the sociopolitical units of the ecosystem include "eight American states, a Canadian province, and thousands of local, regional and special-purpose governing bodies with jurisdiction for management of some aspect of the basin or the lakes. [Hence, c]ooperation is essential."[25] When the outflow via the St. Lawrence River is included, additional area in the province of Quebec must be included in the ecosystem, while the functional economic region that is closely tied to the aquatic system (consisting of the entirety of the eight states, all of Ontario, and lower Quebec) increases the total area impacted by the Great Lakes system to over 1 million square kilometers.[26]

As many as thirty-seven million people live within the drainage basins of the many tributary lower-order watersheds that feed the Great Lakes system itself. Part of the industrial "heartland" and "breadbasket" of North America, this watershed is home to more than 10 percent of the continent's population. For more than a century, this landscape has been thoroughly transformed to support an intensive system of economic production based on highly mechanized forms of extractive and manufacturing activities, together with other aspects of a highly diversified and modern urban economy. These include the world's largest concentration of pulp and paper mills (especially along Lake Michigan) and many metallurgical facilities (especially steel mills). Canada's two largest metropolitan areas are within the basin: Toronto sits squarely on the north shore of Lake Ontario and Montreal sits astride the St. Lawrence River. Two of the ten largest metropolitan areas in the United States (Chicago and Detroit), together with dozens of smaller, yet similarly industrialized, concentrations of urban North America, are located within the Great Lakes ecosystem.

Baltic Sea. Management problems of the Baltic Sea bear similarities to those of the Great Lakes. Each has degraded water resources. While not a freshwater body, the Baltic Sea has low salinity due, in part, to being largely confined by its outlet to the Kattegat and, thence, to the North Sea. Consisting of relatively shallow water passages (some barely reaching 18 meters in depth) through the narrow Danish Belts and the Öresund between Denmark and Sweden, the outlet drastically restricts the circulation of water, ensuring that little mixing with the more saline water in the North Sea occurs. The narrow, shallow outlet to the world's oceans also creates a water-quality problem and makes the shallow Baltic highly vulnerable to pollution. Freshwater discharged from the surrounding catchment area (or, drainage basin) of 1.7 million square kilometers enters the Baltic to become mostly brackish; the Baltic Sea is the largest body of brackish water in the world. Limited mixing with saltwater creates conditions for the sea to become stagnant; customarily, it takes twenty-five to thirty years for the sea to be completely replenished.[27] With an average depth of only 52 meters, the Baltic covers a surface area of 415,266 square kilometers. It is an epicontinental sea, much like the Hudson Bay. Isostatic rebound (the "bouncing back" of the Earth's crust following the removal of glacial ice) following the Würm glacial age (10,000–120,000 years BP) is slowly raising the Baltic Sea floor, exposing new bottom areas to waves and opening up important new nutrient sources for marine species.[28] As a result, the Baltic Sea has evolved into a unique ecosystem, one that has historically provided a rich fishery that helped define the surrounding cultures.

Today, nine countries (Denmark, Estonia, Finland, Germany, Latvia, Lithuania, Poland, Russia, and Sweden) have coastlines on the Baltic Sea, while another five (Belarus, the Czech Republic, Norway, Slovakia, and Ukraine) are partly in the watershed. In all, eighty-five million people live in the catchment area of the Baltic Sea ecosystem, with 15 million living within 10 kilometers of the coast. It is among the most heavily industrialized areas of Europe, with population densities exceeding five hundred persons per square kilometer in the urban regions of the South (along the coast of Poland and Lithuania) and the East (in the Gulf of Finland and the Gulf of Riga). To the North, very low population concentrations make the Gulf of Bothnia more pristine.

The Baltic Sea has long been an important fishery and highway for commerce. Vikings plied its waters by the ninth century; the trading network of the mercantilist Hanseatic League benefited from it as early as the thirteenth century; today more than 130,000 marine vessels travel the Baltic each year.[29] Thus, the Baltic is a heavily used commons, receiving discharges of pollutants from land and seaborne activities; airborne drift brings additional contaminants from outside the region.

Sociopolitical Comparisons

Great Lakes. International cooperation in the interest of water-resource management in the Great Lakes began with the Boundary Waters Treaty of 1909 between the United States and Canada. A principal outcome of this treaty was the creation of the IJC, which serves as a forum to facilitate cooperation and resolve conflicts regarding Great Lakes water quality and water levels, as well as water flows between the United States and Canada.[30] IJC commissioners are appointed by the president or prime minister, respectively. In the United States, they serve at the pleasure of the president; in Canada, the appointments are for fixed terms, determined by the prime minister, and usually last for three years (five years for the chair). Budgeting is provided via Congress and Parliament, with limited additional funding provided via requests that IJC makes to local and state/provincial agencies. The IJC is a unique entity in that it answers only to the two national governments, but does not report to any one agency within either. It is effectively wholly independent, to the extent that the national governments control both the membership and the agenda.

The Boundary Waters Treaty established a unique process for cooperative management of the resource through its organizational structure. Composed of six members—three appointed by each federal government—the IJC is advised by its subsidiary Water Quality Board (composed of representatives from agencies at many jurisdictional levels who promote intergovernmental coordination of Great Lakes programs) and Science Advisory Board (composed of academic and governmental experts). Based in Windsor, Ontario, the IJC administers the original treaty, studies problems of special interest, and arbitrates disputes concerning boundary waters issues. In 1912, among the first problems referred to the IJC for study was water pollution, which led the agency to call for a new treaty to control pollution; such a treaty was not enacted until six decades later.

After five decades of growing industrialization and increased agricultural production in the Great Lakes region, public concern about deteriorating water quality in the system during the 1950s and 1960s led to an increased research agenda focused on environmental problems, particularly those associated with DDT (at one time, a widely used pesticide) and eutrophication (an excessive buildup of algae in aquatic systems, with a consequent reduction in dissolved oxygen). This precipitated a shift from applying reactive management (which long dominated the Great Lakes region, and elsewhere), to the gradual adoption of a proactive management style.[31] A new binational agency, the Great Lakes Fishery Commission, was established in 1955; its first focus was on the parasitic sea lamprey that was causing a decimation of important fisheries throughout the lakes. Coincident with heightened public sentiment for environmental protection and support for environmental legislation in both countries, Canada and the United States signed the first Great Lakes Water Quality Agreement in 1972. The IJC was assigned to monitor progress toward this and successive treaties. The 1972 agreement led to significant reductions in pollutant discharges (principally phosphorous) and demonstrated that environmental-quality enhancements could best be achieved by going beyond localized cleanup campaigns. Specifically, it called for research about Great Lakes problems to be carried out cooperatively by the two countries. Further, it called for surveillance of the system—first of pollution discharges and water chemistry, then of the ecosystem's "health." During the decade that followed, increased monitoring of the lakes and their ecosystem built further evidence to support the growing scientific understanding about how toxic substances biologically accumulate through food chains. By 1978, it became widely acknowledged among the scientific and political community that a systemwide perspective was required to understand the interconnected nature of the Great Lakes and their surrounding landscape. As a result, a second Great Lakes Water Quality Agreement was signed by Canada and the United States in 1978. This called for restoration and ongoing maintenance of "the chemical, physical, and biological integrity of the waters of the Great Lakes Basin Ecosystem," which was defined as "the interacting components of air, land and water and living organisms *including man* within the drainage basin of the St. Lawrence River."[32]

One interesting outcome of the 1978 Great Lakes Water Quality Agreement was the creation of a system for implementing remedial action plans (RAPs) to address serious degradation of the mass balance of toxic substances that enter the Great Lakes system. The mass balance approach—particularly related to phosphorous loading—explicitly recognizes that ecological health is based on a systems perspective; hence, the interconnected ecosystem concept for management became overriding. No longer was it sufficient to impose discharge limits at point sources; now, integrated systemwide approaches were called for. Early success in reducing phosphorous levels in the Great Lakes became an internationally recognized model for binational resource management.[33] Further efforts to achieve the goal of more complete remedial actions would not be so easy.

Baltic Sea. Water-quality degradation in the Baltic began to be reported in the 1950s; prior to that, the sea was still considered healthy. Its condition changed dramatically thereafter, as it became one of the most contaminated seas in the world by the late 1980s.[34] As with the Great Lakes, the problems initially seen were those of eutrophication, which eventually produced toxic algal blooms reaching far out into the Kattegat and affecting the coastlines of Sweden, Denmark, and Norway, as well as the more confined Baltic proper.[35] Phosphorous loading in the Baltic increased eightfold and nitrogen fourfold during the twentieth century.[36] Nutrient loading discharged to the sea from increasingly intensive agricultural production in Sweden, Denmark, Germany, and Poland, together with a gradual exhaustion of oxygen in the deeper Baltic, resulted in a "dead zone" below the halocline throughout nearly one hundred thousand square kilometers of seabed. Reflecting trends that had begun to typify changing diets in the region, eutrophication signaled one of many emergent stresses to the ecosystem.[37] Another stress was the emergence of "toxic stews" of synthetic chemicals discharged into the sea—notably, mercury, DDT, and PCBs—producing a pollution chain alleged to cause dramatic declines in grey and ringed seal populations and widespread deaths of white-tailed eagles.[38] Fears of health impacts on humans led Sweden to ban the use of mercury in 1966; many other Baltic states followed suit by 1972. Around the same time, DDT was banned in many Baltic countries. Since this time, notable declines in DDT and mercury concentrations have signaled some success for water-quality measures taken in the Baltic Sea.

International cooperation for environmental protection in the Baltic region began in the early 1970s when each of the Baltic states signed global conventions meant to limit dumping of industrial wastes (1972 London Dumping Convention) and discharges from marine vessels (1973 MARPOL Protocol). The International Baltic Sea Fisheries Commission (IBSFC) was established in 1973 as a cooperative organization "with a view to preserving and increasing the living resources of the Baltic Sea and the Belts and obtaining the optimum yield";[39] it was a limited regional organization, consisting of Finland, the German Democratic Republic, Poland, Sweden, the Soviet Union, and the European Economic Community (after 1984).

These global (or limited regional) agreements preceded a more significant regionwide pact signed in 1974 at Helsinki. This was the Convention on the Protection of the Marine Environment of the Baltic Sea Area, more commonly known as the Helsinki Convention. Unlike other environmental treaties of the time, it focused on regional (rather than global) cooperation, representing an unusual collaboration between Eastern Bloc and Western European countries.[40] Unique in its focus on a specifically defined region, the Baltic Sea, the Helsinki Convention was wide-ranging in its influence, covering marine- and land-based pollution, dumping, and pollution discharges from waterborne shipping.[41] The seven nations surrounding the sea in 1974 (Denmark, Finland, East Germany, West Germany, Poland, Soviet Union, and Sweden) were the original contracting parties. The convention established a secretariat, the Baltic Marine Environmental Protection Commission, referred to more commonly as the Helsinki Commission, or HELCOM. Originally composed of four committees—Environment, Technological, Maritime, and Combating—its structure bore similarities with the IJC in North America. Remediation strategies and other measures were overseen by the Programme Implementation Task Force (PITF), the real workhorse of the organization. Among the initial targets characterized were wastewater outfalls, particularly from industrial facilities and poor-quality municipal systems in the eastern Baltic. Non–point source pollution was also recognized, and the 1988 Ministerial Declaration targeted a 50 percent reduction of nutrient inputs to the Baltic Sea.

Although initiated due to systemwide water-quality problems, HELCOM is also concerned with air-quality and energy resources as they affect the health of the ecosystem.[42] Automobile ownership in the region is rising rapidly (particularly in the East), along with total vehicle distances driven (greatest in the West). The result is a steady rise in emissions of nitrogen oxides that contribute further to Baltic Sea eutrophication.[43] Combustion of fossil fuels for electrical power and industrial use remains high throughout the region, despite efforts to adopt renewable-energy systems or implement energy-conserving strategies. Hydroelectric power is particularly important in Sweden and Finland, while nuclear power plants supply considerable amounts of electricity in the region (over 50 percent for Sweden, Lithuania, and the St. Petersburg region of Russia).

APPLYING ECOSYSTEM MANAGEMENT IN THE GREAT LAKES AND BALTIC SEA REGIONS

Since their inception through international agreements, remediation efforts in the Great Lakes and Baltic Sea regions have reflected increasing scientific evidence of ecosystem declines, as well as changing political climates for adopting or enforcing environmental policies. The outcomes of these situations have produced different results for the two regions. Despite similar applications of science (a somewhat universal and actively shared body of knowledge), different historically specific sociopolitical contexts constrain the degree to which policy measures have been created and implemented.

The Great Lakes

The use of RAPs was conceived by the IJC to facilitate systemwide environmental improvements by delegating implementation to more localized stakeholders. A follow-up to the 1978 Great Lakes Water Quality Agreement was the designation of forty-three hot spots, or areas of concern (AOCs), throughout the watershed. Thirteen AOCs were in Canada, exclusively, while twenty-five were in the United States; the remaining five were on waterways connecting the lakes and were shared equally by the two countries. Each AOC was delineated by a subwatershed or waterway and was listed on the basis of what IJC calls an "impairment of beneficial use."

Impairments were defined in terms of fourteen criteria in four categories:

1. Ecological health and reproduction (criteria include degradation of fish and wildlife populations; fish tumors or other deformities; reproductive problems or deformities of birds or animals; degradation of phytoplankton and zooplankton populations; degradation of benthic, or floor-dwelling, populations; eutrophication or undesirable algae; loss of fish and wildlife habitat).
2. Human health (criteria include restrictions on fish and wildlife consumption; restrictions on water recreation, essentially beach closings).
3. Human use or welfare (criteria include tainted flavor of fish and wildlife; taste or odor problems in drinking water; restrictions on dredging

activities; degradation of aesthetics; added costs to agriculture or industry).
4. Loss of fish and wildlife habitat.

The Great Lakes Water Quality Agreement was revised in 1987 to include a protocol for RAPs and adoption of a process for reducing pollutant discharges, known as lakewide management plans (LaMPs) for critical pollutants.[44] The 1987 protocol structured the RAP process in three stages. Stage one is primarily a problem-identification or -characterization phase, delimiting the project boundaries and identifying the sources of beneficial-use impairments. Stage two identifies remedial actions, defines an implementation plan, and identifies those agencies with implementation responsibilities. Stage three identifies evaluation, surveillance, and monitoring of the implemented remedial actions. The overall goal of any RAP is to restore beneficial use to the AOC. To date, one AOC (Collingwood Harbour, Ontario) has been delisted from the RAP process. Although coordination of RAPs ultimately falls to the IJC and is administered by one or more lead agencies, top-down actions come from the U.S. EPA and Environment Canada.

Successes and impediments to the RAP process in the Great Lakes illustrate both the promise for this innovative approach and how slowly the ecosystem-management practice is taking hold.[45] In her analysis, Susan Hill MacKenzie[46] proposed a set of standards by which to evaluate the ecosystem approach to resource planning and management, with her specific focus being its use in the Great Lakes. MacKenzie's analysis, similarly reflected in a collection of case studies analyzed by John H. Hartig and Michael A. Zarull,[47] includes three major prerequisites for success: participation of appropriate actors (both intergovernmental and interdisciplinary); development of a mutually agreed-upon decision-making process (including protocols for consensus, development of common visions, and dispute resolution); and legitimacy of the process (including political support, public participation, and funding). Several weaknesses and differences of outcomes became evident through early work on stage one RAPs.

MacKenzie suggests that the Canadian experience at Hamilton Harbour, Ontario, achieved the most successful results for a stage one RAP. Excellent decision making, strong federal and provincial participation, and solid legitimacy enabled the Hamilton Harbour RAP to move smoothly through the stage one process. By contrast, she shows in two cases that American experiences were hampered by weak to inadequate legitimacy, due, in part, to limited federal involvement and limited funding, while stakeholder participation and decision making were weak. Interagency conflict also characterized the American experiences, illustrating the difficulties inherent in adopting a new paradigm for resource management. MacKenzie's conclusions are at odds with those of Hartig and Zarull, who conclude from an analysis of a wider range of case studies that the American examples show evidence of greater success: "[W]hatever the reason, it is evident that Canadian citizens involved in the RAP process will have to learn from their American neighbors."[48]

Recognition of a continuing climate of reduced government funding precipitated the Wingspread Conference at Racine, Wisconsin, in 1996 to address funding strategies for remediation of the AOCs throughout the basin.[49] The IJC was quick to celebrate the notable progress that was being made toward successes in some locations, what later became known as the "beacons of light" AOCs. These included strategic planning for the Black River in Ohio; innovative partnerships for the Ashtabula River in Ohio; cooperative public-private partnerships for the Grand Calumet River/Indiana Harbor Ship Canal in northwestern Indiana; community vision toward sustainable development in Hamilton Harbour, Ontario; pollution permit-trading program for phosphorous at the Bay of Quinte on Lake Ontario; superfund remediation in the Manistique River AOC on Lake Michigan; and creative fund raising for the Muskegon Lake AOCs on Lake Michigan.[50] Against these beacons of light, the IJC recognized that further progress was beginning to stagnate systemwide. This led the IJC to examine progress toward restoration of beneficial uses as an attempt to enhance the restoration process in its 1996 adoption of a new initiative of "status assessments." While not intended as comprehensive environmental audits, status assessments were intended to examine progress toward the restoration and protection of beneficial uses; assess program implementation relative to remedial and preventive actions; and identify and recommend specific actions that could be taken. To date, four status assessments have been completed, and three others are in the process of review.[51]

The IJC recognizes that institutional and financial limitations are impeding progress in the remedial action process. Following the Wingspread Conference, IJC released its "Beacons of Light" report,[52] which identified six major obstacles that need to be overcome in order for the RAP process to go forward effectively. These emphasized lack of planning for major remedial measures; reductions in government funding and staffing; failure to set priorities within and between AOCs; limited efforts to stimulate public participation; random and infrequent transfers of information and technology between AOCs; and failure to quantify the benefits of remediation. The ecosystem approach has been further frustrated by the creation of separate RAPs for single binational AOCs (Niagara River, St. Lawrence River) by U.S. and Canadian authorities.

The Baltic Sea

When HELCOM was established, clear sociopolitical divisions in Europe between the East and West limited the effectiveness of cooperation among the Baltic states. After 1990, as authoritarian governments in Eastern Europe gave way to more democratic regimes, a window for mutual cooperation was opened. A meeting of all Baltic states was convened in September 1990 at Ronneby, Sweden, producing the first Baltic Sea Declaration. This declaration gave teeth to remediation efforts in the region, as signatories agreed to "begin specific initiatives to achieve ecological restoration of the Baltic Sea, and to create the possibility of self-restoration of the marine environment as well as preservation of its ecological balance."[53] In part, these measures were to be paid for by reduced armaments expenditures. The Ronneby meeting paved the way for revisions to the Helsinki Convention in 1992, as the number of its signatories increased to include the nine nations surrounding the Baltic Sea, including newly independent Estonia, Latvia, Lithuania, and Russia, as well as Denmark, Finland, Poland, Sweden, and a reunified Germany.

The spirit of the Baltic Sea Declaration precipitated adoption of the Baltic Sea Joint Comprehensive Environmental Action Programme (JCP), signed into force in 1992. The JCP encouraged policy and regulatory reforms, capacity building, investment in controlling pollution, reducing waste and safely disposing of it, and conserving ecologically sensitive and economically valuable areas. The program "also includes elements to support applied research, environmental awareness, and environmental education."[54] HELCOM was given responsibility for monitoring and implementating the JCP. This "new" HELCOM was recomposed into five groups responsible for monitoring and assessment, land- and sea-based pollution, nature conservation and coastal zone management, and strategy. Membership in PITF was expanded to include each of the newly contracting parties, as well as several other interested governments and intergovernmental organizations.

The new convention extended the original mandate beyond the narrower goals of combating water pollution and management; emphasis now shifted to include habitat conservation, biodiversity, and protection of all ecological processes within a more broadly defined Baltic Sea area (Article 15). The result of this and other components of the JCP (systems linkages, delineation of the watershed as the system boundaries, recognition of people as integral parts of the system) strongly suggest that JCP was adopted with an ecosystem approach in mind.[55]

The JCP established a high-level task force within HELCOM to coordinate remediation efforts. Grants from several sources totaling five million ECU (approximately US$6 million) funded research efforts that led the task force to identify 132 hot spots in the Baltic Sea ecosystem as targets for systemwide remediation and over 2,200 point sources. The hot spots represent concentrations of point sources of pollution, including municipal wastewater outflows and widely distributed industrial facilities. An overwhelming majority of these (98) were located in Eastern Europe; forty-seven of them were designated "priority hot spots."[56] The remaining thirty-four hot spots were in Denmark, Finland, Germany, and Sweden.

Implementation of the JCP was structured in two phases. The first phase (1993–1997) established four priorities: emergency support and systems targeted at municipal water supplies and wastewater facilities in Belarus, Estonia, Latvia, Lithuania, Russia, and Ukraine; improved combined municipal and industrial wastewater treatment systems to reduce organic pollutants by targeting poorly maintained or inoperable facilities; industrial pollution control to replace end-of-pipe measures with process redesign with market-based incentives and increased use of environmental audits and environmental management systems; and reduction of agricultural runoff.[57]

The initial focus of phase one was on twenty-nine priority hot spots in Eastern European states. Within five years, remediation efforts resulted in thirteen hot spots being crossed off the list; as of 2003, nearly fifty hot spots have been delisted. The second phase—expected to last through 2012—targets nonpoint pollution sources, principally agricultural runoff. In all, ECU 9.84 billion (approximately US$11.8 billion) were initially estimated as investments for hot spot remediation, although later estimates were twice that amount systemwide.[58]

It can be expected that diligent efforts by HELCOM to implement the JCP will result in widespread improvements in water quality for the Baltic Sea. Coastal waters are seeing the quickest response to remediation measures, leading to the reopening of many contaminated beaches and helping to reestablish the tourism industry. Fishery resources responded to reduced nutrient loading and algal blooms, eutrophication, and increased levels of dissolved oxygen. Recovery of the deeper water in the Baltic Sea is expected to be much slower, however, particularly due to the inability to control atmospheric emissions from outside the boundaries of the Helsinki Convention. No matter how broadly an ecosystem is defined, it is linked to global processes and policies.

Global environmental policies are pushing a renewed effort for ecosystem management in the Baltic. With heightened focus on regional cooperation, nation states within the Baltic region developed Baltic 21 in response to Agenda 21 and the Rio Declaration adopted by the United Nations Conference on Environment and Development in Rio de Janeiro in 1992. Baltic 21 addresses the question, how can regional cooperation contribute to sustainable development in the Baltic Sea Region? It does so by adopting four major emphases: sustainable fishery; sustainable energy; sustainable industry; and spatial planning for sustainable development. Its long-term policy-making time frame (to 2030) emphasizes regional cooperation via common energy markets and transport policies, as well as coordination of governing activities (including harmonized environmental legislation and taxation). Very forward looking, Baltic 21 acknowledges the fragility of the ecosystem and is grounded in global policy measures (e.g., the polluter-pays and precautionary principles, the Framework Convention on Climate Change, Convention on Biological Diversity, Convention on Long-Range Transboundary Air Pollution, and Convention on the Protection of the Marine Environment of the Baltic Sea Area). Baltic 21 sets the essential objective of Baltic Sea–region cooperation as "the constant improvement of the living and working conditions of their peoples within the framework of sustainable development, sustainable management of natural resources and protection of the environment."[59]

Baltic 21 faces significant barriers to its implementation, including incomplete legislation; weak enforcement of law; custom and certification problems; illegal trade; taxation deficiencies; ineffective administration; and inadequate integration across major sectors. The less integrative, more focused goals of HELCOM have made it more effective at implementing the JCP; Baltic 21 is a more visionary, less successful document.

HELCOM was built on the spirit of the Rio declaration and today is guided by six principles associated with Agenda 21: shared responsibility; adoption of the precautionary principle and the polluter-pays principle; use of best available technologies; avoiding transboundary risks; and monitoring of emissions.[60] Through this principled approach, HELCOM has achieved significant successes with improved water quality basinwide. Reductions were achieved in organic pollutants and nutrients, oxygen-consuming substances, atmospheric nitrogen, and organo-halogen compounds—it is energetically tackling related problems affecting water quality—phasing out leaded gasoline, banning PCBs and DDT, stricter industrial emissions, as well as legislation, monitoring, and enforcement of discharges through maritime commerce. Seal and white-tailed eagle populations have recovered. Importantly, twenty-five hot spots have been cleaned up.

It appears that a recovery of the water quality in the Baltic Sea ecosystem is clearly underway, but what remains to be seen is if the newer provisions for nature conservation and biological diversity of the convention's Article 15 will lead to similar successes for biodiversity and ecological processes, thereby ensuring success for the more extensive and integrated principles of ecosystem management.

INTEGRATING SCIENCE AND POLITICS IN ECOSYSTEM MANAGEMENT

Cooperative international efforts to restore water-based ecosystems in the Great Lakes and Baltic Sea simultaneously illustrate promises for and limitations of employing ecosystem management as an organizing principle.[61] Promises include a dawning acknowledgment that management of natural resources must not artificially compartmentalize ecological systems along jurisdictional lines or agency mandates. Increasingly, holism is replacing reductionism—at least in concept and increasingly in policy. Partnerships among stakeholders are being forged at many levels, and responses to global changes are driving environmental-policy measures at some levels of government. The "new" conceptualization of humans as part of ecosystems is leading to innovative breakthroughs in how to marshal resources toward sustainable development—or at least toward restoration of egregious errors from the past. Despite this changing climate for cooperation, persistent barriers to progress remain. These include competing claims for fiscal resources, inadequate attraction of sufficient stakeholder participation, the legacy of regulatory frameworks that impede success, and inadequacy of ecosystem theory to inform policy making fully.[62] A positive sign amidst times of fiscal uncertainty is the 2003 announcement of a $12-million grant (in part via the World Bank) to support sustainable ecosystem management in the Baltic.[63]

Great Lakes and Baltic Sea Partnerships

Perhaps one of the most significant steps forward is seen when the arenas of science, public policy, and public awareness are all brought together to demonstrate explicitly the potentials for success that come from the mutual collaboration of all stakeholders. Neither the management effort for the Great Lakes ecosystem, nor that for the Baltic Sea ecosystem, exists in a vacuum, and their similarities have attracted considerable attention from outside observers and practitioners. Intercontinental collaboration between stakeholders, using experiences learned from ecosystem management in their respective regions to promote information sharing, resulted in establishing the Great Lakes–Baltic Sea Partnership Program.[64] This collaboration in the interest of water-quality improvements in each region combines scientific and political processes to achieve four goals: sharing information, expertise, and management approaches; strengthening institutional relationships among scientific organizations, professional associations, governments, and NGOs; improving capabilities to manage the two watersheds more effectively; and enhancing environmental decision making through better use of environmental information. This collaboration of scientific and policy-making focus is reflected by efforts to increase the public's interest in ecosystem management. HELCOM and the U.S. EPA facilitate a fellowship and exchange program of scientists and governmental officials in order that both sides may gain additional benefits of multilateral watershed management.

Lessons of partnership may be shared in other ways, too. The Great Lakes and Baltic Sea cases were selected for an international cooperative learning experience during the fall semester 1999, which paired a class in teaching for sustainability at Malmö University (Malmö, Sweden) with a class in integrated resource management from Ball State University (Muncie, Indiana). By sharing their local resource materials and comparing their ideas (from two distinctly different cultures) over the Internet, students from two continents undertook comparative study of the two regions to examine the theoretical principles and practical measures for the ecosystem-management approach to be achieved. Through such means, the "public" may be led to adopt policy measures that require a maturation of thinking that can embrace the scientifically rich focus on ecological systems as being the first-order units for which environmental management efforts must be designed.

IMPLICATIONS FOR GLOBAL ISSUES

The message has become clear: the functional operations of an ecosystem or landscape-scale unit must be addressed clearly and holistically when large-scale restoration or sustainable development is being practiced. This is being recognized for inland ecosystems (such as the Great Lakes), as well as sustainable development of the seas (such as the Baltic).[65] Less clear is how the science behind that message fits within the evolution of the sociopolitical processes of resource management. Although it may be argued that "restoration activities should always be firmly established in a societal framework, that covers legal, administrative, and economic aspects,"[66] R. Edward Grumbine is more emphatic: "Ecosystem management is not just about science nor is it simply an extension of traditional resource management; it offers a fundamental reframing of how humans may work with nature."[67]

John H. Hartig and John R. Vallentyne assert that the "essence of an ecosystem approach is that it relates people to ecosystems that contain them, rather than to environments with which they interact."[68] This may seem like a fine distinction, but it is one with revolutionary implications for the political practice of resource management. It makes emphatic the holistic integration of people as coexisting entities within ecosystems (as embodied in the environmental ethic of ecocentrism), rather than the more conventional distinction (based in anthropocentric ethics) that posits that people and nature are separate.

Against the growing scientific clarity of the ecosystem approach as an operative principle for resource management, the sociopolitical message is clouded. More successful collaboration among European states appears to be leading toward more successful implementation of the ecosystem approach in the Baltic Sea region, whereas the more isolationist histories in North America (at least, within the United States) appear to impede the spirit of cooperation that underlies the fundamental principles of ecosystem management. Again, we may return to one of the central questions: how can environmental policy appropriately address challenges created by politically fragmented spaces? The ecosystem approach to environmental management seems to be a conceptually elegant way forward in its scientific bases, but it must acknowledge the historically rooted sociopolitical constraints to which it must adapt.

Susan Gilbertson of the U.S. EPA acknowledges the sociopolitical complexities that must be overcome:

> You are applying federal authorities, state authorities, local authorities. Sometimes, those authorities are in conflict. And if you think that you can work through this concept and avoid conflict, I think you're kidding yourselves. And I think those of you here recognize that. One of the challenges is: how do you work through that conflict, how do you reach consensus? And in some areas it's easier, and in other areas it's more complex.[69]

Grumbine looks beyond this pragmatic concern. Although he provides many cautions about the barriers to the success of such a new approach, he remains optimistic that the increasing sophistication of ecological understandings will gradually infuse sociopolitical processes in the management of nature. Importantly, he notes that humans must not be discounted as active agents (both positively and negatively) in the transformation process: "[S]uccessful ecosystem management, over time, must nurture both the wildlands at the core of the reserve system and the wildness within human beings."[70] His conclusion sounds an optimistic call:

> Ecosystem management, at root, is an invitation, a call to restorative action that promises a healthy future for the entire biotic enterprise. The choice is ours—a world where the gap between people and nature grows to an incomprehensible chasm, or a world of damaged but recoverable ecological integrity where the operative word is hope.[71]

This call to action represented by ecosystem management sounds a clarion call for cooperation, both across political boundaries and across academic disciplines, as well as in the broader interest of cooperation between science and politics in the international environment.

NOTES

The author wishes to acknowledge participation by Margareta Ekborg and Claes Malmberg and their students in teaching for sustainability at Malmö University, who collaborated on an inquiry-based learning project with students in integrated resource management at Ball State University during the fall semester 1999. Support from the European Teacher Education Network for faculty exchanges between both campuses is also acknowledged. Valuable information about the IJC structure was provided by Jim Chandler, Legal Advisor to the U.S. Section of the IJC.

1. Ludwig Von Bertalanffy, *General System Theory: Foundations, Development, Applications* (New York: George Braziller, 1968).
2. Anthony Giddens, *The Constitution of Society: Outline of a Theory of Structuration* (Cambridge, UK: Polity, 1984).
3. Peter Haggett, *Location Analysis in Human Geography* (London: Edward Arnold, 1965).
4. Brian J. L. Berry, "Cities As Systems within Systems of Cities," *Papers, Regional Science Association* 13 (1964): 147–63.
5. Jan Christian Smuts, *Holism and Evolution* (Westport, Conn.: Greenwood Press, 1973): 336.
6. Timothy F. H. Allen and Thomas W. Hoekstra, *Toward a Unified Ecology* (New York: Columbia University Press, 1992): 44.
7. Malcolm L. Hunter Jr., "The Biological Landscape," in *Creating a Forestry for the 21st Century: The Science of Ecosystem Management,* Kathryn A. Kohm and Jerry F. Franklin, eds. (Washington, D.C.: Island Press, 1997): 57.
8. Cited in R. Edward Grumbine, "What Is Ecosystem Management?" *Conservation Biology* 8(1) (1994): 27–38.
9. Cited in Grumbine, "What Is Ecosystem Management?" 28.
10. Grumbine, "What Is Ecosystem Management?" 28.
11. Preston James, *All Possible Worlds: A History of Geographical Ideas* (Indianapolis: Odyssey, 1972).
12. Lee Botts et al., *The Great Lakes: An Environmental Atlas and Resource Book* (Chicago: U.S. EPA and Toronto, Ont.: Environment Canada, 1987).
13. Allen and Hoekstra, *Toward a Unified Ecology*, 44.
14. WCED, *Our Common Future* (Oxford: Oxford University Press, 1987). See also J. H. Hartig and P. D. Hartig, "Remedial Action Plans: An Opportunity to Implement Sustainable Development at the Grassroots Level in the Great Lakes Basin," *Alternatives* 17 (1990): 26–31.
15. Grumbine, "What Is Ecosystem Management?" 31.
16. Jack Ward Thomas, "Foreword," x, in *Creating a Forestry for the 21st Century: The Science of Ecosystem Management.*
17. S. Bocking, "Visions of Nature and Society: A History of the Ecosystem Concept," *Alternatives* 20(3) (1994): 12.
18. Bruce Mitchell, *Resource and Environmental Management* (Harlow, UK: Longman, 1997): 52.
19. Jack Ward Thomas, "New Directions for the Forest Service," statement before the Subcommittee on National Parks, Forests, and Public Lands and the Subcommittee on Oversight and Investigations, Committee on Natural Resources, U.S. House of Representatives, 3 February 1994, quoted in Steven L. Yaffee, Ali F. Phillips, Irene C. Frentz, Paul W. Hardy, Susanne M. Maleki, and Barbara E. Thorpe, *Ecosystem Management in the United States: An Assessment of Current Practice* (Washington, D.C.: Island Press, 1996): 3.
20. Paul Selman, *Environmental Planning: The Conservation and Development of Biophysical Resources* (London: Paul Chapman, 1992).
21. John B. Loomis, *Integrated Public Lands Management: Principles and Applications to National Forests, Parks, Wildlife Refuges, and BLM Lands* (New York: Columbia University Press, 1993).

22. Loomis, *Integrated Public Lands Management,* 448.
23. Loomis, *Integrated Public Lands Management,* 447.
24. Botts et al., *The Great Lakes: An Environmental Atlas and Resource Book,* 3.
25. Botts et al., *The Great Lakes: An Environmental Atlas and Resource Book,* 5.
26. See Rafal Serafin and Jerzy Zaleski, "Baltic Europe, Great Lakes America and Ecosystem Redevelopment," *Ambio* 17(2) (1988): 99–105.
27. Janusz Kindler and Stephen F. Lintner, "An Action Plan to Clean Up the Baltic," *Environment* 35(8) (1993): 6–15, 28–31.
28. Lars Håkanson, *Physical Geography of the Baltic, The Baltic Sea Environment, Session 1* (Uppsala, Sweden: The Baltic University, 1993).
29. See Ing-Marie Andréasson-Gren, Gabriel Michanek, and Jonas Ebbesson, *Economy and Law—Environmental Protection in the Baltic Region, The Baltic Sea Environment, Session 7* (Uppsala, Sweden: The Baltic University, 1993); Harald Runblom, Mattias Tydén, and Helene Carlbäck-Isotalo, *The Baltic in History, The Baltic Sea Environment, Session 4* (Uppsala, Sweden: The Baltic University, 1993).
30. Susan Hill MacKenzie, *Integrated Resource Planning and Management: The Ecosystem Perspective* (Covelo, Calif.: Island Press, 1996).
31. John H. Hartig and Michael A. Zarull, "A Great Lakes Mission," in *Under RAPs: Toward Grassroots Ecological Democracy in the Great Lakes Basin,* John H. Hartig and Michael A. Zarull, eds. (Ann Arbor: University of Michigan Press, 1992): 5–35.
32. Botts et al., *The Great Lakes: An Environmental Atlas and Resource Book,* 39 (emphasis added).
33. Botts et al., *The Great Lakes: An Environmental Atlas and Resource Book,* 40.
34. Ragnar Elmgren, "Man's Impact on the Ecosystem of the Baltic Sea: Energy Flows Today and at the Turn of the Century," *Ambio* 18(6) (1989): 326–32.
35. Curt Forsberg, *Eutrophication of the Baltic Sea, The Baltic Sea Environment, Session 2* (Uppsala, Sweden: The Baltic University, 1993).
36. Elmgren, "Man's Impact on the Ecosystem."
37. Curt Forsberg, "Conservation of the Baltic Sea Starting at the Kitchen Table," 43–53; and Kerstin Österberg, "A Clean Baltic Sea Requires a New Lifestyle," 3–10 in *A Future for the Baltic? Scientists Discuss the Future of the Baltic Sea,* Kerstin Österberg, ed. (Stockholm: Swedish Council for Planning and Coordination of Research, 1994).
38. Jan Erik Kihlström, *Toxicology—The Environmental Impact of Pollutants, The Baltic Sea Environment, Session 6* (Uppsala, Sweden: The Baltic University, 1993).
39. George R. Francis, "Institutions and Ecosystem Redevelopment in Great Lakes America with Reference to Baltic Europe," *Ambio* 17(2) (1988): 106–11, 107.
40. Gunnel W. Bergström, *Environmental Policy and Cooperation in the Baltic Region, The Baltic Sea Environment, Session 8* (Uppsala, Sweden: The Baltic University, 1993).
41. Francis, "Institutions and Ecosystem Redevelopment."
42. Bergström, *Environmental Policy and Cooperation in the Baltic Region.*
43. Emin Tengström, "Mass Road Transport in the East and the West Threatens the Environment of the Baltic Sea," 23–42, in *A Future for the Baltic? Scientists Discuss the Future of the Baltic Sea.*

44. Great Lakes Commission, *Lakewide Management Plans: An Ecosystem Approach to Protecting the Great Lakes* (Ann Arbor, Mich.: Great Lakes Commission, 2000), at www.glc.org/advisor/, accessed 15 June 2002.

45. John H. Hartig and John R. Vallentyne, "Use of an Ecosystem Approach to Restore Degraded Areas of the Great Lakes," *Ambio* 18(8) (1989): 423–8. John H. Hartig and Michael A. Zarull refer to RAPs as "literally an experiment in grassroots ecological democracy, a once-in-a-lifetime opportunity for concerned citizens, business persons and scientists to substantially influence the management of degraded areas of the Great Lakes and alter the future of this magnificent and unique resource"; see Hartig and Zarull, "Keystones for Success," 275, in *Under RAPs*.

46. MacKenzie, *Integrated Resource Planning and Management.*

47. Hartig and Zarull, "Keystones for Success." These authors note the need to judge success on the following factors: continued public involvement; effective communication and cooperation; resource commitments; research needs; and building a record of success.

48. Hartig and Zarull, "Keystones for Success," 268.

49. IJC, "Wingspread Conference: Funding Strategies for Restoration of Areas of Concern in the Great Lakes Basin," summary report from a conference sponsored by the IJC and the Johnson Foundation at the Wingspread Conference Center, Racine, Wisconsin, July 23–25, 1996, at www.ijc.org/boards/annex2/wingrap.html, accessed 15 June 2002.

50. IJC, "Beacons of Light," Special Report on Successful Strategies Toward Restoration in Areas of Concern under the Great Lakes Water Quality Agreement, 1998, at www.ijc.org/boards/annex2/beacon/beacon.html, accessed 1 February 2002. For further insights into these AOCs, see the following chapters in *Under RAPs:* Julie A. Letterhos, "Dredging Up the Past: The Challenge of the Ashtabula River Remedial Action Plan," 121–38; Michael O. Holowaty, Mark Reshkin, Michael J. Mikulka, and Robert D. Tolpa, "An Overview of the Modeling and Public Consultation Processes Used to Develop the Bay of Quinte Remedial Action Plan," 161–83; G. Keith Rodgers, "Hamilton Harbour Remedial Action Planning," 59–72; and Frederick Stride, Murray German, Donald Hurley, Scott Millard, Kenneth Minns, Kenneth Nicholls, Glenn Owen, Donald Poulton, and Nellie de Geus, "An Overview of the Modeling and Public Consultation Processes Used to Develop the Bay of Quinte Remedial Action Plan," 161–83.

51. IJC, "Status Assessments," 2001, at www.ijc.org/boards/annex2/status.html, accessed 30 September 2001.

52. IJC, "Beacons of Light."

53. Mitchell, *Resource and Environmental Management,* 65.

54. Kindler and Lintner, "An Action Plan to Clean Up the Baltic," 12.

55. Mitchell, *Resource and Environmental Management*, 66.

56. Bergström, *Environmental Policy and Cooperation in the Baltic Region*, 24.

57. Bergström, *Environmental Policy and Cooperation in the Baltic Region*, 24.

58. Kindler and Lintner, "An Action Plan to Clean Up the Baltic," 15.

59. Baltic 21, "Baltic 21: Background, Objectives, Goals, Organisation," 2000, at www.ee/baltic21, accessed 28 January 2002.

60. HELCOM, "Convention on the Protection of the Marine Environment of the

Baltic Sea Area, 1992 (entered into force on 17 January 2000)," (Helsinki: Helsinki Commission, 2000).

61. See Francis, "Institutions and Ecosystem Redevelopment," 111; Hallett J. Harris, Victoria A. Harris, Henry A. Regier, and David J. Rapport, "Importance of the Nearshore Area for Sustainable Redevelopment in the Great Lakes with Observations on the Baltic Sea," *Ambio* 17(2) (1988): 112–20; AnnMari Jansson and Bengt-Owe Jansson, "Energy Analysis Approach to Ecosystem Redevelopment in the Baltic Sea and Great Lakes," *Ambio* 17(2) (1988): 131–6; Henry A. Regier, Pekka Tuuainen, Zdzislaw Russek, and Lars-Erik Persson, "Rehabilitative Redevelopment of the Fish and Fisheries of the Baltic Sea and the Great Lakes," *Ambio* 17(2) (1988): 121–30; and Serafin and Zaleski, "Baltic Europe," 99–105.

62. IJC, "Evaluating Successful Strategies for Great Lakes Remedial Action Plans," Proceedings from a roundtable discussion sponsored by the IJC and the Johnson Foundation at the Wingspread Conference Center, Racine, Wisconsin, 25–27 July 1995, at www.ijc.org/boards/annex2/rapconf.html, accessed 1 February 2002; Mitchell, *Resource and Environmental Management.*

63. HELCOM, Press Release, 17 March 2003, at www.helcom.fi/helcom/news/206.html, accessed 4 April 2003.

64. USEPA, Great Lakes-Baltic Sea Partnership. Framework Plan, 1999, at www.epa.gov/grtlakes/baltic/plan.html, accessed 1 February 2002.

65. See Nikolaus Gelpke, "Regional Cooperation in Science: The Helsinki Convention for the Baltic," and Hiran W. Jayewardene, "The Indian Ocean Marine Affairs Cooperation (IOMAC)," in *Ocean Governance: Sustainable Development of the Seas,* ed., Peter Bautista Payoyo, (Tokyo: United Nations University Press, 1994), at www.unu.edu/unupress/unupbooks/uu15oe/uu15oe00.htm, accessed 4 April 2003.

66. Per Brinck, Lars M. Nilsson, and Uno Svedin, "Ecosystem Redevelopment," *Ambio* 17(2) (1988): 84–9.

67. Grumbine, "What Is Ecosystem Management?" 27.

68. Hartig and Vallentyne, "Use of an Ecosystem Approach," 424.

69. U.S. EPA/Environment Canada, *Practical Steps to Implement an Ecosystem Approach in Great Lakes Management*, Proceedings of a workshop cosponsored by U.S. EPA and Environment Canada, in cooperation with the IJC and Wayne State University, Detroit, Michigan, 1995, at www.ijc.org/boards/wqb/toc.html, accessed 30 January 2002.

70. Grumbine, "What Is Ecosystem Management?" 35.

71. Grumbine, "What Is Ecosystem Management?" 35.

V

BEYOND CASE STUDIES

12

Toward Theory

Neil E. Harrison and Gary C. Bryner

In environmental issues science is intimately related to politics and policy making. Policy makers believe in its importance and call for science to assess and measure environmental problems as a basis for political negotiation of response strategies. Scientists believe that the knowledge they produce is and should be relied upon in policy making and sometimes that it should directly determine policy. But these case studies show that the influence of science on politics is varied and sporadic, and they question the importance of the role that science has played in a wide variety of environmental issues. Some of the cases show that rather than the seer and guide of politics, science is its handmaiden, subordinate to political need.

That science is affected by politics should be no surprise. Science, especially "big" basic science with little future for commercialization, is always subject to political whim. Particle physics and astronomy, for example, are funded by governments for the economic benefits of construction and operation, as much as for knowledge, the purpose of pure science. Because science funding is limited and always insufficient, policy makers and their appointees choose between research proposals, indirectly selecting what knowledge is to be pursued and how.[1] Most science for the environment is instituted and, thus, designed, to answer questions of interest to policy makers that may not be of interest to pure science. Sometimes the political environment may demand conclusions from scientists ahead of sufficient scientific understanding of the relevant processes.

In this chapter we draw some common themes out of the case studies, critique the theories outlined in the introduction, and offer some alternate theories for understanding the relationship between science and politics and its effect on international environmental policy.

ANALYZING THE CASES

We begin with the two questions raised in chapter 1: how do science and politics relate in international environmental issues? and how does the interplay of science and politics influence international environmental policy? Because of the richness of the cases, readers' answers to these questions will differ, interpretations will vary, and different lessons will be emphasized. In this section, we assess the connections, similarities, and differences between the cases to identify some general themes and invite readers to do the same. To illustrate how the case studies in this book may be aggregated into a foundation for theory building, we identify three principal themes, each of which has several components that are common to all. First, the context and conditions surrounding each case are important. The second principal lesson is that processes vary in structure and participants. Third, the role of ideas is different in each case.

Conditions and the Context

The environmental problems studied in the case studies differ as much in their social contexts as in their essential natures. This makes the success of a broad explanatory or predictive theory unlikely. Most of the cases identify uncertainty in scientific knowledge and policy outcomes as a primary influence affecting both political processes and the uses of science. They also note that these issues are very complex, with many causative factors, myriad and diverse effects, and differential impacts on participants. Understanding each of the problems seems to demand specific science that is adapted not only to the characteristics of the environmental problem, but also to its political meaning. Finally, the use of defined rules is sometimes counter-productive.

Uncertainty has been defined as "the certainty that one of several outcomes will occur; but which one is unknown and unknowable."[2] It is only possible to prove that all swans are white by observing every swan that ever lived. If every one of a large sample of swans is white, we still cannot logically conclude that swans are always white: the next observed swan might be black. As more and more comparable events are observed, certainty (confidence in our hypothesis) increases. But novelty guarantees uncertainty, and environmental issues are invariably complex, novel, and interdisciplinary science problems. Dioxin drift, acid rain, climate change, BH, and BSE were novel problems that stretched scientific knowledge and methods. But well-studied phenomena also may generate uncertainty. Although fish reproduction and migration have been researched for years, the sustainable harvest of tuna is uncertain.

Uncertainty is everywhere; we are rarely absolutely certain of a relationship between a cause and an effect. Uncertainty about the causes and consequences of a problem creates the risk that policy will be wrong. This is particularly problematic in international environmental issues. Hierarchical systems, such as firms or even domestic political systems, handle risk better because they are more effectively able to monitor and modify decisions and implement new measures. Firms are rewarded through the market for taking risk by making decisions under uncertainty, and national governments have the legitimate power to control risk by acting and reacting as evidence changes.

Uncertain problems (like the environmental problems studied in this book) are especially difficult to handle in the anarchic international system, because the nature of the problem must be negotiated through contending national preferences and interpretations. Each socially constructed problem definition implies a different bundle of possible solutions that comprise national obligations and benefits. And once a policy is agreed on and an international treaty ratified, it is implemented by sovereign countries that interpret the agreement into national and local policies through their interests and political processes.

International environmental issues are complex. The depth of detail in the case studies shows that several scientific, cultural, political, social, and economic factors played some role in some part of the process of science, politics, and their interaction. In none of the cases does a single simple explanation suffice. Ethics, personal beliefs, and economics in climate change, reputation and values in dioxin drift, political access and scientific serendipity in acid rain, or international political structures and institutional interests in BSE are only some of the apparently causative factors. Adding a further complication to the international policy process, environmental issues are usually less important to most political actors than economic growth, social issues, and other calls on the attention and pockets of governments.

Science is specific to place, time, and context and not, as usually is assumed, a monolithic universal methodology. This is most clearly seen in the different approaches to acid rain between Europe, North America, and Asia. But it also can be seen in the different uses of science and experimentation in the Bluefin Tuna case and in selection of GCMs in climate change. Some of this specificity reflects the peculiar characteristics of, and the disciplines used in, each issue, and some of it reflects policy needs. But if the methods and goals of science change between issues, the relationship with politics is likely to be unique to each case. And sometimes, as in the tuna case, experimental science may threaten the resource to be protected, limiting the uses of empirical natural science in policy making.

Although many of the case studies focus on international science projects and their reports, in every international environmental issue, countries use their own scientific programs to understand the effects of the issue on their specific interests. Their scientific institutions and structures apply their individual strengths and weaknesses and their own abilities and methodological preferences to interpret scientific data for their national policy makers. But many developing countries have few resources for creating such localized knowledge and are bystanders to the cut and thrust of international science. Chapter 8, by Kenneth Wilkening, which compares the European, Asian, and North American approaches to acid rain, offers the clearest example of the different uses of science and individual choices in scientific methods. For example, the European RAINS model, which was effective in the EU, was not adapted for use in the other regions. In climate change the United States, the UK and other EU members, Brazil, India, and China all funded substantial scientific projects for understanding their relationship to the issue, each with a different national purpose and emphasis that employs different disciplinary techniques. For example, India is most concerned with population effects and land-use constraints on rice farming, Brazil with tropical deforestation, and the United States with fossil fuel consumption.

Rules are the heart of international environmental policy. In conventional wisdom, rules are made to be broken, but in international relations they are made to be manipulated and interpreted. Recently, scholars of international relations have been greatly agitated by issues of institutional effectiveness.[3] Research has tried to show that some international institutions (primarily treaties) are more effective than others in order to identify the structures that seem most effective. However, this research overlooks one important aspect of institutional effectiveness: that overly detailed rules may allow signatories significant room for interpretation.

International treaties comprise statements of principles that enshrine norms of international behavior in a specific issue area and extensive detailed and legalistic rules that translate norms and principles into definitions of right and wrong behaviors. Because there is no international government to enforce treaties, enforcement is believed to require detailed rules. In this view a transgression must be clearly and narrowly defined before sanctions may be administered by the other signatories. However, as the Bluefin Tuna case shows, rules cannot be defined for every eventuality and every condition; rules may be manipulated as countries' preferences or environmental conditions change, and rules established by regional and international institutions may conflict.[4] Japan's increased fish take in the guise of scientific experimentation was within the letter of the law, but outside its spirit. Its behavior did not conform to principles established by UNCLOS and was probably not intended to be permitted by the framers of any of the relevant treaties, though it was not explicitly excluded. Regarding climate change, countries have haggled endlessly over rules for funding, reporting, and administration, but have not agreed on the central problem of defining the meaning of "dangerous anthropogenic interference" with the climate. Where science cannot define the problem, countries routinely seek to make implementation rules open to several interpretations. Because of the withdrawal of the United States, the Marrakesh accords are riddled with special provisions favoring select countries and industries.

Processes

Factors that were important in the development in one case are insignificant or absent in others. The differences in communication of scientific findings cannot be explained merely by the characteristics of the scientific problem. Science in environmental issues is rarely pure, but in some issues, politics demands that it be more policy-relevant than in others. Some issues permit openings for the individual influence of scientists while others do not. The public interest rarely appears unless there is a strong image motivating public concern. And chance or serendipity plays a variable role among the cases. Communication of scientific results is an important factor in many cases. Aside from personal relations between (usually government) scientists and policy makers that may influence the latter's understanding of the problem, science communicates with politics through formal, often international, reports. In many cases, scientific reports had little influence on political interactions. If science is the basis of effective international environmental policy, it is curious that political discourse has often taken little account of apparently authoritative scientific reports.

The cases suggest that in addition to being scientifically authoritative, science reports influence political discourse when they are expressed in terms that speak to the needs of policy makers. But these are conflicting goals. A report is authoritative in a novel and developing science if it is comprehensive, boasts broad participation from eminent scientists in the relevant fields, and clearly defines the areas and degrees of uncertainty. But it is more useful to policy makers if it minimizes uncertainty and expresses greater conviction about the parameters of the problem and the costs and benefits of alternate solutions. For example, EU interest in solving the acid rain problem accelerated once the scientists created the concept of critical loads and a predictive model—the RAINS model—both of which imply a substantial certainty of knowledge about environmental processes.

Individual knowledge brokers gain their influence from their ability to make sense of uncertainty, give it a meaning, and take a position on it, even if that is implicitly policy prescriptive. Sir Crispin Tickell and David Fisk have influenced UK climate policy with such skill; John Sununu similarly influenced President Bush Senior. But international scientific reports cannot play that role without attracting legitimate complaints that its conclusions are not supported by the science.

Individuals outside government may influence environmental policy making. Although there is little evidence in the cases that epistemic communities exist or have influence in policy making, several cases spotlight the influence of individual scientists or policy makers. In acid rain, an individual scientist, later helped by his colleagues, used the popular media to make acid rain a political issue in Sweden and then among European states. Individual staff within Environment Canada organized a meeting that serendipitously influenced climate discourse. Not unexpectedly, leading policy makers and heads of government from Europe to North America also have influenced political discourse. Because of their positions or their abilities, their personal views can lead or stall international policy making.

The public interest is often lost in the muddy waters of policy making. The concept of a public interest may be useful, but it is rarely an accurate representation of the policy positions favored by the public: there may be many publics with an interest in an issue. It is reasonable to assume that most of the public would be interested in a safe food supply, though concerned about its cost. But in the face of BSE, feed suppliers were concerned about government regulation that might increase their costs, and government ministries worried about public perceptions and panic.

The public image of an issue directly affects its political fortunes. If the public interest is not clearly stated, it may be ignored by government. Those environmental issues that have motivated the greatest public reaction and advanced the most rapidly on the political agenda have enjoyed the clearest and simplest image of a complex environmental problem. The average member of the public is unfamiliar with scientific niceties and is too concerned with personal and financial security to fully comprehend the implications of climate change, overfishing, or acid rain. Local debate over dioxins or destruction of nearby fishing grounds may excite their interest, but long-term international problems are usually too obscure. Effective motivating images bring home to the layperson the dangers to them of not immediately addressing the environmental problem. And reports of scientific conferences that step beyond the data to express environmental dangers with colorful images like nuclear winter have more impact than carefully worded scientific reports from intergovernmental boundary organizations that also effectively muzzle policy advocacy by individual technical experts.

The coincidence of relatively insignificant events may have a disproportionate effect. For example, the accidental meeting of a Canadian and a Swedish scientist helped to advance Canadian acid rain research and pinpoint U.S. sources as a potential cause of Ontarian acid deposition. Unaware of research published in Swedish journals, most Canadian scientists had looked no further than the Sudbury smelters when they observed acid deposition. Swedish scientists were themselves aided by the convening of the CSCE in Helsinki in 1975. They were able to make a political meeting shrouded in Cold War rhetoric aware of, and responsive to, their scientific concerns. The 1988 conference in Toronto was designed to debate climate change and publicize scientific data. It unexpectedly grew as national leaders signed on. This attracted the media and the conference became a cause celèbre when its message was connected to the North American drought.

Ideas

Although environmental problems may be conceptualized in several ways, policy makers often attach themselves to one understanding, ignoring others that are, in scientific terms, equally valid. A single environmental problem that is part of the global environment is selected, usually for political, but sometimes for scientific, reasons. But the physical or political locus of the problem may be misunderstood. That some issues attract more reliance on beliefs is not solely explained by their natural complexity.

The locus of the problem is not always self-evident, and policy makers may look to the wrong part of the system for causes and solutions. That is, they may frame the problem incorrectly. In this book we have defined "international environmental" to include those problems with effects across borders and those that may appear in several countries out of a common international or global cause. However, the case studies reflect the principles of international law and diplomatic practice: those environmental issues that have transboundary effects are more readily embraced by scientists and policy makers than those with local effects from international causes.

Effects are more clearly identified and international institutions aimed at remediation are more easily constructed, but the more pervasive environmental effects of industrialization, agricultural practices, and international trade are routinely ignored—as in forests—or given special, inviolate status—as in climate change. For example, when trade-induced logging is identified as a common cause of deforestation in many countries, it is ignored by both scientific reports and political forums. The UN FCCC gives priority to existing trade agreements rather than environmental protection. But systems of trade, finance, industrial production, and capital flows are so endemic to international relations and so prized by the powerful developed countries that they are rarely questioned and never changed, even in the face of serious environmental threats. Early in the preparation for the 1992 Earth Summit, the United States forced climate change to be separated out of the catch-all "sustainable development" UNCED process as an issue amenable to institution building. The United States then believed that, as in the case of stratospheric ozone depletion, a separate international convention could be effective.

In just one case—dioxin in Ames—the cause dominated the effect: the incinerator's ability to cause a transboundary environmental problem. It is curious that those who believed Commoner and wanted to close down or at least fully test the plant never brought the victims (in Northern Canada) into the dialogue. When the cause became the focus of attention, transboundary effects were ignored.

Beliefs are influential in international environmental issues, even as policy makers tout science as central to framing environmental problems. Although beliefs are a primary motivator of domestic political behavior, they are even more salient in international environmental issues, because uncertainty is greater and because public interest is usually slight. The absence of a public voice grants the government (including its bureaucrats) more leeway in international negotiations. Scientific uncertainty means that the problem is not clearly mapped and the effects of alternative policies are poorly understood. In the absence of knowledge, it is easy for policy makers to eschew scientific skepticism and resort to their beliefs to make choices. Believing that BSE would not cross the species barrier into humans, even after it had crossed to pigs and cats, many scientists and most policy makers downplayed and underestimated the threat to food security. This belief seems to have reflected more hope than knowledge, but for government officials it only may have been a calculated risk. Some EU scientists believed animal welfare should be considered in assessing the effects of BH. History, culture, and public sentiment made EU scientists more cautious in embracing scientific innovation than their U.S. colleagues.

Beliefs about the veracity and skill of scientists were important in the Ames debate between a recognized authority on dioxin and a local engineering professor. Yet, the unknown engineering professor "won." The objective strength of their respective arguments was only one of many factors that influenced the outcome of the "contest." At least as important were various subjective factors by which public groups assigned value to their arguments. Commoner, an environmentalist, was supported by progressive citizens, and the engineering professor—one of "us"—by the city establishment. Illustrating how public environmental choices emerge from the mass of citizens, values shared among the members of opposing groups largely determined whom they would believe.

Beliefs influence to whom policy makers listen, what they hear, and the meaning they assign to what they hear. Often they choose to trust advisors who hold comparable values to their own. The first President Bush relied on his economic advisors for the interpretation of climate science. Their value was enhanced by a domestic economic slowdown, but his choice also reflects his greater trust in them than in senior EPA administrators. His son, who in practice has much less interest in conserving the environment than he professes, has ridiculed the EPA's scientific interpretations as products of the bureaucracy, which in his mind is strong language indeed. UK agriculture officials believed that preserving ministerial integrity was more valuable than warning the public about the possible dangers in eating beef. Beliefs may be as simple as a conservative government's opposition to regulation, which delayed an effective response to BSE and underpinned U.S. support for BH. Similarly, Ontario bureaucrats long believed that acid deposition was caused by the Sudbury smelters and that regulating them would damage the province's economy. And differences in beliefs may cause seemingly unbridgeable divides. Japan believes that its experimentation will not irreparably reduce tuna populations, while, using the same data, Australia and New Zealand believe that it will.

Absent NGOs

There is one notable absence from the case studies: all but one of them accord little weight to NGOs. There are two general types of NGOs: those that operate from the top down and those that work from the bottom up. The large NGOs based in the developed countries that are active at international negotiations operate like professional lobbyists. They are as diligent in marketing their interests as any business and have substantial technical and financial resources. Often their delegates to international meetings have stepped through the revolving door to government more than once. They effectively disseminate research results, especially to delegates from developing countries, and promote favored policies, such as the precautionary principle. The majority of NGOs, however, focuses on an environmental, development, or aid agenda at the community level. They rely on volunteers and activists whose commitment and energy overcome their inevitable lack of resources and effectively grow self-reliant responses to local problems. Such NGOs were effective in pushing the Sonoran Desert biosphere reserve and helping local communities adapt to environmental and regulatory changes.

Scholars have found evidence that NGOs have significant influence on environmental policy.[5] Lobbyist NGOs talk directly to policy makers, but they are used or ignored according to the interests of countries and national policy makers. The scholarly consensus is now moving to the view that long overlooked grassroots NGOs are creating effective governance processes outside formal governmental and international institutions that may have more influence over social behaviors than top-down government policies.[6]

Perhaps the issues in these cases were not conducive to NGO influence because they were in some sense too big; in climate change and growth hormones, NGOs have been active, but not very influential. Even the largest NGOs rarely perform big, cutting edge research themselves, but in climate change they have helped delegates interpret scientific and critical reports. Yet, with a few exceptions (for example, China, India, and Brazil), developing countries are neither a large part of the problem nor of its solution, and the developed countries have all the scientists and policy analysts they need. Or it may be that these issues are "under the radar" for lobbyist NGOs because they would not excite popular interest, as in the tuna case, for example, and there are no local victims to mobilize grassroots NGOs. In Ames's dioxin issue, local issue-specific groups formed, but they were barely organized, and external NGOs were not attracted. Dry scientific arguments and legalese would excite few Australians, and esoteric and speculative science does not sell well in Iowa. The absence of NGOs from Ontario's acid rain issue is surprising, but may be explained by the pervasive assumption that the Sudbury smelters were necessary to economic growth.

It is possible that the absence of NGOs is the result of the case study authors' implicit theories. If they did not look very hard for NGO influence, they would not find it. But it is more likely that NGOs neither have nor expect much influence in the relatively less tangible environmental issues that are the subject of intergovernmental debates. NGOs earn a greater return in media attention and have a greater impact in issues that are highly emotive or very local.

EVALUATING CURRENT THEORIES

The unique context and conditions, processes, and ideas surrounding each case makes for interesting examples of how science and politics intersect in environmental policy making. How can we integrate our findings into broader themes and theories?

Theories of the relationship between science and politics are theories of social action. Relations between humans, whether within science or politics or across the divide, are properly the sphere of social science. What, then, is social science theory? What does it comprise and how is it tested? There is not space here to address these questions in detail: a brief summary will have to suffice. Readers can examine and test other theories.

Theory is an empirically testable guess intended to describe, explain, or predict events. Theories shape the way we frame problems and look for answers. They provide perspective on how to assess what is being studied, define what data is relevant, and suggest how events and other observations should be related and interpreted. Theory provides a framework for identifying multiple examples of a phenomenon, finding common themes across them, and deducing from them predictions about future events. A theory is an abstraction from the massive detail of innumerable events in search of patterns expected to reoccur in the future.

Because theory tells us where to look and what data to seek, it is not practical to analyze an environmental issue without using some kind of theory. Without theory, every piece of data, great or small, would be equally interesting. But if a theory is poorly developed or fails to describe some process or explain its outcome, it can misdirect researchers from what is really important. Thus, analysis of each international environmental policy—including, as in the global forest issue, the choice to do nothing—could be guided by a comprehensive analytical methodology, as much as by a comprehensive theory that aspires to prediction.

Theory may be limited to categories of actors who are likely to be important in an issue, directing attention to their actions and expressed beliefs, and avoiding elaborate contextual assumptions and imputations of actor goals, motives, and other intangibles. For example, in environmental issues the developed countries are disproportionately more influential than the developing countries. The G-77 may be able to modify international policy at the margins, but without the developed countries, and increasingly the United States, there are insufficient resources to solve most international environmental problems. Despite the UN's norm of consensus and one nation one vote, it is widely recognized that the United States is more important than St. Kitts and Nevis in any environmental issue.

Among individuals, government leaders often have power over environmental issues through political processes and access to popular media. Although leading scientists and research administrators also may have media access, they generally use it to popularize their research findings without advocating particular policies, though they may be effective if they did.

As these cases have shown, however, chance events and nonpublic individuals and groups may have a disproportionate influence on outcomes. The thick description of events in the case studies may reveal unexpected relationships between science and politics and unpredictable determinants of international environmental policy.[7] One conclusion, then, that readers might draw from these case studies is that the study of international environmental politics and policies should focus on thick descriptions, rather than the testing of hypotheses generated from theories that are likely to have only limited explanatory and predictive value. But proponents of a methodological, rather than a theoretical, approach face the difficult task of how to aggregate the lessons from different case studies and develop ways of improving the policy-making process and the policies they produce.

How did the four common theories that we outlined in the introduction describe or explain the interplay of science and politics in the cases? We conclude that the cases illustrate some of the limitations of each theory. None described or explained the whole pattern of events described in the cases. We expect that none would be able to predict outcomes. Part of the problem is that they are somewhat incompatible. For example, mutual construction is a structural theory in which agents operate in accordance with formal institutional structures, but discursive politics is poststructuralist, meaning that the structures are invisible ideas that regulate and direct behaviors no less effectively. In epistemic communities, ideas (beliefs and values) seem only to infect technical experts who pursue a political agenda by manipulating scientific uncertainty, and influence is unidirectional, from science to politics. This latter aspect accords more closely with rational theory, but that theory minimizes the impact of ideas, except in the diluted sense of interests that are largely objective (that is, knowable and calculable) desires. Since no one theory is likely to be able to account for the complexity found in these and other case studies of international environmental policy making, integrating and synthesizing theories is essential in constructing a more sophisticated understanding of these cases.

These four theories all fall short in accounting for the interaction of science and politics. How can existing theories be adjusted, developed, and refined in order to better account for the interaction of science and politics in international environmental policy making, and how does that interaction shape policies that governments pursue? We examine the four theories introduced in the introduction in light of the main themes that emerge from the case studies written by our colleagues and suggest a modest synthesis that focuses on the political construction of knowledge—the way in which politics affects the creation of scientific understanding. We then suggest two more radical conceptions of the interplay of science and politics. One explains international environmental policies as emerging from a complex process of interactions among interdependent agents at multiple levels of analysis. The second argues that all international environmental policy is the reflection of entrenched economic structures. We urge readers to do the same—to imagine new and creative ways to use the insights from existing theories and the evidence from the case studies to develop new explanations of how science and politics interact and influence international environmental policy and from there to suggest how policy making can be improved.

Rational Choice

There is no evidence that science and politics are discrete and that their relationship is orderly and sequential as this theory predicts. In the rational model, politics calls on science to frame the problem; science collects data, defines the problem, and outlines the costs and benefits of a range of alternatives; politics selects among the alternatives. In the case studies, the interactions vary, but they are always muddled and confused, with little clear evidence of causality between science and politics. In Ontario, politics was ahead of the science in defining the problem as transboundary and only then financing a research program to "prove" it. More than once, politics ignored the science showing the effects of acid deposition in Ontario. In other cases, science had little or no effect on politics. President Bush ignored the evidence of the TAR in setting U.S. policy, and senior officials in the Major government worked to prevent the Health Ministry's warning on BSE. Though the rational theory model may yet reside in the minds of some policy makers and many laypersons, it is too simplistic to be a useful analytical tool.

Epistemic Communities

This theory argues that individuals, primarily technical experts, influence politics through their ability to authoritatively interpret scientific uncertainty. In doing so they use their values and beliefs to reach conclusions out of incomplete or contradictory information. In climate change, several individual scientists were able to influence political decision making. But beyond the early (1985–1990) climate policy advocacy of scientific meetings, there is little evidence of a conscious or accidental congruence of scientists' beliefs through which they interpreted uncertainty and influenced politics. Indeed, the assumptions and beliefs of scientists may inhibit scientific inquiry as in the early interpretations of lake acidification in Ontario, which concluded, perhaps too quickly, that the lead smelters were the only source of the observed pollution. Similarly, disagreement between scientists from Australia and New Zealand and their Japanese counterparts' over the meaning of incomplete fisheries information probably inhibited effective management. The absence of a common definition of "forest" contributed to the absence of a global forest convention.

Although not conclusive, the case studies suggest that epistemic communities are likely to be more influential in raising an issue than in making policy. It is in the nascent stages of problem definition that exaggerated interpretations of incomplete scientific data may be most influential.

Discursive Practices

This theory predicts that ideas discipline both scientists and policy makers, influencing processes in both science and politics, and defining what can be thought and what is debated. Such practices can only be identified when sought out using a detailed hermeneutical methodology. Although the case studies in this book entail thick description, they were not designed to ferret out discursive practices. However, there are a few instances where the available data suggests that discursive practices may have played a role. For example, the idea of critical loads calmed political concerns about appropriate policy in the European acid rain debate. Once this idea and the RAINS model were in place, policy became the routine adjustment of regulations to advancing scientific knowledge. But this also suggests an alternate explanation: that policy makers were happy to set scientifically based standards and allow regulations to adjust relatively automatically, taking themselves out of the loop. We consider in more detail below the related issue of how science is stated.

In the Great Lakes and the Baltic Sea, holistic notions of ecosystems that crossed borders played a role in establishing cross-border environmental management arrangements. Conceptualizing ecosystems does not come easily to delegates trained in the pursuit of national interests. In climate change the idea that action was urgently necessary and possible took temporary hold between 1988 and 1991, only to fade once the United States opposed defined targets and timetables. In international forest policy the presumed international effects of deforestation obscured its international causes.

Simplifying and structuring common ideas do not occur in other cases. This raises the question, why not? What this theory does not address is the causes or conditions under which structuring ideas arise. Are they merely serendipitous? Or do they reflect a cooperative undercurrent in an issue—like regulation of ozone depleting substances—in which a collective solution is tantalizingly close? This yet undeveloped theory needs to be extended to define the conditions under which ideas take hold and discipline both science and politics.

Mutual Construction

Politics always influences environmental science, if only because much research is funded by governments. Politics chooses among alternate research plans on the basis of scientific and other criteria. In Ontario policy makers quickly funded a large research project once they accepted the idea that acid deposition may have originated outside the province. Until then, they had ignored scientific research and blocked research or narrowed its scope.

But this theory goes further than arguing that science chooses methodologies that may not reflect pure science, but that will provide policy makers with the information they need for making informed choices. In climate change GCMs were chosen, despite the many problems inherent in forecasting with computer models of complex processes, to help encapsulate scientific knowledge and predict the effects of alternate policies. These models became a target of much criticism, and policy makers seem not to have crafted policy in response to their scenarios of impending harm. But the RAINS model of acid rain was accepted by European policy makers and relied on, greatly simplifying policy and local regulation. International forest policy was inhibited by scientists' inability to define forests, but in the climate change sinks argument, this problem was avoided by politics defining the concept for science.

Such modeling may cause more problems than it solves. In the dioxin case, did Commoner intentionally include several estimates (that he at first refused to acknowledge or adjust) in order to strengthen his interpretation of the science and influence policy? Or was the design of his model a response to a need to simplify political comprehension and policy selection?

The other cases do not show a similar pattern of political influence on scientific methodology. The question not answered is why politics influenced scientific models in some issues and not others.

ALTERNATE THEORIES

Current theories provide a sufficient grounding for a more elaborate and detailed theory of the relationship between science and politics and the effect of that relationship on international environmental policy. They already provide for the effects of beliefs on scientific interpretation, the power of ideas on both political and scientific behavior, and the influence of policy needs on scientific methods. Although parsimony is valued in theory building—ceteris paribus the simplest theory is the best—combining and extending current theories is now necessary. If a theory is 50 percent more complex and has more than 50 percent greater explanatory power, it would be preferable to a simpler theory. What other theories might be helpful in exploring the intersection of science and politics in the cases examined here?

Political Construction

Building on and integrating the available theories in light of the themes noted in the case studies might result in a theory of the political construction of knowledge. Science is part of a system structured to produce knowledge understood as a consensual statement of the environmental problem expressed in terms relevant to policy formation. Inputs into this system are scientific data, the beliefs of experts, and the beliefs of political actors included in the system. The IPCC was deliberately structured to allow nonscience actors from business, NGOs, and governments to help define "knowledge." Political intervention in knowledge formation may be less formal, as when political actors decide on the extent and form of scientific research. In Ontario, government officials first opposed and then supported a large-scale research effort that led directly to U.S. acknowledgement of its role in acid deposition in Ontario.

This knowledge production system interacts with a system structured to create a knowledge-based policy. Power determines the system structure: the most powerful in an issue area are most able to influence outcomes. In environmental issues, power reflects contribution to the problem in terms of pollution or resource use or to its solution through financial or technical resources.

An important contextual variable is the political milieu, or the environment, in which scientific and politics actors are operating. What policy makers believe significantly influences both the use of knowledge (and, thus, of science) and the design of the final policy. These beliefs may reflect their upbringing, their education, or their political allegiances and change only rarely. As noted above, policy makers' beliefs played a role in several of the environmental issues studied in this book.

The political environment includes the domestic political pressures on this and other issues and the other international problems they are facing.[8] Policy in Ontario changed after the political environment shifted as the public became alarmed about the potential damage to their vacation lakes. Changes in knowledge were a consequence, not a cause, of this change. The political environment was an insignificant factor in the tuna case, but important in the dioxin debate. The political environment can change with little warning and in unpredictable ways, but in so doing it may open a space for policy action that knowledge cannot. Activist scientists, who may be considered an epistemic community, briefly achieved this feat by aggressively interpreting knowledge in the Toronto conference report during a long drought in North America.

Knowledge brokers, who may be individuals or intergovernmental or nongovernmental organizations, provide a two-way interaction between the two systems. They interpret knowledge for the policy system and convey knowledge requirements from the policy system to the knowledge system. Knowledge brokers gain a hearing in the political system through their technical qualifications, their access, and their ability to explain complex technical issues in policy-relevant and accessible language. International science panels in climate change or forests were designed to have (and accepted as having) the first two qualifications. Skill in explaining technical issues in simple language is aided by beliefs that can fill gaps in knowledge to form a coherent lay explanation. Scientific scruples prevent formal science panels from doing this, as the IPCC has found, and individual (usually government) scientists often have been very influential with individual policy makers in this role.

Knowledge brokers also ensure that system structures are designed to produce policy-relevant knowledge. Scientists like Bert Bolin (who participated in the Swedish acid rain issue and the IPCC), who are experienced participants in, and administrators of, large scale and cross-national research projects, understand well what knowledge policy makers need. They structure the knowledge system accordingly, with appropriate science plans, review panels, and reporting committees. The large-scale government-supported acid rain research was carefully designed to answers certain questions of interest to Ontario's policy makers. The comparable U.S. project was designed to address U.S. policy concerns.

Although the political construction blending of current theories is quite comprehensive, it is perhaps more structured than the evidence from the cases suggests is appropriate and perhaps privileges politics over science in the relationship. It also lacks a central statement of its generative idea.

Emergence Processes

The relationship between science and politics and production of international environmental policy can be though of in terms of "emergence processes" drawn from the science of complex systems. As there is not space here to develop this theory completely, the following is a sketch of the main ideas.[9]

Complex systems are emergent (system behavior and structure emerge from the interaction of its parts), dynamic, and ever changing and potentially nonlinear (that is, disproportionately sensitive to small changes in internal and external conditions). The decisions, policies, and international institutions of the international environmental system seen in the case studies can be said to dynamically "emerge" from the rule-ordered interaction among interdependent agents.[10] In a simple system, each part has a defined purpose, and its role in the system and its relations to every other part is defined. For example, in a bicycle the wheels are held within a frame, and the pedal crank turns the rear wheel through a chain. Change the relationship of the parts—exchange a wheel and the handlebars—and the system ceases to be a bicycle. In a complex system, structure is but one state of evolving relations between the system agents, like a single frame in a movie: in an ecosystem, trees and animals grow and die, but the system is still recognizably a forest.

Each agent—an individual or, at higher levels of aggregation, an organization (e.g., firm or NGO) or a nation-state—chooses its behavior in accordance with an internal model of the world around it within the constraints of rules that govern interactions with other agents. An internal model is a set of desires (and beliefs about how to satisfy those desires) that change with learning, which, as in social construction theories, results from interactions with other agents.[11] Rules govern interactions among agents, providing some order and preventing chaos or anarchy.[12] Without a global government to make and enforce rules, nation-states individually choose their behaviors and the international system moves between chaos (for example, World War II) and substantial order (as in empires or the Cold War). But there is extensive governance through formal rules, such as the treaties and agreements observed in the case studies, and informal rules as in customary international law.[13]

What are the implications of this approach? There are several. First, this approach focuses attention on how policy emerges from bottom-up processes. International environmental policy is not only the creation of states, but it is influenced in many ways by interactions and choices at lower levels of aggregation. National policy making is a complex process through which the nation-state constructs its internal model of the issue: what it wants in an international agreement and how it expects to get it. The national internal model evolves out of the interaction of many related processes, including the negotiated conclusions of authoritative scientific reports, international discourse between states, the emergent demands of interest groups and the public through domestic political processes, and the beliefs and preferences of governments and leaders (for example, President George W. Bush's rejection of Kyoto).[14]

Second, because decision making is distributed, change in international policy may emerge through social interactions from small beginnings. Individual agents may influence international policy by changing the internal models of influential policy makers much as Sir Crispin Tickell influenced Mrs. Thatcher's preferences on climate change and how to achieve them, or as John Sununu influenced President Bush Senior. A scientist may propose a new interpretation of complex data, as when the language on a "discernible" anthropogenic effect strengthened the SAR and increased its influence. Or an individual may introduce a new idea to the public, changing many internal models, as a scientist did in Sweden and a reporter did in Ontario, which in turn influenced (through domestic political processes) the internal models of larger aggregations: the nation of Sweden and the Province of Ontario. Often, as in these examples, scientific evidence is a primary tool of persuasion and learning.

Third, the international policy process is dynamic: problems are never solved. As scientific understanding accumulates, agent models evolve, and political processes change; international environmental problems are continuously redefined, and new possible solutions created. Although IPCC reports have made clear that the full implementation of the Kyoto Protocol will not prevent the climate from changing, this is the current political solution to the problem, one that has evolved considerably since it was drafted in 1997. Despite evidence of global deforestation, its contribution to global warming, and its deleterious local effects, policy makers who once supported an international treaty to limit it have now rejected one.

Fourth, conflict between science and politics is inevitable, and the tension between the rules of the two systems affects international policy. Both science and politics are social activities governed by rules of behavior, but the rules in the two social systems are different, causing conflict between them. Science is the process of continuously testing contingent hypotheses and refining them: knowledge in the sense of "proof" never occurs. But policy makers want to begin with scientific knowledge and "solve" the social and political problem of distributing the solution's consequences. In the scientific process, scientists are constrained by rules of peer review and replication of empirical results from concluding too generously on incomplete data. Such rules are absent from politics: desires and beliefs guide behavior, only constrained by the need for consensus among sovereign states with individual internal models. Scientists (tentatively) accept information as knowledge only when a majority concurs on the method of collection and analysis of the data—when Kuhn's normal science is constructed. But politics is complete when policy makers agree on a collective interpretation of the problem and a collective solution, not when objective data shows that the problem no longer exists. Despite the efforts of boundary organizations, reports in regulatory science are the negotiated product between scientific rigor and policy relevance.

Fifth, this approach suggests a different research methodology. Analytical focus should be pointed at processes of social interaction, as is already advocated by the constructivist literature. But this needs to be extended beyond the international level to domestic processes and even issues of the authority (influence over others' behavior) of individuals and organizations within those processes. Fortunately, this approach often lends itself to computer simulations to identify the interaction rules from which system behaviors emerge.[15] And as argued in the appendix, the absence of theories successfully defining cause-effect relations (which are not possible in complex systems) suggests that the case studies in this volume should be approached through a rigorous analytical method emphasizing agents and the rules of their interactions.

Finally, if policy makers and activists recognized that international environmental policy is an emergent property of complex and dynamic processes, they may change their behavior. Policy makers should emphasize general rules, principles, and norm building instead of overly precisely defined obligations. If both natural and political processes are dynamic, a proliferation of narrow and pedantic institutions and treaties quickly becomes a barrier to intelligent policy responses, as the tuna case showed. Although most international environmental agreements include provision for changes as the science evolves, this is more honored in the breach, and environmental treaties remain subordinate to global international economic institutions, limiting opportunities for adaptation to changing political and scientific conditions. The emergence processes perspective also suggests that NGOs and activists should put more resources into changing internal models (through learning) at the grassroots level. In the long term, the substantial process changes necessary to better mitigate international environmental challenges come from pressures from below, from domestic processes.

The Critical Theory Option

The theories discussed above are largely aimed at explaining how environmental politics works and why it produces the policy outcomes it does, how to predict political and policy outcomes, and how to improve the policy-making process and policies themselves. If one begins with the agenda of explaining, predicting, and improving, then the challenge here is to create better, more powerful theories that account for more of the factors that are central to parsimonious explanations, predictions, and improvements. Readers may have already begun to think of other ways to combine and extend these theories into a much richer, more useful way of thinking about the case studies discussed here and found elsewhere.

However, another way in which we think about theories is from a critical, transformative perspective, where the task is not to explain or to make marginal improvements or solve problems, but to fundamentally reshape environmental politics and the policies they produce. Critical theory seeks to identify the core problems with and shortcomings of environmental politics and policies, calls into question their legitimacy, and explores alternative paradigms and terms of discourse.[16]

The case studies presented above, for example, can be examined in light of the extent to which environmental policy making is constrained by powerful economic and political interests that are able to protect their interests even when they produce environmentally damaging outcomes. From this perspective, the results of scientific research on environmental harms will never be as potent as demands to protect important economic interests from regulation and ensure that their capacity to produce wealth remains unfettered. President Bush, who with some of his cabinet colleagues is from the "oil patch," has preferred energy production (especially of fossil fuels) to demanding reductions that would reduce GHG emissions. ExxonMobil, the world's largest corporation, is a strong supporter of his policies and his rejection of the Kyoto Protocol. Logging interests support international trade in timber and usually reject efforts to conserve forests. Japan's experimental fishing was a response to popular demand for a much-prized fish. Even when scientific findings about environmental harms are clear and unambiguous, if they challenge the prerogatives of powerful economic actors, they will not lead to effective remedies. Urban opposition in the United States to the required increase in water flow to the Colorado River delta has prevented conservation of all of Mexico's biosphere reserve ecosystem, even though the effects of the river's overallocation are evident.

Most of the case studies demonstrate that science influences politics. But no case provides an example of scientific findings dominating political calculations. Even where there is strong scientific consensus, such as for acid rain, economic interests determine how governments respond. If advocates of ecological sustainability are correct, the world cannot continue to support policies that are rooted in short-term economic profit-seeking, rather than sustaining ecological systems and natural resources. Scientific knowledge that identifies the prerequisites for ecological sustainability must be given priority in public policy making, or human life itself will be threatened.[17]

Ecological theories, such as ecological sustainability, deep ecology, and global ecology, begin with the proposition that the health of the biosphere is the primary public value, because it makes all other activities possible. The real test of policies is whether they preserve the critical ecological services that ecosystems provide, not whether they contribute to short-term economic growth. Ecological science, not politics or economics, is the primary focus of theory. If ecologists can explain convincingly how life is dependent on preserving ecological systems and natural resources, identify threats to those systems and resources, and suggest what actions need to be taken to preserve them, then that knowledge must set the policy agenda, rather than the demands of companies to gain access to natural resources or to be able to reduce their costs by polluting their environs. Science and politics do not interact as equals, but the former shapes and directs the latter. The Baltic Sea and Great Lakes cases suggest that scientific recognition of the interaction between environmental issues does influence political thinking. The acid rain chapters show how scientific knowledge can shape policy, and that has even occurred in some nations in response to research on climate change. But the test here is whether scientific arguments about sustainability dominate political considerations. If these ecologists are right, then the analytic task is to figure out how to ensure that political and economic interests are subservient to the demands of protecting ecosystem health and how to design policy-making processes and policies to ensure that they are consonant with these fundamental values.

Ecological sustainability does not exhaust the possibilities of critical theories. Others may argue that democratic values of deep participation, consensus building, and participation in self-government are the primary values by which international environmental policy making must be assessed. A radical transformation of politics that empowers all global citizens to protect their interests and share in decisions that affect their quality of life, one might argue, is the imperative that should guide explanations of how policy making occurs and how it should take place. Science here clearly plays a subservient role to politics. Political values of equality, justice, opportunity, and self-government shape the search for scientific knowledge to ensure that it contributes to the achievement of political goals.[18]

CONCLUSION

This book presents extensive detailed data showing that the relationship between science and politics is more varied and more complex than current theories predict. Neither science nor politics is monolithic; neither can be studied alone. But separating science and politics for analytical purposes (as is done in this book) allows greater clarity of thought and analysis, and illuminates the many interactions between science and politics out of which emerges international environmental policy. As a foundation for a richer and more effective theory of international environmental policy, we have suggested three alternate approaches to theory, supported by the data in the cases. The relationship between science and politics needs much further study if international environmental policy is to be understood and improved. We hope that we have created a learning text (for both students and scholars) that opens many opportunities for further empirical research and the construction of new theories to explore how science and politics come together in the making of international environmental policies, and how the contributions each bring to the policy-making process can be enhanced.

NOTES

1. The demand for knowledge may not conform to the rules of neoclassical economics in which marginal demand falls as the cost of marginal supply increases (and demand and supply converge on an equilibrium).

2. George T. Chacko, *Decision-Making under Uncertainty: An Applied Statistics Approach* (New York: Praeger, 1991): 5.

3. Oran R. Young, ed., *The Effectiveness of International Environmental Regimes: Causal Connections and Behavioral Mechanisms* (Cambridge, Mass.: MIT Press, 1999); David G. Victor, Kal Raustiala, and Eugene B. Skolnikoff, eds., *The Implementation and Effectiveness of International Environmental Commitments: Theory and Practice* (Cambridge, Mass.: MIT Press, 1998).

4. The accounting scandals of 2001 and 2002 in which major corporations (e.g., Enron, WorldCom) misstated their earnings were, in part, the result of the U.S. preference for detailed accounting rules. "Generally accepted accounting principles" are based on principles, but constitute a body of highly specific rules. Enron, for example, structured its transactions around the rules, but was not in compliance with several principles of accounting. In Europe and in the international accounting standards, principles are emphasized over detailed rules. Some international environmental-politics scholars argue that the norms represented by treaties are more important than their rules, an interpretation most international lawyers would contend. See Martha Finnemore, "Norms, Culture, and World Politics: Insights from Sociology's Institutionalism," *International Organization* 50(2) (Spring 1996): 325–47; Audie Klotz, "Norms Reconstituting Interests: Global Racial Equality and U.S. Sanctions against South Africa," *International Organization* 49(3) (Summer 1995): 451–78. In the U.S. policy literature, it is accepted that policy is not implemented as designed, and it should be no surprise that slippage occurs when sovereign countries are required to implement the provisions of complex international treaties. For example, see Jeffrey L. Pressman and Aaron Wildavsky, *Implementation* (Berkeley: University

of California Press, 1973). Many EU members fail to implement many of the directives that they participated in creating, suggesting that they accept the rules in principal, but are unwilling or unable to implement them when required to.

5. Thomas Princen and Matthias Finger, *Environmental NGOs in World Politics: Linking the Local and the Global* (London: Routledge, 1994); Sheldon Kamieniecki, ed., *Environmental Politics in the International Arena: Movements, Parties, Organizations, and Policy* (Albany: State University of New York Press, 1993).

6. Paul Wapner argues in *Environmental Activism and World Civic Politics* (Albany: State University of New York Press, 1996) that NGO influence is now primarily through civic, rather than formal, politics.

7. Clifford Geertz, "Deep Play: Notes on the Balinese Cockfight," *Daedalus* (Winter): 1–38; and *The Interpretation of Cultures: Selected Essays* (New York: Basic Books, 1973).

8. John W. Kingdon in *Agendas, Alternatives, and Public Policies* (Boston: Little, Brown and Company, 1984) develops the concept of streams of events, one of which is the political stream. A convergence of the three streams (with problems and policy—or available solutions—streams) opens a window of opportunity for policy action. Though developed for U.S. domestic policy, it is an attractive model that has only rarely been applied in an international political context (see Neil E. Harrison, "Heads in the Clouds, Feet in the Sand: Multilateral Policy Coordination in Global Environmental Issues," Ph.D. Dissertation, University of Denver, Denver, Colorado, 1994).

9. There are three broad categories of theories of complexity, all of which model emergent systems. Models from physics that do not exhibit adaptive characteristics and models from cybernetics—see John H. Holland, *Hidden Order: How Adaptation Builds Complexity* (Reading, Mass.: Perseus Books, 1995)—and biology that do—see Humberto Maturana and Francisco Varela, *The Tree of Knowledge: The Biological Roots of Human Understanding* (Boston: Shambhala, 1992); Stuart Kauffman, *At Home in the Universe: The Search for the Laws of Self-Organization and Complexity* (New York: Oxford University Press, 1995). For a highly readable overview of the development of complexity theories, see M. Mitchell Waldrop, *Complexity: The Emerging Science at the Edge of Order and Chaos* (New York: Simon & Schuster, 1992). A book on the use of complexity and agent-based models in international relations is in preparation: Neil E. Harrison, ed., *International Complexity: Agent-Based Models in Global and International Studies* (Albany, N.Y.: SUNY Press, forthcoming).

10. Membership in a group, nation, or sovereign country; continual interaction; and ethnicity and identity are among the other factors that make agents "interdependent." The interdependence of countries that interact continually on multiple issues is well recognized. See Robert O. Keohane and Joseph S. Nye, *Power and Interdependence*, 2nd ed. (Boston: Little, Brown, 1989); and Robert O. Keohane and Robert Axelrod, "Achieving Cooperation under Anarchy: Strategies and Institutions," in *Cooperation under Anarchy*, Kenneth A. Oye, ed. (Princeton, N.J.: Princeton University Press, 1986).

11. John Gerard Ruggie, *Constructing World Polity: Essays on International Institutionalization* (London: Routledge, 1998); Vendulka Kubálková, Nicholas Onuf,

and Paul Kowert, "Constructing Constructivism," in *International Relations in a Constructed World*, ed. Vendulka Kubálková, Nicholas Onuf, and Paul Kowert (Armonk, N.Y.: M. E. Sharpe, 1998): 3–21; Mark Neufeld, "Reflexivity and International Relations Theory," *Millennium: Journal of International Studies* 22(1) (1993): 53–76.

12. Relations between states have been called "anarchical" meaning "absence of government." But there is governance provided by a network of organizations (formalized institutions), institutions, and less formal rules and practices of international law and diplomacy (see e.g., Robert O. Keohane, "International Institutions: Two Approaches," *International Studies Quarterly* 32(4) [December 1988]: 379–96; Oran R. Young, *International Governance: Protecting the Environment in a Stateless Society* [Ithaca: Cornell University Press, 1994]; Oran R. Young, *International Cooperation: Building Regimes for Natural Resources and the Environment* [Ithaca: Cornell University Press, 1989]).

13. Customary laws are those laws recognized by states that have emerged from practice, such as the inviolate sovereignty of states. These may be referred to in formal international agreements, but are not defined by them.

14. Peter B. Evans, Harold K. Jacobson, and Robert D. Putnam, eds., *Double-Edged Diplomacy: Institutional Bargaining and Domestic Politics* (Berkeley: University of California Press, 1993); Robert D. Putnam, "Diplomacy and Domestic Politics: The Logic of Two-Level Games," *International Organization* 42(3) (Summer 1988): 427–60.

15. Joshua M. Epstein and Robert Axtell, *Growing Artificial Societies: Social Science from the Bottom Up* (Washington, D.C.: Brookings Institution Press, 1996).

16. Scott Burchill, "Introduction," in Burchill and Andrew Linklater, *Theories of International Relations* (New York: St Martin's Press, 1996): 18–24.

17. For more on this view, see Gary C. Bryner, *Gaia's Wager: Environmental Movements and the Challenge of Sustainability* (Lanham, Md.: Rowman & Littlefield, 2001).

18. For more on this issue, see Gary Minda, *Postmodern Legal Movements* (New York: NYU Press, 1995).

Appendix

Teaching with Case Studies

Neil E. Harrison

I hear and I forget.
I see and I remember.
I do and I understand.
—Confucius

The design of case studies varies with their purpose, which may be either as a research or teaching tool. For either purpose they assemble data from multiple sources to give a detailed understanding of a complex issue within a defined context. In research case studies, primarily used to test the application of ideas and theories, data is collected for the variables that the theory defines as significant. Theory is a short cut to comprehension. If theory is well designed, it defines what relationships are important and what data may be ignored. It helps to distinguish useful from extraneous information. But if it is misguided, it misdirects the analyst from what is truly important, acting as blinders that prevent observation of reality. Didactic pedagogy is appropriate and effective when there is a fixed and definable body of knowledge to be conveyed, theories are well defined and empirically tested, and extensive data is available that clearly measures theoretical variables.

Case studies are often used in the preparation of theories and for initial empirical testing of theory-derived hypotheses of cause-effect relations. In the social sciences, case studies are a rich source of data, but that data may be tainted by factors not theorized as important or not measurable. Despite these limitations, they are a valuable tool in theory building and a favorite of doctoral candidates. Edited collections of case studies also are designed to test theories and hypotheses and, therefore, have little teaching value.

Teaching case studies are designed to encourage in-class debate among students through problem-solving scenarios, theory-building exercises, or simulations. To be effective they are relatively theory agnostic and include all data that may be relevant to their purpose. The case studies in this book were custom written to give students a slice of reality with which they can begin to argue from the specific to the general. Thus, they are theory-building cases that enhance debate across the diversity of knowledge and experience of the student discussants. A mix of students with natural and social scientific backgrounds is likely to create a richer range of suggested explanations for events.

LEARNING WITH CASE STUDIES

Case studies are a proven means to encourage independent analytical thinking. They have been used for many years in business schools to stimulate the complex, innovative, and adaptive modes of thinking that can be successfully applied to a wide range of real problems. Case studies are less well accepted in the natural and social sciences, which have been more concerned with theory building than practical analysis of complex and context-dependent phenomena.

The case study method is most valuable for illustrating the art of analyzing and solving complex problems, such as international environmental policy, that are poorly understood.

> Cases bring to the classroom a slice of reality—often incomplete, undigested, unanalyzed—to provide a measure of simulated or vicarious experience which is to be searched for meaning. Case teaching and learning is inductive. The student learns to reason from the specific to the general, linking disparate events and bits of data into patterns for analysis.[1]

In this way case studies stimulate independent thought in the students, interaction in the classroom, and collective construction of the dominant patterns in the data. Participation in case studies is active learning from which comes understanding.

Even without an explicit theory, perceived patterns in historical data can help analysis and comprehension of new problems as they arise. Allowing for evident contextual differences—time, place, and culture and scientific knowledge, for example—the sort of data that is helpful in understanding the interplay of science and politics in stratospheric ozone depletion and the effect of that interaction on policy should be useful in climate change. By beginning with a search for patterns of similarity, differences can be highlighted through classroom debate and the beginnings of theory discovered.

The skepticism that a well-designed case study engenders is healthy and the first step in understanding the complexity of reality. In designing this book, we sought case studies that foster informed debate and do not conclude it. Unlike a chapter in a collection intended to support a research perspective, in this book a case study should, as far as practical, be neutral. They have been designed to describe the relevant issues and provide the data to fuel classroom and scholarly debate; they do not "solve" the problem so much as define the parameters within which a solution may be found. The authors' ideas about the most important factors in the interaction of science and politics are implicit in the text—in the choice of what data they exclude to keep within the publisher-imposed word limit—rather than explicit in pursuit of a theoretical explanation. We believe that a good teaching case study reads like a novel: there are clues throughout the story, there is a message in the text, but no categorical answer.

An apocryphal story illustrates our thinking here: a fan wrote to playwright Samuel Beckett with a long and complete explanation of the meaning of his successful, but enigmatic, play "Waiting for Godot." Beckett wrote back that if he had known what it meant, he wouldn't have written it. Like good literature, a well-written case study similarly includes rich (or thick) description, consideration of context, and depth, but does not attempt a simple explanation for a complex reality. Such cases may be dismantled and understood in many ways and many patterns identified in the data. Through instructor-guided classroom debate of each case, different meanings can be compared and contrasted, modified and reworked, until agreement coalesces around dominant patterns. As the semester progresses, the patterns identified in each case may be compared for commonalities, consistencies, and inconsistencies that identify important differences between cases. But always, the data drives ideas, inductively creating patterns that may be fitted to other apparently similar cases. This is the reverse of most case studies of international environmental issues where the ideas and theories drive the data.[2]

GETTING STARTED

The case studies in this book are intentionally rich in data and short on theoretical explanations. It is reasonable to presume that in social environments the actions of social actors are important. Thus, when faced with much data and few explanations, a useful initial cut can be made by asking, who wanted what, when, and how did he or she go about getting it? This focuses attention on actors, those who act in the science or politics processes (who may be individuals, groups, organizations, or states).[3]

Although they may not themselves clearly understand what they want, an approximation of what actors value may be inferred from their actions and statements. Rational consideration of what they should want, given their position (e.g., business organizations usually value access to resources) may also be useful.[4] However, imputed values or preferences can only be educated guesses and should be handled with caution.

Was the structure of the process influential? It is reasonable to assume that the rules that define who is involved and how they are required to relate may affect the availability of ideas and policy alternatives, how they are evaluated and compared, and which choices and compromises are made. In science the sources and constraints of funding influence research objectives. What criteria do funding agencies use to select among proposals? If funding is by governmental institutions, does a political or policy purpose dominate?

Theory can never anticipate every condition: context may be a vital variable in explaining outcomes and must also be considered. For example, the general feeling of the times (often called the zeitgeist) may have some impact and should be assessed. Ideas have greater currency at some times rather than others. The environment was rarely considered by U.S. business or government institutions in the 1960s, but forty years later it had become central to most decision making. The effect of the currency of certain ideas is clear in the effect that the Enlightenment has had on how we view the world.[5] Before the Enlightenment it was common for Europeans to believe that insects had rights comparable to humans, and where the interests of humans and insects were opposed, judicial arbitration was appropriate. The idea that animals are our moral equals seems absurd to the modern mind, but it was the norm for much of human history and still may be found in some traditional communities.

Group psychology or "herd mentality" is much studied in finance and economics.[6] Investors often wrongly believe that the investing style that was profitable yesterday will always be profitable. Keynes emphasized the importance of capitalists' "animal spirits"; Simon noted that rationality is usually limited and unable to process all the available data; and Janis warned of the danger of "groupthink" in policy making.[7] Is there any evidence that ideas captured the minds of the participants and limited the range of alternate policies considered?

TEACHING THE CASES

Teaching case studies are commonly used in three ways: by lecturing a case, illustrating it, or theorizing it. In the first instance, the instructor takes students through the case step by step and draws the conclusion. In our view this is a waste of a well-written teaching case study as it minimizes student participation and, as Confucius commented, they will soon forget the lesson. Illustrating a case means that the instructor will use the case as an example of a specific lesson that he or she wants students to learn. Students are passive observers of the instructor's brilliant and incisive analysis. In theorizing a case, the instructor uses the cases to demonstrate the efficacy of a particular theory. A good instructor is able to identify which parts of the case support the theory and where events in the case tend to oppose the theory.

These cases were written for a different approach, which has been called "choreographing a case class" in which students actively participate in discerning patterns within and between the cases and proposing possible explanations for events. This approach has been described as follows:

> An instructor . . . leads students through the key conceptual and decision issues in the case without necessarily prejudging the correctness of their students' contributions . . . if the case is an instrument to stimulate inductive frameworking, then vigorous debate on the merits of an argument are an absolute cornerstone of a good case discussion . . . [and] the instructor should not presuppose that his or her point of view is the most accurate. Neither do [we] suggest that "wrong" analyses should be pardoned but that patient discussion, rather than [the] instructor, first should expose the faulty premise. . . . A good case instructor does not intellectually dominate the case discussion.[8]

Each class can begin with a discussion of who the major actors were and what they were doing. Students may be assigned to write a short paper on the actions of an individual country or principal individual actor prior to the class. They can represent their assigned actor in class and explain what they did.

Next the instructor can offer several (four or five may be a useful limit) themes within the case. Some useful themes can be found in the section introductions, in each case, and in the concluding chapter. The instructor summarizes discussion when moving from one theme to the next. By listening (the primary instructor skill in the case method; the instructor "guides" rather than "teaches" a case), the instructor can distill students' meanings from their class interventions and find connections between them. Structuring the class carefully around summaries (on a board or flipchart) of students' comments permits a useful concluding section. This is where the discussion of each theme is summarized and any broad generalizations are extracted for use later in creating an inductive theoretical framework.

At the end of the semester (or the section in which the cases are used), the instructor can lead a session discussing the commonalities and differences between the cases as a step toward building a theory of the interplay of science and politics in the international environment. Using the concluding chapter as the assigned reading, the instructor might lead discussion on the significant factors we identified. Are there other examples of these factors? Did any of these factors not apply in any of the cases? Is there contrary evidence in any of the cases suggesting that any of these factors are not important? Are there other common factors that we overlooked? What are their theoretical implications?

I have successfully used this approach in courses on foreign policy and international environmental policy. The students enjoy speculating about what factors might be important, what motivates actors, which decision practices they follow, and the effect of the structure of their interactions. I found my principal task was to keep class discussion moving and accumulating, and to prevent it from getting too heated as everyone chipped in. Students who at the beginning of the course could not understand what a theory was, by the end were actively theorizing.

TEACHING NOTES

In a text of this kind it is not practical to attach teaching notes that help the instructor define each case's objectives, understand the context, and set up a class plan. However, we provide much of what would be in teaching notes throughout the book. For example, we include:

- The objective of each case. In general this is to explore the interplay between science and politics and understand the effect of that interaction on international environmental policy. This strategic objective is the same in all cases, but more limited intermediate goals are suggested in the section introductions and the cases themselves.
- The scientific and social context as a part of each case.
- Guiding questions on each case. These are appropriate for both teachers and students and can be found in the introduction to each section and in the cases themselves.

The four theories of science-politics interaction that we describe in the introduction can be used as the core of class plans. Each case can be compared to these theories to see which works and how they fail to describe or explain events. Is there any evidence of an epistemic community? How influential are individual scientists? Do any ideas take hold of the scientific or political discourse and limit policy alternatives? How do policy makers' needs direct scientific methods or influence scientific reports? And so on. Based on the class response to the questions in the section introductions and the cases, the instructor can move discussion toward testing the individual theories and, ultimately, to constructing an inductive theoretical framework.

NOTES

1. Peter B. Zimmerman, "Case Development and Teaching: Communicating the Results," In *Notes on the Case Method* (M18–9701368.0) President and Fellows of Harvard College.

2. One well-known example from among many such collections of case studies is Peter M. Haas, Robert O. Keohane, and Marc A. Levy, eds., *Institutions for the Earth: Sources of Effective International Environmental Protection* (Cambridge, Mass.: The MIT Press, 1993). Such books are designed as part of a theoretical debate and are less effective in the classroom.

3. There is debate in the technical literature about the definition of "state." In this book we use the term to mean the governing structure within a country that includes official government institutions and those individuals and organizations outside formal government that may influence government decision making.

4. Not all businesses always want to desecrate natural ecosystems for profit. Some corporations consciously try to reduce their ecological impact, and regarding climate change, insurance companies have forcefully supported strong international policy, where oil companies and auto manufacturers oppose them.

5. Luc Ferry, *The New Ecological Order*, trans. Carol Volk (Chicago: University of Chicago Press, 1995). Ferry argues that the Enlightenment created the bifurcation in the human mind between humanity and its environment. Before then, French villagers, for example, would consider all God's creation as one and assign comparable rights to nonhumans, including bugs and weevils.

6. An early and entertaining example is Charles Mackay, *Extraordinary Popular Delusions and the Madness of Crowds* (New York: Noonday Press, 1932).

7. John Maynard Keynes, *The General Theory of Employment, Interest, and Money* (New York: Harcourt, Brace, & World Inc., 1964); Herbert A. Simon, *Administrative Behavior,* 3rd ed. (New York: Free Press, 1976); Irving Janis, *Groupthink* (Boston: Houghton Mifflin, 1982).

8. V. Kasturi Rangan, "Choreographing a Case Class," Note 9–595–074 on Case Method, Harvard Business School (Boston, Mass.: Harvard Business School Publishing).

Index

About the Contributors

Michael M. Bell is associate professor of rural sociology at the University of Wisconsin, Madison, and a part-time composer of folk and classical music. He is particularly interested in dialogue, democracy, and unfinalizability in social, ecological and musical life. He is the author, along with Gregory Peter, Susan Jarnagin, and Donna Bauer, of the forthcoming book *Farming for Us All: Practical Agriculture and the Cultivation of Sustainability* (2004). The second edition of his *An Invitation to Environmental Sociology* (1998) will also appear in 2004.

M. Leann Brown is an associate professor of political science/international political economy at the University of Florida. Previously, she was a Fulbright European Union Research Fellow and spent a semester in Brussels affiliated with the Environmental Committee of the European Parliament. She was also program coordinator for the International Studies Association. Her current research focuses on how scientific uncertainty and organizational learning affect environmental policymaking in the EU.

Richard C. Brusca is executive director of the Arizona–Sonora Desert Museum in Tucson, Arizona. He is the author of over 100 books and scholarly papers in the fields of invertebrate zoology, marine biology, biodiversity, and the natural history of the Sonoran Desert region. He has been working in the Gulf of California for over 30 years.

Gary C. Bryner is a professor in the Public Policy Program at Brigham Young University and a research associate in the Natural Resources Law Center at the University of Colorado School of Law. He has a J.D. from Brigham Young University, a Ph.D. in government from Cornell University, and a B.S. and M.A. in economics from the University of Utah. He has been a guest fellow at the Natural Resources Defense Council, the Brookings Institution, and the National Academy of Public Administration.

Michael S. Carolan is an assistant professor of Sociology and Environmental Studies at Whitman College. He received his Ph.D. in sociology at Iowa State University in 2002. Before coming to Whitman, he was a visiting Fellow at Wageningen University in the Netherlands. His research interests include environmental sociology, agriculture and food systems, the sociology of knowledge, and the sociology of trust.

Radoslav Dimitrov is an assistant professor at the Department of Political Science of the University of Utah. He has published research on environmental regime formation; the connection between knowledge, power and interests in global environmental politics; and the concept of environmental security in relation to water issues. His articles appear in *International Studies Quarterly, Journal of Environment and Development,* and *Society and Natural Resources.* In addition to his academic work, he works as an analyst for the *Earth Negotiations Bulletin* and attends international meetings on environmental issues.

James Eflin is a geographer with degrees from the University of Colorado (B.A., M.A.) and the University of Washington (Ph.D.). His research and teaching are at the interface between environment and society, with an underlying theme of sustainability. He is an associate professor in natural resources and environmental management at Ball State University, where his teaching includes energy resources, integrated resource management, environmental ethics, and sustainable futures.

Jeremy Firestone is an assistant professor of marine policy, College of Marine Studies, University of Delaware. He holds a J.D. from the University of Michigan Law School and a Ph.D. in public policy from the University of North Carolina. He specializes in international and domestic ocean and coastal law, policy, governance, and regulation, with an emphasis on fish and wildlife resources, coastal resource protection, and marine pollution.

Neil E. Harrison, Ph.D., is executive director of the Sustainable Development Institute (SDI) and is a member of the adjunct faculty at the University of Wyoming. He has published *Constructing Sustainable Development* (2000) and authored several journal articles and research papers. At SDI he heads up research projects that use complex systems theories to help implement sustainable development. He is editing a book for SUNY Press on the uses of complex systems theories in international studies.

Don Munton is professor and former chair of international studies at the University of Northern British Columbia. He teaches environmental policy, Canadian foreign policy, research methods, and international security. His research has focused both on international environmental issues, particularly water pollution, acid rain, and hazardous wastes, and on arms control and disarmament questions, particularly changing public attitudes about nuclear weapons over the Cold War period.

Tom Polacheck is a fishery scientist whose research has focused on the population dynamics of marine resources and who has been involved in the provision of scientific advice to several international management organizations. Since 1991 he has had primary responsibility for the Australian assessments of Southern Bluefin Tuna (SBT) and was the principal Australian scientific adviser in the international SBT dispute.

Marvin S. Soroos is a professor of political science and public administration at North Carolina State University, where he teaches courses in international environmental politics, law, and policy. His books include *Beyond Sovereignty: The Challenge of Global Policy* (1986) and *The Endangered Atmosphere: Preserving a Global Commons* (1997). He has recently been a visiting professor at Williams College.

Kenneth E. Wilkening is an assistant professor in the International Studies Program at the University of Northern British Columbia. He specializes in international environmental policy in the Asia and Pacific region, and is particularly interested in the science-policy interface of emerging issues such as trans-Pacific air pollution. He has a B.S. in physics, an M.S. in environmental and systems engineering, and a Ph.D. in international environmental policy from the Institute for Environmental Studies at the University of Wisconsin, Madison.